Topics in Stochastic Processes

This is Volume 27 in
PROBABILITY AND MATHEMATICAL STATISTICS

A Series of Monographs and Textbooks

Editors: Z. W. Birnbaum and E. Lukacs

A complete list of titles in this series appears at the end of this volume.

Topics in Stochastic Processes

ROBERT B. ASH

Department of Mathematics
University of Illinois
Urbana, Illinois

MELVIN F. GARDNER

Department of Mathematics
University of Toronto
Toronto, Ontario, Canada

ACADEMIC PRESS   New York   San Francisco   London   1975

A Subsidiary of Harcourt Brace Jovanovich, Publishers

996811

ACADEMIC PRESS, INC.
111 Fifth Avenue, New York, New York 10003

United Kingdom Edition published by
ACADEMIC PRESS, INC. (LONDON) LTD.
24/28 Oval Road, London NW1

Library of Congress Cataloging in Publication Data

Ash, Robert B
 Topics in stochastic processes.

 (Probability and mathematical statistics series ;
 Bibliography: p.
 Includes index.
 1. Stochastic processes. I. Gardner, Melvin F.,
joint author. II. Title.
QA274.A79   519.2   74-17991
ISBN 0-12-065270-6
AMS (MOS) 1970 Subject Classification: 60-62

PRINTED IN THE UNITED STATES OF AMERICA

Contents

PREFACE   vii

Chapter 1 L^2 **Stochastic Processes**

1.1	Introduction	1
1.2	Covariance Functions	14
1.3	Second Order Calculus	30
1.4	Karhunen–Loève Expansion	37
1.5	Estimation Problems	43
1.6	Notes	49

Chapter 2   **Spectral Theory and Prediction**

2.1	Introduction; L^2 Stochastic Integrals	50
2.2	Decomposition of Stationary Processes	57
2.3	Examples of Discrete Parameter Processes	69
2.4	Discrete Parameter Prediction: Special Cases	77
2.5	Discrete Parameter Prediction: General Solution	81
2.6	Examples of Continuous Parameter Processes	88
2.7	Continuous Parameter Prediction in Special Cases; Yaglom's Method	97
2.8	Some Stochastic Differential Equations	104
2.9	Continuous Parameter Prediction: Remarks on the General Solution	111
2.10	Notes	112

Chapter 3 **Ergodic Theory**

3.1	Introduction	113
3.2	Ergodicity and Mixing	117
3.3	The Pointwise Ergodic Theorem	124
3.4	Applications to Real Analysis	136
3.5	Applications to Markov Chains	141
3.6	The Shannon–McMillan Theorem	148
3.7	Notes	159

Chapter 4 **Sample Function Analysis of Continuous Parameter Stochastic Processes**

4.1	Separability	161
4.2	Measurability	169
4.3	One-Dimensional Brownian Motion	175
4.4	Law of the Iterated Logarithm	181
4.5	Markov Processes	186
4.6	Processes with Independent Increments	195
4.7	Continuous Parameter Martingales	200
4.8	The Strong Markov Property	205
4.9	Notes	209

Chapter 5 **The Itô Integral and Stochastic Differential Equations**

5.1	Definition of the Itô Integral	210
5.2	Existence and Uniqueness Theorems for Stochastic Differential Equations	219
5.3	Stochastic Differentials: A Chain Rule	226
5.4	Notes	233

Appendix 1	**Some Results from Complex Analysis**	234
Appendix 2	**Fourier Transforms on the Real Line**	242
References		248
Solutions to Problems		250
INDEX		319

Preface

This book contains selected topics in stochastic processes that we believe can be studied profitably by a reader familiar with basic measure-theoretic probability. The background is given in "Real Analysis and Probability" by Robert B. Ash, Academic Press, 1972. A student who has learned this material from other sources will be in good shape if he feels reasonably comfortable with infinite sequences of random variables. In particular, a reader who has studied versions of the strong law of large numbers and the central limit theorem, as well as basic properties of martingale sequences, should find our presentation accessible.

We should comment on our choice of topics. In using the tools of measure-theoretic probability, one is unavoidably operating at a high level of abstraction. Within this limitation, we have tried to emphasize processes that have a definite physical interpretation and for which explicit numerical results can be obtained, if desired. Thus we begin (Chapters 1 and 2) with L^2 stochastic processes and prediction theory. Once the underlying mathematical foundation has been built, results which have been used for many years by engineers and physicists are obtained. The main result of Chapter 3, the ergodic theorem, may be regarded as a version of the strong law of large numbers for stationary stochastic processes. We describe several interesting applications to real analysis, Markov chains, and information theory.

In Chapter 4 we discuss the sample function behavior of continuous parameter processes. General properties of martingales and Markov processes are given, and one-dimensional Brownian motion is analyzed in detail. The purpose is to illustrate those concepts and constructions that are basic in any discussion of continuous parameter processes, and to open the gate to allow the reader to proceed to more advanced material on Markov processes and potential theory. In Chapter 5 we use the theory of continuous parameter processes to develop the Itô stochastic integral and to discuss the solution of stochastic differential equations. The results are of current interest in communication and control theory.

The text has essentially three independent units: Chapters 1 and 2; Chapter 3; and Chapters 4 and 5. The system of notation is standard; for example, 2.3.1 means Chapter 2, Section 3, Part 1. A reference to "Real Analysis and Probability" is denoted by RAP.

Problems are given at the end of each section. Fairly detailed solutions are given to many problems.

We are indebted to Mary Ellen Bock and Ed Perkins for reading the manuscript and offering many helpful suggestions.

Once again we thank Mrs. Dee Keel for her superb typing, and the staff of Academic Press for their constant support and encouragement.

Chapter 1

L^2 Stochastic Processes

1.1 Introduction

We shall begin our work in stochastic processes by considering a down-to-earth class of such processes, those whose random variables have finite second moments. The objective of this chapter is to develop some intuition in handling stochastic processes, and to prepare for the study of spectral theory and prediction in Chapter 2.

First we must recall some notation from probability and measure theory.

1.1.1 Terminology

A *measurable space* $(S, \mathscr{S})$ is a set S with a σ-field $\mathscr{S}$ of subsets of S. A *probability space* $(\Omega, \mathscr{F}, P)$ is a measure space $(\Omega, \mathscr{F})$ with a probability measure P on $\mathscr{F}$. If X is a function from Ω to S, X is said to be *measurable* (notation $X: (\Omega, \mathscr{F}) \to (S, \mathscr{S})$) iff $X^{-1}(A) \in \mathscr{F}$ for each $A \in \mathscr{S}$. If S is the set R of reals and $\mathscr{S} = \mathscr{B}(R)$, the class of Borel sets of R, X is called a *random variable* or, for emphasis, a *real random variable*; if $S = \bar{R}$ (the extended reals) and $\mathscr{S} = \mathscr{B}(\bar{R})$, X is said to be an *extended random variable*; if $S = C$ (the complex numbers) and $\mathscr{S} = \mathscr{B}(C)$; X is called a *complex random variable*. More generally, if S is the set R^n of n-tuples of real numbers, and $\mathscr{S} = \mathscr{B}(R^n)$, X is called an *n-dimensional random vector*; X may be regarded as an n-tuple $(X_1, \ldots, X_n)$ of random variables (RAP, p. 212). If $S = R^\infty$, the set of all sequences of real numbers, and $\mathscr{S} = \mathscr{B}(R^\infty)$, X is said to be a *random sequence*; X may be regarded as a sequence $(X_1, X_2, \ldots)$ of random variables (RAP, 5.11.3, p. 233).

We now give the general definition of a stochastic process, and the probability student will see that he has already encountered many examples of such processes.

1.1.2   Definitions and Comments

Let $(\Omega, \mathscr{F}, P)$ be a probability space, $(S, \mathscr{S})$ an arbitrary measurable space, and T an arbitrary set. A *stochastic process* on $(\Omega, \mathscr{F}, P)$ with *state space* $(S, \mathscr{S})$ and *index set* T is a family of measurable functions X_t: $(\Omega, \mathscr{F}) \to (S, \mathscr{S})$, $t \in T$.

Note that if T is the set of positive integers and $S = R$, $\mathscr{S} = \mathscr{B}(R)$, the stochastic process $\{X_t, t \in T\}$ is a sequence of random variables. Similarly, we can obtain n-dimensional random vectors $(T = \{1, \ldots, n\}$, $S = R$, $\mathscr{S} = \mathscr{B}(R))$ and even single random variables $(n = 1)$.

A synonym for stochastic process is *random function*; let us try to explain this terminology. Let S^T be the collection of all functions from T to S, and let $\mathscr{S}^T$ be the *product σ-field* on S^T. (Recall that $\mathscr{S}^T$ is the smallest σ-field containing all measurable rectangles

$$\{\omega \in S^T : \omega(t_1) \in B_1, \ldots, \omega(t_n) \in B_n\},$$

$t_1, \ldots, t_n \in T$, $B_1, \ldots, B_n \in \mathscr{S}$, $n = 1, 2, \ldots$; see RAP, 4.4.1, p. 189.)

Now suppose that for each $t \in T$ we have a function $X_t \colon \Omega \to S$; define $X \colon \Omega \to S^T$ by

$$X(\omega) = (X_t(\omega), t \in T).$$

Then X is a measurable map of $(\Omega, \mathscr{F})$ into $(S^T, \mathscr{S}^T)$ iff each X_t is a measurable map of $(\Omega, \mathscr{F})$ into $(S, \mathscr{S})$ (see Problem 1).

Thus a stochastic process is a measurable mapping X from Ω into the function space S^T. Intuitively, the performance of the experiment produces a sample point ω, and this in turn determines a collection of elements $X(t, \omega) = X_t(\omega)$, $t \in T$. In other words, the outcome of the experiment determines a function from T to S, called the *sample function* corresponding to the point ω. If T is an interval of reals, it is often helpful to visualize t as a time parameter, and to think of a stochastic process as a random time function.

In this chapter we shall be concerned with L^2 processes, defined as follows. An L^2 *stochastic process* is a family of real or complex random variables X_t, $t \in T$, such that $\|X_t\|^2 = E(|X_t|^2) < \infty$ for all $t \in T$. Thus the state space is R or C, and we have a second moment restriction.

1.1.3 Examples

(a) Let θ be a random variable, uniformly distributed between 0 and 2π. (We may take $\Omega = [0, 2\pi)$, $\mathscr{F} = \mathscr{B}[0, 2\pi)$, $P(B) = \int_B (1/2\pi)\, dx$, $\theta(\omega) = \omega$.)

We define a stochastic process by $X_t = \sin\,(at + \theta)$ where a is fixed and t ranges over all real numbers. Explicitly, $X_t(\omega) = \sin\,(at + \theta(\omega))$, $\omega \in \Omega$. Thus the process represents a sine wave with a random phase angle (see Figure 1.1).

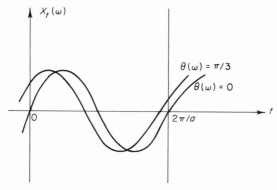

FIGURE 1.1

(b) Let V and W be independent random variables with distribution functions F and G (take $\Omega = R^2$, $\mathscr{F} = \mathscr{B}(R^2)$, $P(B) = \iint_B dF(x)\, dG(y)$, $V(x, y) = x$, $W(x, y) = y$). For t real, let $X_t = 0$ for $t < V$, $X_t = W$ for $t \geq V$, that is, $X_t(\omega) = 0$ for $t < V(\omega)$, $X_t(\omega) = W$ for all $t \geq V(\omega)$. The process represents a step function with random starting time and random amplitude (see Figure 1.2).

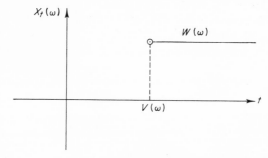

FIGURE 1.2

We now describe a basic approach to the construction of stochastic processes.

1.1.4   The Kolmogorov Construction

If $\{X_t, t \in T\}$ is a stochastic process, we may compute the distribution of $(X_{t_1}, \ldots, X_{t_n})$ for each $t_1, \ldots, t_n \in T, t_1 < \cdots < t_n$. (If T is not a subset of the reals, a fixed total ordering is assumed on T. This avoids the problem of having to deal with all permutations of $t_1, \ldots, t_n$.) For example, in 1.1.3(b) we have

$$P\{X_2 > 1, 2 \le X_5 \le 4, X_{27} > 3\} = P\{V \le 2, 3 < W \le 4\}$$
$$= F(2)(G(4) - G(3)).$$

In fact if we specify in a consistent manner the distribution of $(X_{t_1}, \ldots, X_{t_n})$ for all finite subsets $\{t_1, \ldots, t_n\}$ of T, it is possible to construct a stochastic process with the given finite-dimensional distributions. This is a consequence of the Kolmogorov extension theorem (RAP, 4.4.3, p. 191). To see how the theorem is applied, let $\Omega = S^T$, $\mathscr{F} = \mathscr{S}^T$; we assume throughout that S is a complete, separable metric space and $\mathscr{S} = \mathscr{B}(S)$, the class of Borel sets of S. For each finite subset $v = \{t_1, \ldots, t_n\}$ of T, suppose that we specify a probability measure P_v on $\mathscr{S}^n$; $P_v(B)$ is to represent $P\{\omega \in \Omega : (X_{t_1}(\omega), \ldots, X_{t_n}(\omega)) \in B\}$, which will equal $P\{\omega \in \Omega : (\omega(t_1), \ldots, \omega(t_n)) \in B\}$ if we take $X_t(\omega) = \omega(t), t \in T$.

The hypothesis of the Kolmogorov extension theorem requires that the P_v be consistent, that is, if $u = \{\tau_1, \ldots, \tau_k\} \subset v = \{t_1, \ldots, t_n\}$, the projection $\pi_u(P_v)$ (RAP, p. 190) must coincide with P_v. Now

$$[\pi_u(P_v)](B) = P_v\{y = (y(t_1), \ldots, y(t_n)) \in S^n : (y(\tau_1), \ldots, y(\tau_k)) \in B\}.$$

In terms of finite-dimensional distributions, $\pi_u(P_v)$ gives the distribution of $(X_{\tau_1}, \ldots, X_{\tau_k})$ *as calculated from the distribution of the larger family* $(X_{t_1}, \ldots, X_{t_n})$, whereas P_u gives the distribution of $(X_{\tau_1}, \ldots, X_{\tau_k})$ as originally specified. Thus the consistency requirement is quite natural. For example, if we want X_1, X_2, and X_3 to be independent, normally distributed random variables, we cannot at the same time demand that X_1 be uniformly distributed; it must be normal.

Now let us recall the precise statement of the Kolmogorov extension theorem:

Assume that for each finite nonempty subset v of T we are given a probability measure P_v on $\mathscr{S}^n$, where n is the number of elements in v; assume also that the P_v are consistent. Then there is a unique probability measure P on $\mathscr{S}^T$ such that $\pi_v(P) = P_v$ for all v, that is, if $v = \{t_1, \ldots, t_n\}$, $B \in \mathscr{S}^n$, we have

$$P\{\omega : (\omega(t_1), \ldots, \omega(t_n)) \in B\} = P_v(B).$$

Thus if we set $X_t(\omega) = \omega(t)$, $t \in T$ (or equivalently, if we take X to be the identity map on S^T), we have produced a stochastic process with the given finite-dimensional distributions.

Now suppose that $\{Y_t, t \in T\}$ is another process (on a possibly different probability space $(\Omega', \mathscr{F}', P')$) with the same finite-dimensional distributions. Define the mapping $Y: \Omega' \to S^T$ by $Y(\omega) = (Y_t(\omega), t \in T)$; then (Problem 1) Y is a measurable function from $(\Omega', \mathscr{F}')$ to $(S^T, \mathscr{S}^T)$. If $B \in \mathscr{S}^T$, we assert that $P_X(B) = P_Y(B)$; this holds when B is a measurable cylinder since the two processes have the same finite-dimensional distributions, and the general result follows from the Carathéodory extension theorem. The statement that $P_X = P_Y$ is the precise version of the intuitive statement that the finite-dimensional distributions determine the distribution of the entire process.

The above discussion applies with only minor notational changes if the fixed state space $(S, \mathscr{S})$ is replaced by a family of spaces $(S_t, \mathscr{S}_t)$ where each S_t is a complete separable metric space and $\mathscr{S}_t$ is the class of Borel sets.

We shall often use the Kolmogorov extension theorem to construct stochastic processes in situations where the initial data consist of the finite-dimensional distributions. In particular, if we specify that $X_{t_1}, \ldots, X_{t_n}$ be independent whenever $t_1 < \cdots < t_n$, the consistency requirement is satisfied, so that it is possible to construct a family of independent random variables X_t with arbitrary distribution functions F_t, $t \in T$. In some cases, more complicated stochastic processes are constructed using the independent random variables X_t; the following example illustrates this idea.

1.1.5 Example

Suppose that "customers" arrive at times T_1, $T_1 + T_2$, ..., $T_1 + \cdots + T_n$, ..., where the T_i are independent, strictly positive random variables, T_1 having distribution function F_1 and the T_i, $i > 1$, each having distribution function F.

Let $Y_n = T_1 + \cdots + T_n$ be the arrival time of the nth customer, and let $N(t)$ be the number of customers arriving in the interval $(0, t]$. Formally,

$$N(t) = k \quad \text{if} \quad Y_k \le t < Y_{k+1}, \quad k = 0, 1, \ldots \quad \text{(define } Y_0 \equiv 0).$$

For any $t \ge 0$, let Z_t be the waiting time from t to the arrival of the next customer. We claim that

$$\{Z_t \le x\} \overset{\text{a.e.}}{=} \bigcup_{n=1}^{\infty} \{t < Y_n \le t + x < Y_{n+1}\}. \tag{1}$$

(We say that the sets A and B are equal a.e. iff $P(A \triangle B) = 0$.) For if $t < Y_n \le t + x < Y_{n+1}$, the nth customer arrives in $(t, t + x]$, so $Z_t \le x$. Conversely, if $Z_t \le x$, then some customer arrives in $(t, t + x]$, and hence (a.e.) there is a last customer to arrive in this interval. (If not, then $\sum_{n=1}^{\infty} T_n < \infty$, and this has probability zero by the strong law of large numbers.) If customer n is the last to arrive in $(t, t + x]$, then $t < Y_n \leq t + x < Y_{n+1}$.

We now find the distribution of Z_t. By (1),

$$P\{Z_t \leq x\} = \sum_{n=1}^{\infty} P\{t < Y_n \leq t + x < Y_{n+1}\}. \tag{2}$$

But

$$P\{t < Y_n \leq t + x < Y_{n+1}\} = P\{t < Y_n \leq t + x, \; Y_n + T_{n+1} > t + x\}$$

$$= \iint_{\substack{t < u \leq t+x \\ u+v > t+x}} dF_{Y_n}(u) \, dF_{T_{n+1}}(v)$$

$$= \int_{t < u \leq t+x} P\{T_{n+1} > t + x - u\} \, dF_{Y_n}(u),$$

as would be expected by a conditioning argument. Thus

$$F_{Z_t}(x) = \sum_{n=1}^{\infty} \int_{(t,\, t+x]} [1 - F(t + x - u)] \, dF_{Y_n}(u). \tag{3}$$

We now assume that the T_n, $n > 1$, have finite expectation m, and

$$F_1(x) = \frac{1}{m} \int_0^x [1 - F(y)] \, dy. \tag{4}$$

(This forces the arrival rate, that is, the average number of customers arriving per second, to be constant; see Problem 4. We make the assumption in order to obtain F_{Z_t} explicitly.)

By RAP (p. 280, Problem 2), $F_1(\infty) = 1$, and hence F_1 is the distribution function of a random variable.

Now if X is a random variable with distribution function G, the *generalized characteristic function* of X (or of G) is defined by

$$M(s) = \int_R e^{-sx} \, dG(x). \tag{5}$$

$M(s)$ is defined for those complex numbers s for which the integral is finite. If we set $s = -iu$, we obtain the "ordinary" characteristic function. The proof given in RAP (8.1.2, p. 322) shows that if $X_1, \ldots, X_n$ are independent random variables whose generalized characteristic functions $M_1, \ldots, M_n$ are defined at s, then the generalized characteristic function M of $X_1 + \cdots + X_n$ is defined at s, and

$$M(s) = \prod_{i=1}^{n} M_i(s).$$ (6)

Returning to the original problem, let M_1 and M be the generalized characteristic functions of T_1 and T_n, $n > 1$; we have, for Re $s > 0$,

$$M_1(s) = \int_0^\infty e^{-sx}\, dF_1(x) = \frac{1}{m}\int_0^\infty e^{-sx}(1 - F(x))\, dx.$$ (7)

We now assume that the T_n, $n > 1$, have a density f (this assumption can be avoided, but it simplifies the calculations). Note that by (4), T_1 has density $f_1(x) = m^{-1}(1 - F(x))$, $x \geq 0$. Then (7) becomes

$$M_1(s) = \frac{1}{ms} - \frac{1}{m}\int_0^\infty e^{-sx}\int_0^x f(y)\, dy\, dx$$

$$= \frac{1}{ms} - \frac{1}{m}\int_0^\infty f(y)\int_y^\infty e^{-sx}\, dx\, dy$$

$$= \frac{1}{ms} - \frac{1}{ms}\int_0^\infty f(y)e^{-sy}\, dy.$$

Thus

$$M_1(s) = \frac{1}{ms}(1 - M(s)),$$

or

$$\frac{1}{ms} = \frac{M_1(s)}{1 - M(s)} = M_1(s)\sum_{n=0}^{\infty}[M(s)]^n.$$

But the generalized characteristic function of Y_n is $M_1(s)[M(s)]^{n-1}$, hence

$$\frac{1}{ms} = \sum_{n=1}^{\infty} M_{Y_n}(s).$$ (8)

Since $T_1, \ldots, T_n$ have densities, so does Y_n. Thus

$$\sum_{n=1}^{\infty} M_{Y_n}(s) = \sum_{n=1}^{\infty}\int_0^\infty e^{-sx}f_{Y_n}(x)\, dx,$$

hence by (8),

$$\frac{1}{ms} = \int_0^\infty e^{-sx}\left[\sum_{n=1}^{\infty} f_{Y_n}(x)\right] dx.$$ (9)

But $(ms)^{-1} = \int_0^\infty e^{-sx}m^{-1}\,dx$, and it follows from the uniqueness theorem for Laplace transforms (Problem 7) that

$$\sum_{n=1}^\infty f_{Y_n}(x) = \frac{1}{m} \qquad \text{a.e. on} \quad [0, \infty) \quad [\text{Lebesgue measure}]. \tag{10}$$

By (3),

$$F_{Z_t}(x) = \frac{1}{m} \int_t^{t+x} [1 - F(t + x - u)]\,du$$

$$= \frac{1}{m} \int_0^x [1 - F(y)]\,dy \qquad (\text{set } y = t + x - u)$$

$$= F_1(x).$$

Thus if we start counting at any time t, the waiting time to the next customer has exactly the same distribution as the initial waiting time starting at $t = 0$.

(If we do not assume that the T_n, $n > 1$, have a density, the above analysis may be carried through by replacing expressions of the form $f_{Y_n}(x)\,dx$ by $dF_{Y_n}(x)$. Equation (10) becomes $\sum_{n=1}^\infty F_{Y_n}(x) = x/m$ a.e.; this is proved using the uniqueness theorem for the Laplace transform of a distribution function (see, for example, Widder, 1941).)

In fact if W_1, W_2, ... are the successive waiting times starting at t (so $W_1 = Z_t$), then W_1, W_2, ... are independent, and W_i has the same distribution as T_i for all i. To see this, note that

$$\{W_1 \le x_1, \ldots, W_k \le x_k\} = \bigcup_{n=0}^\infty \{Y_n \le t < Y_{n+1} \le t + x_1,$$

$$T_{n+2} \le x_2, \ldots, T_{n+k} \le x_k\}. \tag{11}$$

For it is clear that the set on the right-hand side of (11) is a subset of the set on the left. Conversely, if $W_1 \le x_1, \ldots, W_k \le x_k$, then a customer arrives in $(t, t + x_1]$, hence there is a first customer in this interval, say customer $n + 1$. Then $Y_n \le t < Y_{n+1} \le t + x_1$, and also $W_i = T_{n+i}$, $i = 2, \ldots, k$, as desired. Therefore

$$P\{W_1 \le x_1, \ldots, W_k \le x_k\} = \left[\sum_{n=0}^\infty P\{Y_n \le t < Y_{n+1} \le t + x_1\}\right] \prod_{i=2}^k F(x_i)$$

$$= P\{Z_t \le x_1\} \prod_{i=2}^k F(x_i).$$

Fix j and let $x_i \to \infty$, $i \neq j$, to conclude that the W_i are independent and W_i has the same distribution as T_i for all i. In particular, $N(t + h) - N(t)$ has the same distribution as $N(h)$.

The stochastic process $\{N(t), t \geq 0\}$ is sometimes called a *delayed renewal process*. (The word *delayed* refers to the fact that T_1 has a different distribution from the T_n, $n > 1$; if the distributions are the same, *delayed* is omitted.)

Physically, the T_i can be regarded as the lifetimes of a succession of products such as light bulbs. If $T_1 + \cdots + T_n = t$, then bulb n has burned out at time t, and the light must be renewed by placing bulb $n + 1$ in position.

If F_1 and F are related as in (4), then $\{N(t), t \geq 0\}$ is called a *uniform* (delayed) renewal process. The most important example is the *Poisson process*, obtained by specifying that the T_n, $n > 1$, have the exponential density $f(x) = \lambda e^{-\lambda x}$, $x \geq 0$ (where $\lambda > 0$). Then $m = 1/\lambda$, so by (4),

$$F_1(x) = \lambda \int_0^x [1 - (1 - e^{-\lambda y})]\, dy = 1 - e^{-\lambda x}, \qquad x \geq 0.$$

Therefore $F_1 = F$.

The reason for the use of the name *Poisson* is that for each t, $N(t)$ has a Poisson distribution with parameter λt, that is,

$$P\{N(t) = k\} = \frac{e^{-\lambda t}(\lambda t)^k}{k!}, \qquad k = 0, 1, \ldots.$$

To see this, observe that

$$P\{N(t) \leq k\} = P\{T_1 + \cdots + T_{k+1} > t\},$$

and $T_1 + \cdots + T_{k+1}$ has density $\lambda^{k+1} x^k e^{-\lambda x}/k!$, $x \geq 0$. (This is an exercise in basic probability theory; see Ash, 1970, p. 197 for details.) Thus

$$P\{N(t) \leq k\} = \int_t^\infty \frac{1}{k!} \lambda^{k+1} x^k e^{-\lambda x}\, dx.$$

Successive integration by parts (again see Ash, 1970, p. 197) yields

$$P\{N(t) \leq k\} = \sum_{i=0}^k e^{-\lambda t} \frac{(\lambda t)^i}{i!},$$

as desired.

Since $E[N(t)] = \lambda t$, $\lambda = 1/E(T_i)$ may be interpreted as the average number of customers per second. Furthermore, since $\mathrm{Var}\, N(t) = \lambda t$, $\{N(t), t \geq 0\}$ is an L^2 process.

A key property of the Poisson process is the following.

1.1.6   Theorem

Let $\{N(t),\ t \geq 0\}$ be a Poisson process. The process has *independent increments*, that is, if $0 < t_1 < \cdots < t_m$, then $N(t_1) - N(0)$, $N(t_2) - N(t_1)$, ..., $N(t_m) - N(t_{m-1})$ are independent; equivalently, if

$$0 \leq t_1 < t_2 \leq t_3 < t_4 \leq \cdots \leq t_{2n-1} < t_{2n},$$

then $N(t_2) - N(t_1)$, $N(t_4) - N(t_3)$, ..., $N(t_{2n}) - N(t_{2n-1})$ are independent.

PROOF   We break the argument into several parts. Fix $t_1 > 0$, and let W_1, W_2, ... be the successive waiting times starting from t_1, as in 1.1.5.

(a) $N(t_1)$ and W_1 are independent.
We do this by an induction argument. First note that

$$P\{N(t_1) = 0,\ W_1 \leq x_1\} = P\{t_1 < T_1 \leq t_1 + x_1\}$$

$$= e^{-\lambda t_1} - e^{-\lambda(t_1 + x_1)} = e^{-\lambda t_1}(1 - e^{-\lambda x_1})$$

$$= P\{N(t_1) = 0\}P\{W_1 \leq x_1\}.$$

If $P\{N(t_1) = n - 1,\ W_1 \leq x_1\} = P\{N(t_1) = n - 1\}P\{W_1 \leq x_1\}$ for   all $t_1 > 0$, $x_1 \geq 0$, then

$$P\{N(t_1) = n,\ W_1 \leq x_1\} = P\{T_1 + \cdots + T_n \leq t_1$$

$$< T_1 + \cdots + T_{n+1} \leq t_1 + x_1\}$$

$$= \int \cdots \int_A \lambda^{n+1} e^{-\lambda(y_1 + \cdots + y_{n+1})}\, dy_1 \cdots dy_{n+1}$$

where

$$A = \{(y_1, \ldots, y_{n+1}): 0 \leq y_1 \leq t_1,\ y_2, \ldots, y_{n+1} \geq 0,$$

$$y_2 + \cdots + y_n \leq t_1 - y_1 < y_2 + \cdots + y_{n+1} \leq t_1 + x_1 - y_1\}.$$

Thus

$$P\{N(t_1) = n,\ W_1 \leq x_1\}$$

$$= \int_0^{t_1} \lambda e^{-\lambda y_1} P\{T_2 + \cdots + T_n \leq t_1 - y_1$$

$$< T_2 + \cdots + T_{n+1} \leq t_1 + x_1 - y_1\}\, dy_1$$

$$= \int_0^{t_1} \lambda e^{-\lambda y_1} P\{N(t_1 - y_1) = n - 1,\ W_1^* \leq x_1\}\, dy_1$$

where W_1^* is the waiting time starting from $t_1 - y_1$. By the induction hypothesis and the fact that W_1^* and W_1 have the same distribution (see 1.1.5),

$$P\{N(t_1) = n, W_1 \leq x_1\}$$

$$= P\{W_1 \leq x_1\} \int_0^{t_1} \lambda e^{-\lambda y_1} P\{N(t_1 - y_1) = n - 1\} \, dy_1$$

$$= P\{W_1 \leq x_1\} \int_0^{t_1} \lambda e^{-\lambda y_1} e^{-\lambda(t_1 - y_1)} \frac{[\lambda(t_1 - y_1)]^{n-1}}{(n - 1)!} \, dy_1$$

$$= P\{W_1 \leq x_1\} e^{-\lambda t_1} \frac{(\lambda t_1)^n}{n!} = P\{W_1 \leq x_1\} P\{N(t_1) = n\}.$$

(b) $N(t_1), W_1, W_2, \ldots$ are independent.
To prove this, we compute

$$P\{N(t_1) = n, W_1 \leq x_1, \ldots, W_k \leq x_k\}$$

$$= P\{T_1 + \cdots + T_n \leq t_1 < T_1 + \cdots + T_{n+1} \leq t_1 + x_1,$$

$$T_{n+2} \leq x_2, \ldots, T_{n+k} \leq x_k\}$$

$$= P\{T_1 + \cdots + T_n \leq t_1 < T_1 + \cdots + T_{n+1} \leq t_1 + x_1\}$$

$$\times \prod_{i=2}^{k} P\{T_{n+i} \leq x_i\}$$

$$= P\{N(t_1) = n, W_1 \leq x_1\} \prod_{i=2}^{k} P\{W_i \leq x_i\}$$

since the W_i and the T_i all have the same distribution

$$= P\{N(t_1) = n\} \prod_{i=1}^{k} P\{W_i \leq x_i\} \qquad \text{by (a).}$$

We may now prove the theorem. Assume the result holds through the integer $m - 1$. Then

$$P\{N(t_1) = k_1, N(t_2) - N(t_1) = k_2, \ldots, N(t_m) - N(t_{m-1}) = k_m\}$$

$$= P\{N(t_1) = k_1, W_1 + \cdots + W_{k_2} \leq t_2 - t_1 < W_1 + \cdots + W_{k_2+1},$$

$$W_1 + \cdots + W_{k_2+k_3} \leq t_3 - t_1 < W_1 + \cdots + W_{k_2+k_3+1}, \ldots,$$

$$W_1 + \cdots + W_{k_2+\cdots+k_m} \leq t_m - t_1 < W_1 + \cdots + W_{k_2+\cdots+k_m+1}\}$$

$$= P\{N(t_1) = k_1\} P\{N(t_2 - t_1) = k_2, N(t_3 - t_1) - N(t_2 - t_1) = k_3, \ldots,$$

$$N(t_m - t_1) - N(t_{m-1} - t_1) = k_m\}$$

by (b) and the fact that the W_i have the same distribution as the T_i

$$= P\{N(t_1) = k_1\} \prod_{i=2}^{m} P\{N(t_i - t_1) - N(t_{i-1} - t_1) = k_i\}$$

by the induction hypothesis

$$= P\{N(t_1) = k_1\} \prod_{i=2}^{m} P\{N(t_i) - N(t_{i-1}) = k_i\}$$

since $N(t + h) - N(t)$ has the same distribution as $N(h)$. ‖

Let us look at a basic application of 1.1.6. Suppose that $0 \le t_1 < \cdots < t_{n+1}$, and we wish to compute

$$P\{N(t_{n+1}) = a_{n+1} \mid N(t_1) = a_1, \ldots, N(t_n) = a_n\}$$

where $a_1, \ldots, a_n$ are nonnegative integers and $a_1 \le \cdots \le a_n$. Theorem 1.1.6 will be used to show that this coincides with

$$P\{N(t_{n+1}) = a_{n+1} \mid N(t_n) = a_n\}$$

so that the distribution of the state of the process at the "future" time t_{n+1}, given the behavior at "past" times $t_1, \ldots, t_{n-1}$ and the "present" time t_n depends only on the present state. This is the *Markov property*, which will be examined systematically in Chapter 4. To verify it, we write

$$P\{N(t_{n+1}) = a_{n+1} \mid N(t_1) = a_1, \ldots, N(t_n) = a_n\}$$
$$= P\{N(t_{n+1}) - N(t_n) = a_{n+1} - a_n \mid N(t_1) - N(0) = a_1,$$
$$N(t_2) - N(t_1) = a_2 - a_1, \ldots,$$
$$N(t_n) - N(t_{n-1}) = a_n - a_{n-1}\}$$
$$= P\{N(t_{n+1}) - N(t_n) = a_{n+1} - a_n\} \qquad \text{by 1.1.6}$$
$$= P\{N(t_{n+1}) - N(t_n) = a_{n+1} - a_n \mid N(t_n) - N(0) = a_n\}$$

again by 1.1.6

$$= P\{N(t_{n+1}) = a_{n+1} \mid N(t_n) = a_n\}.$$

We may compute this probability explicitly:

$$P\{N(t_{n+1}) - N(t_n) = a_{n+1} - a_n\} = P\{N(t_{n+1} - t_n) = a_{n+1} - a_n\}$$
$$= e^{-\lambda(t_{n+1} - t_n)} \frac{[\lambda(t_{n+1} - t_n)]^{a_{n+1} - a_n}}{(a_{n+1} - a_n)!}.$$

Problems

1. Let $(\Omega, \mathcal{F})$ and $(\Omega_t, \mathcal{F}_t)$, $t \in T$, be measurable spaces, and let $\Omega' = \prod_{t \in T} \Omega_t$, $\mathcal{F}' = \prod_{t \in T} \mathcal{F}_t$ (see RAP, p. 189). If $Y_t: \Omega \to \Omega_t$, $t \in T$, and we define $Y: \Omega \to \Omega'$ by $Y(\omega) = (Y_t(\omega), t \in T)$, show that Y is measurable iff each Y_t is measurable.

2. In the uniform renewal process, find the distribution of the time U_t from the arrival of the *last* customer on or before time t to the present time t (take $U_t = t$ if no customer arrives in $(0, t]$).

3. In the Poisson process, let $V_t = Z_t + U_t$ (U_t as in Problem 2). Thus in terms of light bulbs, V_t is the lifetime of the bulb in service at time t. Find the distribution and the expectation of V_t.

Note that V_t and T_1 do not have the same distribution, although each is the lifetime of a single bulb. This is the so-called waiting time paradox; an explanation is that the length of time that the bulb in service at t has already lasted has no effect on the distribution of its future life because of the "memoryless" feature of the exponential density, that is,

$$P\{T_n \le r + x \mid T_n > r\} = P\{T_n \le x\}, \qquad r, x \ge 0.$$

4. (a) Show that $E[N(t)] = \sum_{n=1}^{\infty} P\{Y_n \le t\}$ for any renewal process, delayed or not.

(b) In the uniform renewal process, show that $E[N(t)] = t/m$.

5. Let $\{N(t), t \ge 0\}$ be a Poisson process. If $Y_n = T_1 + \cdots + T_n$ is the arrival time of the nth customer, show that $(Y_1, \ldots, Y_n)$ has density $f(y_1, \ldots, y_n) = \lambda^n e^{-\lambda y_n}$, $0 \le y_1 \le y_2 \le \cdots \le y_n$; $f = 0$ elsewhere.

6. In Problem 5, show that given $N(t) = n$, the joint distribution of $Y_1, \ldots, Y_n$ is the same as if the arrival time were chosen independently with uniform density over $[0, t]$, and then ordered (so that Y_k is the kth smallest of the arrival times). In other words, the conditional distribution of $Y_1, \ldots, Y_n$ given $N(t) = n$ is the same as the distribution of the order statistics of n independent, uniformly distributed random variables.

7. Let f be a Borel measurable function from $[0, \infty)$ to R. If $\int_0^{\infty} e^{-\alpha t} f(t) \, dt = 0$ for all $\alpha > 0$, show that $f = 0$ a.e. (Lebesgue measure).

8. There is of course a natural metric associated with L^2 convergence; we can also introduce a metric corresponding to convergence in probability, as follows. Let $g: [0, \infty) \to [0, \infty)$ be Borel measurable, bounded, increasing, continuous at 0, with $g(x + y) \le g(x) + g(y)$ for all x, y, $g(0) = 0$, $g(x) > 0$ for $x > 0$ (for example, $g(x) = x/(1 + x)$ or min $(x, 1)$). If X and Y are random variables on $(\Omega, \mathcal{F}, P)$, define

$$d(X, Y) = E[g(|X - Y|)].$$

It may be verified that d is a metric on the space M of random variables on $(\Omega, \mathscr{F}, P)$ if functions in M that agree almost everywhere are identified.

Show that d-convergence is equivalent to convergence in probability.

9. In general, there is no metric corresponding to almost everywhere convergence. In fact, let $\mathscr{C}$ be any collection of random variables on $(\Omega, \mathscr{F}, P)$ such that convergence in probability does not imply convergence a.e. for sequences in $\mathscr{C}$. Show that there is no function $d = d(X, Y)$, X, $Y \in \mathscr{C}$, such that $d \geq 0$ and, for $X, X_1, X_2, \ldots \in \mathscr{C}$,

$$d(X_n, X) \to 0 \qquad \text{iff} \qquad X_n \to X \quad \text{a.e.}$$

10. Show that there is no universal probability space on which all possible stochastic processes can be defined. (*Hint:* Consider a very large family of independent random variables.) Note that there is a universal probability space for all possible random variables; see RAP (p. 222, Problem 7).

11. In 1.1.5, we used Problem 2 of RAP (p. 280) to show that $m^{-1} \int_0^x [1 - F(y)] \, dy \to 1$ as $x \to \infty$. The following result is a related application. If X is a nonnegative random variable such that $E(X^r) < \infty$ for some $r > 0$, show that $x^r[1 - F(x)] \to 0$ as $x \to \infty$.

1.2   Covariance Functions

When the term L^2 *theory* is used in connection with stochastic processes, it refers to the properties of an L^2 process that can be deduced from its covariance function, defined as follows.

1.2.1   Definitions and Comments

Let $\{X(t), t \in T\}$ be an L^2 process. The *covariance function* of the process is defined by

$$K(s, t) = \text{Cov} \, [X(s), X(t)].$$

Since we are dealing with complex random variables, we must extend the earlier definition (RAP, 5.10.10, p. 229); we define

$$\text{Cov} \, [X(s), X(t)] = E\{[X(s) - m(s)]\overline{[X(t) - m(t)]}\}$$

where $m(t) = E[X(t)]$. Thus $K(s, t)$ is the inner product of $X(s) - m(s)$ and $X(t) - m(t)$ in the space $L^2(\Omega, \mathscr{F}, P)$, so that every L^2 process has a covariance function. Furthermore, we may write

$$K(s, t) = E[X(s)\overline{X(t)}] - m(s)\overline{m(t)}.$$

Note that by the Cauchy–Schwarz inequality,

$$| K(s, t) |^2 \leq \| X(s) - m(s) \|_2^2 \, \| X(t) - m(t) \|_2^2 = K(s, s) K(t, t).$$

If T is an interval of reals, the L^2 process $\{X(t), \, t \in T\}$ is said to be *stationary in the wide sense* iff $m(t)$ is constant for all t and $K(s, t) = K(s + h, t + h)$ for all s, t, h such that $s, t, s + h, t + h \in T$. In other words, $K(s, t)$ depends only on the difference between s and t, so that $K(s + t, s)$ may be expressed as a function of t alone. Abusing the notation slightly, we write $K(s + t, s) = K(t)$; that is,

$$K(t) = \text{Cov} \, [X(s + t), X(s)] = E\{[X(s + t) - m][\overline{X(s) - m}]\};$$

$K(t)$ is defined for those t expressible as $u - v$ for $u, v \in T$. The Cauchy–Schwarz inequality yields

$$| K(t) | \leq K(0) = E[\, | X(s) - m |^2].$$

An arbitrary process $\{X(t), \, t \in T\}$, is called *strictly stationary* iff the joint distribution functions $F_{t_1 \cdots t_n}(x_1, \, \ldots, \, x_n) = P\{X(t_1) \leq x_1, \, \ldots, \, X(t_n) \leq x_n\}$ have the property that

$$F_{t_1 \cdots t_n} = F_{t_1 + h, \, \ldots, \, t_n + h}$$

for all $n = 1, 2, \ldots$ and all $t_1, \ldots, t_n, h$ such that $t_1 < \cdots < t_n$ and all t_i, $t_i + h \in T$. In other words the joint distribution function of $X(t_1), \ldots, X(t_n)$ depends only on the differences $t_2 - t_1, t_3 - t_2, \ldots, t_n - t_{n-1}$. Thus if the "observation points" $t_1, \ldots, t_n$ are translated rigidly, the joint distribution is not changed. (Strictly speaking, in the complex-valued case, with $X(t) = Y(t) + iZ(t)$, $F_{t_1 \cdots t_n}$ should be replaced by the joint distribution function of $Y(t_1), \ldots, Y(t_n), Z(t_1), \ldots, Z(t_n)$.)

Strict stationarity implies wide sense stationarity for an L^2 process since

$$E[X(s)\overline{X(t)}] = \iint x\bar{y} \, dF_{st}(x, y) = \iint x\bar{y} \, dF_{s+h, \, t+h}(x, y)$$

$$= E[X(s + h)\overline{X(t + h)}],$$

and similarly

$$E[X(t)] = E[X(t + h)].$$

The converse is not true however. For example, let the $X(t)$ be independent random variables, each with mean 0 and variance 1. Then $E[X(s + t)X(s)] = K(t) = 0$ for $t \neq 0$, and $K(0) = 1$. But the process need not be strictly stationary; for example, we might take $X(t)$ to be normally distributed for $t \geq 0$ and uniformly distributed for $t < 0$. Thus $F_t(x) = P\{X(t) \leq x\}$ may depend on t.

In Chapters 1 and 2, *stationary* will mean "stationary in the wide sense" unless otherwise specified.

Note that a covariance function always satisfies $K(t, s) = \overline{K(s, t)}$; thus in the stationary case, $K(-t) = \overline{K(t)}$. Since $\overline{K(t, t)} = K(t, t)$, $K(t, t)$ is real, as is $K(0)$ in the stationary case.

One advantage of concentrating on the L^2 theory of stochastic processes is that complicated questions concerning measurability and continuity of the sample functions can be avoided. Such questions, which require us to assemble a formidable apparatus, will be considered in Chapter 4.

We are going to consider the problem of realizing a stochastic process with a specified covariance function. To do this we need the basic properties of the multidimensional Gaussian distribution. Since complete proofs are given in Ash (1970, pp. 279–284), we shall only state the results here.

1.2.2 The Multidimensional Gaussian Distribution

Let $X = (X_1, \ldots, X_n)$ be a (real) random vector. The *characteristic function* of X (or the *joint characteristic function* of $X_1, \ldots, X_n$) is defined by

$$h(u_1, \ldots, u_n) = E[e^{i(u_1 X_1 + \cdots + u_n X_n)}], \qquad u_1, \ldots, u_n \text{ real}$$

$$= \int_{-\infty}^{\infty} \cdots \int_{-\infty}^{\infty} \exp\left(i \sum_{k=1}^{n} u_k x_k\right) dF(x_1, \ldots, x_n)$$

where F is the distribution function of X. It will be convenient to use a vector–matrix notation. If $u = (u_1, \ldots, u_n) \in R^n$, $\mathbf{u}$ will denote the column vector with components $u_1, \ldots, u_n$. Similarly we write $\mathbf{x}$ for col $(x_1, \ldots, x_n)$ and $\mathbf{X}$ for col $(X_1, \ldots, X_n)$. A superscript t will indicate the transpose of a matrix.

Just as in one dimension, the characteristic function determines the distribution function uniquely (see RAP, p. 327, Problem 1).

The random vector $X = (X_1, \ldots, X_n)$ is said to be *Gaussian* (or $X_1, \ldots, X_n$ are said to be *jointly Gaussian*) iff the characteristic function of X is

$$h(u_1, \ldots, u_n) = \exp(i\mathbf{u}^t\mathbf{b}) \exp\left(-\tfrac{1}{2}\mathbf{u}^t K \mathbf{u}\right)$$

$$= \exp\left[i \sum_{r=1}^{n} u_r b_r - \frac{1}{2} \sum_{r,s=1}^{n} u_r K_{rs} u_s\right] \tag{1}$$

where $b_1, \ldots, b_n$ are arbitrary real numbers and K is an arbitrary real symmetric nonnegative definite n by n matrix. (Nonnegative definite means that $\sum_{r,s=1}^{n} a_r K_{rs} a_s$ is real and nonnegative for all real numbers $a_1, \ldots, a_n$.)

(a)   Let X be a random n-vector. Then X is Gaussian iff $\mathbf{X}$ can be expressed as $W\mathbf{Y} + \mathbf{b}$ where $b = (b_1, \ldots, b_n) \in R^n$, W is an n by n matrix, and $Y_1, \ldots, Y_n$ are independent normal random variables with 0 mean.

The matrix K of (1) is given by WDW^t where $D = \text{diag}\,(\lambda_1, \ldots, \lambda_n)$ is a diagonal matrix with entries $\lambda_j = \text{Var}\,Y_j$, $j = 1, \ldots, n$. (To avoid having to treat the case $\lambda_j = 0$ separately, we agree that normal with expectation m and variance 0 will mean constant at m.) The matrix W can be taken as orthogonal.

Furthermore, it is always possible to construct a Gaussian random vector corresponding to a prescribed symmetric nonnegative definite matrix K and a point $b \in R^n$.

(b)   In (a) we have $E(\mathbf{X}) = \mathbf{b}$, that is, $E(X_j) = b_j$, $j = 1, \ldots, n$, and K is the covariance matrix of the X_j, that is, $K_{rs} = \text{Cov}\,(X_r, X_s)$, $r, s = 1, \ldots, n$.

(c)   Let X be Gaussian with representation $\mathbf{X} = W\mathbf{Y} + \mathbf{b}$, W orthogonal, as in (a).

If K is nonsingular, then the random variables $X_j^* = X_j - b_j$ are *linearly independent*, that is, if $\sum_{j=1}^n a_j X_j^* = 0$ a.e., then all $a_j = 0$. In this case X has a density given by

$$f(\mathbf{x}) = (2\pi)^{-n/2}(\det K)^{-1/2}\exp\left[-\tfrac{1}{2}(\mathbf{x} - \mathbf{b})^t K^{-1}(\mathbf{x} - \mathbf{b})\right].$$

If K is singular, the X_j^* are linearly dependent. If, say, $\{X_1^*, \ldots, X_r^*\}$ is a maximal linearly independent subset of $\{X_1^*, \ldots, X_n^*\}$, then $(X_1, \ldots, X_r)$ has a density of the above form, with K replaced by K_r, the matrix consisting of the first r rows and columns of K; $X_{r+1}^*, \ldots, X_n^*$ can be expressed (a.e.) as linear combinations of $X_1^*, \ldots, X_r^*$.

The result that K is singular iff the $X_j^* = X_j - E(X_j)$ are linearly dependent is true for arbitrary random variables in L^2.

(d)   If X is a Gaussian n-vector and $\mathbf{Z} = A\mathbf{X}$ where A is an m by n matrix, then Z is a Gaussian m-vector. In particular:

(e)   If $X_1, \ldots, X_n$ are jointly Gaussian, so are $X_1, \ldots, X_m$, $m \leq n$.

(f)   If $X_1, \ldots, X_n$ are jointly Gaussian, then $a_1 X_1 + \cdots + a_n X_n$ is a Gaussian random variable.

(g)   If $X_1, \ldots, X_n$ are jointly Gaussian and uncorrelated (Cov $(X_r, X_s) = 0$ for $r \neq s$), they are independent.

We now return to the problem of constructing a process with a given covariance function. Let $\{X(t),\ t \in T\}$ be an L^2 process with covariance function $K = K(s, t)$. From the definition of covariance it follows that K is

(Hermitian) *symmetric*, that is, $K(t, s) = \overline{K(s, t)}$, $s, t \in T$; K is also *nonnegative definite*, that is, for all $t_1, \ldots, t_n \in T$ $(n = 1, 2, \ldots)$ and all complex numbers $a_1, \ldots, a_n$,

$$\sum_{r,s=1}^{n} a_r K(t_r, t_s)\bar{a}_s \quad \text{is real and nonnegative.}$$

(In other words, the matrix $[K(t_r, t_s)]$, $r, s = 1, \ldots, n$, is nonnegative definite.) To see this, note that if $X^*(t) = X(t) - E[X(t)]$,

$$\sum_{r,s=1}^{n} a_r K(t_r, t_s)\bar{a}_s = E\left[\sum_{r,s=1}^{n} a_r X^*(t_r)\overline{a_s X^*(t_s)}\right]$$

$$= E\left[\left|\sum_{r=1}^{n} a_r X^*(t_r)\right|^2\right] \geq 0.$$

We now show that symmetry and nonnegative definiteness are sufficient conditions for there to exist an L^2 process with a specified covariance function.

1.2.3 Theorem

Let $K = K(s, t)$, $s, t \in T$, be a complex-valued function on $T \times T$ that is symmetric and nonnegative definite. Then there is an L^2 process $\{X(t), t \in T\}$ whose covariance function is precisely K. (The index set T is arbitrary here; it need not be a subset of R.)

PROOF   First assume K real valued. Given $t_1, \ldots, t_n \in T$, $t_1 < t_2 < \cdots < t_n$, we specify that $X(t_1), \ldots, X(t_n)$ be jointly Gaussian with covariance matrix $[K(t_r, t_s)]$, $r, s = 1, \ldots, n$. By 1.2.2(e), $X(t_{i_1}), \ldots, X(t_{i_j})$ must be jointly Gaussian with covariance matrix $[K(t_{i_r}, t_{i_s})]$, $r, s = 1, \ldots, j$. But this agrees with the original specification of the joint distribution of $X(t_{i_1}), \ldots, X(t_{i_j})$. Therefore the Kolmogorov consistency requirement is satisfied, and the result follows.

In the complex case, let $K = K_1 + iK_2$. If $c_r = a_r + ib_r$, then

$$\sum_{r,s=1}^{n} c_r K(t_r, t_s)\bar{c}_s = \sum_{r,s=1}^{n} K_1(t_r, t_s)(a_r a_s + b_r b_s)$$

$$+ \sum_{r,s=1}^{n} K_2(t_r, t_s)(a_r b_s - a_s b_r)$$

(There are no imaginary terms since K is nonnegative definite.)

$$= \mathbf{d}^t L \mathbf{d} \qquad \text{where } d_r = a_r, \quad r = 1, \ldots, n$$

$$= b_{r-n}, \quad r = n+1, \ldots, 2n$$

and

$$
L = \begin{matrix} \\ \end{matrix}
\begin{array}{c}
a_1 \\ \vdots \\ a_n \\ b_1 \\ \vdots \\ b_n
\end{array}
\overset{\displaystyle a_1 \cdots a_n \quad\quad b_1 \cdots b_n}{
\left[
\begin{array}{cc}
K_1(t_r, t_s) & K_2(t_r, t_s) \\[2ex]
-K_2(t_r, t_s) & K_1(t_r, t_s)
\end{array}
\right]}.
$$

Note that the element in row r, column $n + s$ of L contributes $a_r K_2(t_r, t_s)b_s$ to the quadratic form; the element in row $n + r$, column s contributes $-b_r K_2(t_r, t_s)a_s$. The element in row $n + s$, column r of L is $-K_2(t_s, t_r) = K_2(t_r, t_s)$ by (Hermitian) symmetry of K, and it follows that L is symmetric. Since K is nonnegative definite, so is L.

Now let $Y(t_1), \ldots, Y(t_n), Z(t_1), \ldots, Z(t_n)$ be jointly Gaussian random variables with zero mean and covariance matrix $\frac{1}{2}L$. Define $X(t_r) = Y(t_r) - iZ(t_r)$. Then

$$
\begin{aligned}
E[X(t_r)\overline{X(t_s)}] &= E[Y(t_r)Y(t_s) + Z(t_r)Z(t_s)] \\
&\quad + iE[Y(t_r)Z(t_s) - Y(t_s)Z(t_r)] \\
&= K_1(t_r, t_s) + iK_2(t_r, t_s) = K(t_r, t_s).
\end{aligned}
$$

Thus the $X(t_r)$ are *complex jointly Gaussian random variables* (that is, the real and imaginary parts $Y(t_1), \ldots, Y(t_n), Z(t_1), \ldots, Z(t_n)$ are jointly Gaussian) with covariance matrix $[K(t_r, t_s)]$, $r, s = 1, \ldots, n$. Again, 1.2.2(e) shows that the Kolmogorov consistency conditions hold for the $X(t)$, $t \in T$, and the result follows. ‖

1.2.4   Comments

(a)   The stochastic process $\{X(t), t \in T\}$ is said to be a *Gaussian process* iff for each $t_1, \ldots, t_n \in T$, $n = 1, 2, \ldots$, $(X(t_1), \ldots, X(t_n))$ is jointly Gaussian. (By convention, a Gaussian random vector is assumed to consist of real rather than complex random variables, unless otherwise specified.) For a Gaussian process, the covariance function completely determines all finite-dimensional distributions, hence stationarity in the wide sense is equivalent to strict stationarity.

(b)   If $\{X(t), t \in T\}$ has covariance function K and $f: T \to C$, then $\{X(t) + f(t), t \in T\}$ also has covariance function K; thus changing the means of the individual random variables does not change the covariance, and we need not take the $X(t)$ to have zero mean in 1.2.3.

(c)   No uniqueness assertion can be made in 1.2.3. In fact the proof shows that for any L^2 process there is a complex Gaussian process with the same covariance.

(d)   If T is a real interval and $\{X(t), t \in T\}$ is a stationary L^2 process with covariance $K = K(t) = \text{Cov}\,[X(s + t),\ X(s)]$, then K is symmetric $(K(-t) = \overline{K(t)})$ and nonnegative definite $(\sum_{r,\,s=1}^{n} a_r K(t_r - t_s)\bar{a}_s \geq 0$ for all $t_1, \ldots, t_n \in T$ and all complex numbers $a_1, \ldots, a_n, n = 1, 2, \ldots)$. Conversely, let T be a real interval and let $I = \{u - v : u, v \in T\}$. If K is a symmetric, nonnegative definite complex-valued function on I, there is a stationary L^2 process $\{X(t), t \in T\}$ with covariance function K.

For if $K^*(s, t) = K(s - t)$, $s, t \in T$, then K^* is symmetric and nonnegative definite, hence there is an L^2 process $\{X(t), t \in T\}$ with

$$\text{Cov}\,[X(s + t),\ X(s)] = K^*(s + t, s) = K(t).$$

For the remainder of this section, T will be either the set of integers (the "discrete parameter case") or the set of reals (the "continuous parameter case"). We are going to derive an analytic characterization of stationary covariance functions; namely, that the class of continuous covariance functions of stationary L^2 processes is exactly the class of Fourier transforms of finite measures, in other words, the class of characteristic functions (see RAP, 8.1.1, p. 322). The measures are defined on $\mathscr{B}([-\pi, \pi])$ in the discrete parameter case and on $\mathscr{B}(R)$ in the continuous parameter use.

First, we need a few properties of nonnegative definite functions.

1.2.5   Lemma

If K is any nonnegative definite function on T, then:

(a) $K(0) \geq 0$;

(b) $K(-u) = \overline{K(u)}$; thus K is automatically symmetric;

(c) $|K(u)| \leq K(0)$; and

(d) $|K(u) - K(v)|^2 \leq 2K(0)[K(0) - \text{Re}\,K(u - v)]$;

thus if $T = R$ and K is continuous at 0, then K is uniformly continuous on R.

PROOF   Recall that nonnegative definite means that

$$\sum_{j,\,k=1}^{n} z_j \bar{z}_k K(t_j - t_k) \geq 0 \qquad \text{for all} \quad z_1, \ldots, z_n \in C,$$

$$t_1, \ldots, t_n \in T, \quad n \geq 1.$$

(a)   Take $n = 1$, $z_1 = 1$, $t_1 = 0$.

(b) Take $n = 2$, $z_1 = z_2 = i$, $t_1 = 0$, $t_2 = u$; then $2K(0) + K(-u) + K(u) \geq 0$, in particular (using (a)) $K(-u) + K(u)$ is real. Therefore $\operatorname{Im} K(-u) = -\operatorname{Im} K(u)$.

(c) The result is clear if $K(u) = 0$. If $K(u) \neq 0$, take $n = 2$, $z_1 = 1$, $z_2 = -x/\overline{K(u)}$, $t_1 = u$, $t_2 = 0$, where x is an arbitrary real number. Then (using (b))

$$K(0) - 2x + \frac{K(0)x^2}{|K(u)|^2} \geq 0.$$

Since x is arbitrary, the above quadratic in x must have a nonpositive discriminant, and the result follows.

(d) Take $n = 3$, $z_1 = 1$, $z_2 = z$, $z_3 = -z$, $t_1 = 0$, $t_2 = u$, $t_3 = v$, where z is an arbitrary complex number. Then

$$0 \leq K(0) + \bar{z}K(-u) - \bar{z}K(-v) + zK(u) + |z|^2 K(0)$$

$$- |z|^2 K(u - v) - zK(v) - |z|^2 K(v - u) + |z|^2 K(0)$$

$$= K(0) + 2\operatorname{Re}(z[K(u) - K(v)]) + 2|z|^2[K(0) - \operatorname{Re} K(u - v)].$$

If $K(u) - K(v) = |K(u) - K(v)|e^{i\theta}$, set $z = xe^{-i\theta}$, x real. Then

$$0 \leq K(0) + 2x|K(u) - K(v)| + 2x^2[K(0) - \operatorname{Re} K(u - v)].$$

Since this holds for all $x \in R$, the discriminant cannot be positive, and the desired inequality is obtained. ‖

We may now give the characterization in the discrete case.

1.2.6 Herglotz's Theorem

A complex-valued function K on the integers is the covariance function of a stationary L^2 process if and only if there is a finite measure μ on $\mathscr{B}[-\pi, \pi]$ such that

$$K(n) = \int_{-\pi}^{\pi} e^{inu} \, d\mu(u) \qquad \text{for all integers } n.$$

PROOF If K is a covariance function it is nonnegative definite, so that for all $N \geq 1$ and $x \in R$,

$$G_N(x) = \frac{1}{2\pi N} \sum_{j,k=1}^{N} e^{-ijx} \overline{e^{-ikx}} K(j - k) \geq 0.$$

Since the number of values of (j, k) in the sum such that $j - k = m$ is $N - |m|$, we get

$$G_N(x) = \frac{1}{2\pi N} \sum_{|m| < N} (N - |m|) e^{-imx} K(m) \geq 0.$$

If μ_N is the measure on $\mathscr{B}[-\pi, \pi]$ with density G_N, then

$$\int_{-\pi}^{\pi} e^{inx} \, d\mu_N(x) = \frac{1}{2\pi N} \sum_{|m| < N} (N - |m|) K(m) \int_{-\pi}^{\pi} e^{i(n-m)x} \, dx$$

$$= \begin{cases} (1 - |n|/N) K(n) & \text{if } |n| < N \\ 0 & \text{otherwise.} \end{cases} \tag{1}$$

Since the measures μ_N are concentrated on a bounded interval, and $\mu_N[-\pi, \pi] = K(0)$, the family $\{\mu_N\}$ is tight, so that there is a subsequence $\{\mu_{N_k}\}$ converging weakly to a measure μ on $\mathscr{B}[-\pi, \pi]$ (see RAP, 8.2.4, p. 330). Then (RAP, 4.5.1, p. 196), letting N approach infinity in (1) along the sequence $N_1, N_2, \ldots$, we obtain

$$\int_{-\pi}^{\pi} e^{inx} \, d\mu(x) = K(n).$$

Conversely, if $K(n) = \int_{-\pi}^{\pi} e^{inu} \, d\mu(u)$, then

$$\sum_{j, k=1}^{n} z_j \bar{z}_k K(n_j - n_k) = \int_{-\pi}^{\pi} \left| \sum_{j=1}^{n} z_j \exp[in_j u] \right|^2 d\mu(u) \geq 0.$$

Thus K is nonnegative definite; by 1.2.4(d), K is the covariance function of stationary L^2 process. ‖

Herglotz's theorem will now be used in the analysis of the continuous parameter case.

1.2.7 Bochner's Theorem

A complex-valued function K on R that is continuous at the origin is the covariance function of a stationary L^2 process if and only if there is a finite measure μ on $\mathscr{B}(R)$ such that

$$K(t) = \int_R e^{itu} \, d\mu(u) \qquad \text{for all } t \in R.$$

PROOF If K is a covariance function, it is nonnegative definite. Thus for each n, the function $K(\cdot/2^n)$ is nonnegative definite on the integers; so by 1.2.6, there is a finite measure μ_n on $\mathscr{B}[-\pi, \pi]$ such that for each integer k,

$$K(k/2^n) = \int_{-\pi}^{\pi} e^{ikx} \, d\mu_n(x).$$

Define

$$f_n(u) = \int_{-\pi}^{\pi} \exp[i2^n ux] \, d\mu_n(x), \qquad u \in R.$$

Then f_n is the characteristic function of a measure concentrated on $[-2^n\pi, 2^n\pi]$. (An easy way to see this is to note that if f is the characteristic function of a random variable X and $g(u) = f(2^n u)$, then $g(u) = E[\exp(iu2^n X)]$, so that g is the characteristic function of $2^n X$.)

In particular, f_n is nonnegative definite and, since $k2^{-n} = k2^{m-n}2^{-m}$,

$$f_m(k2^{-n}) = K(k2^{-n}) \qquad \text{for } m \geq n \text{ and all integers } k. \qquad (1)$$

We are going to show that $\{f_m\}$ converges pointwise to K; the key step is to show that the f_m form a uniformly equicontinuous family on R. Assuming this for the moment, the Arzela–Ascoli theorem (RAP, A8.5, pp. 396–397) guarantees the existence of a subsequence $\{f_{n_i}\}$ converging pointwise to a continuous limit f. (Use the theorem on the compact interval $[-r, r]$ for each $r = 1, 2, \ldots$, and construct the f_{n_i} by a diagonalization procedure.) By Lévy's theorem (RAP, 8.2.8, p. 332), f is a characteristic function. But by (1), $f = K$ on the dyadic rationals, and, by 1.2.5(d), K is continuous on R; hence $f = K$ on R.

Thus K is a characteristic function, so that we may write $K(t) = \int_R e^{itu} \, d\mu(u)$ for some finite measure μ. (The converse is proved as in 1.2.6.)

To begin the proof of uniform equicontinuity, we note that for any u, $v \in R$ we may write $u - v = (k + \theta)2^{-m}$ where k is an integer, $|k2^{-m}| \leq |u - v|$ and $|\theta| \leq 1$. Then by 1.2.5(d) and the triangle inequality,

$$|f_m(u) - f_m(v)|^2 \leq 2f_m(0)[f_m(0) - \mathrm{Re}\, f_m(u - v)]$$

$$\leq 2f_m(0)\,|f_m(0) - \mathrm{Re}\, f_m(k2^{-m})|$$

$$+ 2f_m(0)\,|\mathrm{Re}\, f_m(k2^{-m}) - \mathrm{Re}\, f_m(u - v)|. \qquad (2)$$

By (1), the first term of (2) equals $2K(0)\,|K(0) - \mathrm{Re}\, K(k2^{-m})|$; since K is continuous at 0 and $|k2^{-m}| \leq |u - v|$, this term can be made less than $\varepsilon^2/2$ (ε a given positive number) provided $|u - v|$ is small enough, say $|u - v| < \delta$. The square of the second term is at most

$$4K^2(0)\,|f_m(k2^{-m}) - f_m(u - v)|^2$$

$$\leq 8K^3(0)[f_m(0) - \mathrm{Re}\, f_m(u - v - k2^{-m})] \qquad \text{by 1.2.5(d)}$$

$$= 8K^3(0)[f_m(0) - \mathrm{Re}\, f_m(\theta 2^{-m})]$$

$$= 8K^3(0)\int_{-\pi}^{\pi} [1 - \cos\theta x]\, d\mu_m(x) \qquad \text{by definition of } f_m$$

$$\leq 8K^3(0)\int_{-\pi}^{\pi} [1 - \cos x]\, d\mu_m(x) \qquad \begin{array}{l}\text{since } \cos\theta x \geq \cos x \text{ for}\\ |\theta| \leq 1 \text{ and } |x| \leq \pi\\ \text{(draw a picture)}\end{array}$$

$$= 8K^3(0)[f_m(0) - \mathrm{Re}\, f_m(2^{-m})] \qquad \text{by definition of } f_m$$

$$= 8K^3(0)[K(0) - \mathrm{Re}\, K(2^{-m})] \qquad \text{by (1)}.$$

Since K is continuous at 0, it follows that the second term of (2) will be less than $\varepsilon^2/2$ for large m, say $m \geq M$. Thus if $m \geq M$ and $|u - v| < \delta$, we have $|f_m(u) - f_m(v)| < \varepsilon$. By 1.2.5(d), each $f_j, j < M$, is uniformly continuous on R. Consequently, the f_m form a uniformly equicontinuous family.   ∥

1.2.8   Comments

To summarize: We have now established that if K is a complex-valued function on T, the following conditions are equivalent:

(a) K is nonnegative definite (hence symmetric by 1.2.5(b)) and continuous at 0 (this is automatically satisfied when T is the set of integers).

(b) K is the covariance function of a stationary L^2 process, and K is continuous at 0.

(c) K is the characteristic function of a finite measure, defined on $\mathscr{B}[-\pi, \pi]$ in the discrete case, and $\mathscr{B}(R)$ in the continuous case.

The equivalence of (a) and (b) was proved in 1.2.4(d); (b) and (c) are equivalent by 1.2.6 and 1.2.7.

Note also that in (c), the measure μ is determined by K. In the continuous case, this was proved in (RAP, 8.1.3, p. 323); the discrete case result can then be derived by extending the given measures to $\mathscr{B}(R)$ by defining $\mu([-\pi, \pi]^c) = 0$.

We now consider Brownian motion, one of the most important L^2 processes.

1.2.9   Brownian Motion

Let $\{B(t), t \geq 0\}$ be a Gaussian process with $E[B(t)] = 0$ for all t, and

$$K(s, t) = \sigma^2 \min (s, t)$$

where $\sigma^2 > 0$.

First, we verify that K is a realizable covariance function. The following auxiliary result is helpful; if $\{X(t)\}$ is any L^2 process with independent increments, then for $s \leq t$,

$$\text{Cov } [X(s), X(t)] = E[\,|X(s) - m(s)|^2]$$
$$(= \text{Var } X(s) \text{ if the } X(s) \text{ are real-valued})$$

where $m(s) = E[X(s)]$.

To prove this, write

$$[X(s) - m(s)][\overline{X(t) - m(t)}] = [X(s) - m(s)][\overline{X(s) - m(s)}$$
$$+ \{\overline{X(t) - m(t)} - \overline{(X(s) - m(s))}\}],$$

and use the independence of increments.

In particular, a Poisson process with parameter λ has covariance function $K(s, t) = \operatorname{Var} X(s) = \lambda s$, $s \leq t$, so that $K(s, t) = \lambda \min(s, t)$. It follows that the covariance function of Brownian motion is realizable (this may also be verified directly; see Problem 1). This illustrates the point that the covariance functions of two processes may be the same and yet the probability distributions may be quite different.

Note that $E[B^2(0)] = K(0, 0) = 0$, hence $B(0) = 0$ a.e. (This does not contradict the fact that the process is Gaussian; recall our convention of 1.2.2(a) that a constant random variable is regarded as normal with variance 0.)

Now if $0 \leq t_1 < t_2 \leq t_3 < t_4$,

$$E[(B(t_2) - B(t_1))(B(t_4) - B(t_3))] = t_2 - t_2 - t_1 + t_1 = 0.$$

Thus if $0 \leq t_1 < t_2 \leq t_3 < t_4 \leq \cdots \leq t_{2n-1} < t_{2n}$, the random variables $B(t_2) - B(t_1)$, $B(t_4) - B(t_3)$, $\ldots$, $B(t_{2n}) - B(t_{2n-1})$ are uncorrelated. But the random variables are jointly Gaussian by 1.2.2(d), hence they are independent by 1.2.2(g). Therefore the process has independent increments. Furthermore, $B(t + h) - B(t)$ is normally distributed with mean 0 and variance

$$E[(B(t + h) - B(t))^2] = K(t + h, t + h) - 2K(t, t + h) + K(t, t)$$

$$= \sigma^2(t + h) - 2\sigma^2 t + \sigma^2 t$$

$$= \sigma^2 h.$$

Thus the distribution of $B(t + h) - B(t)$ does not depend on t, that is, the process has *stationary increments*.

The process $\{B(t), t \geq 0\}$ is called *Brownian motion*; it has been used as a model for the movement of a microscopic particle undergoing molecular bombardment in a liquid. Some insight as to why the model is appropriate is gained by considering a random walk with a very large number of steps and a very small step size. Such a walk is a reasonable approximation for the movement of a particle undergoing random bombardment, and we can justify, at least intuitively, that the random walk approximates Brownian motion.

Suppose that the particle starts at the origin and jumps every Δt seconds, moving to the right a distance Δx with probability $\frac{1}{2}$, or to the left a distance Δx with probability $\frac{1}{2}$ (we consider only one-dimensional motion). If $X_n(t)$ is the position of the particle at time $t = n \Delta t$, then $X_n(t)$ is the sum of independent random variables $Y_1, \ldots, Y_n$, where $P\{Y_i = \Delta x\} = P\{Y_i = -\Delta x\} = \frac{1}{2}$ for all i. Now $\operatorname{Var} X_n(t) = n(\Delta x)^2$, and hence

$$X_n(t) = \frac{(Y_1 + \cdots + Y_n)}{\sqrt{n}\, \Delta x} \sqrt{n}\, \Delta x = Z_n \sqrt{n}\, \Delta x$$

where Z_n has mean 0 and variance 1. We may write $\sqrt{n}\,\Delta x = \sqrt{t}\,\Delta x/\sqrt{\Delta t}$, so if we assume that

$$\frac{(\Delta x)^2}{\Delta t} \to \sigma^2 > 0 \qquad \text{as} \quad \Delta t \to 0,$$

the central limit theorem implies that $X_n(t)$ converges in distribution to a random variable that is normal with mean 0 and variance $\sigma^2 t$, in other words, $X_n(t) \xrightarrow{\mathrm{d}} B(t)$.

If $0 \le t_1 < \cdots < t_k$, similar results may be obtained for the convergence in distribution of the random vector $(X_n(t_1), \ldots, X_n(t_k))$ to $(B(t_1), \ldots, B(t_k))$; see Problem 9.

Throughout the book, the notation $B(t)$ will always indicate that the Brownian motion process is being considered.

The analysis of 1.2.9 shows that another way of defining Brownian motion is to require that $\{B(t),\, t \ge 0\}$ be a process with independent increments such that $B(0) \equiv 0$ and for $s,\, t \ge 0$, $B(s) - B(t)$ is normally distributed with mean 0 and variance $\sigma^2 \left| s - t \right|$.

Problems

1. Show directly that $\min(s, t)$, $s, t \ge 0$, is nonnegative definite.

2. In the uniform renewal process, let $V_1, V_2, \ldots$ be a sequence of independent, identically distributed random variables with 0 mean and variance σ^2; assume also that $(T_1, T_2, \ldots)$ and $(V_1, V_2, \ldots)$ are independent. Define a stochastic process $\{X(t),\, t \ge 0\}$ as follows: If $0 \le t < T_1$ let $X(t) = V_1$; if

$$T_1 + \cdots + T_n \le t < T_1 + \cdots + T_{n+1}$$

let $X(t) = V_{n+1}$ ($n = 1, 2, \ldots$) (see Figure P1.1). Show that $\{X(t)\}$ is stationary with covariance function $K(t) = \sigma^2(1 - F_1(\left| t \right|))$.

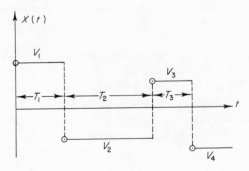

FIGURE P1.1

3. Let $X_1, \ldots, X_n$ be random variables. What conditions on the joint characteristic function of $X_1, \ldots, X_n$ are equivalent to the independence of the X_i?

4. Let (X_1, X_2) be Gaussian with zero mean and covariance matrix

$$K = \begin{bmatrix} \sigma^2 & \sigma^2 \\ \sigma^2 & \sigma^2 \end{bmatrix}.$$

Express $\binom{X_1}{X_2}$ as $W\binom{Y_1}{Y_2}$ where W is an orthogonal 2 by 2 matrix and Y_1, Y_2 are independent normal random variables.

5. (a) Give an example of Gaussian random variables X and Y such that (X, Y) is not Gaussian and $X + Y$ is not Gaussian.

(b) Give an example of Gaussian random variables X and Y such that X and Y are uncorrelated but not independent. (Note that (X, Y) cannot be Gaussian.)

6. Give an example of two complex jointly Gaussian random variables that are uncorrelated but not independent.

7. (a) If $K_1(s, t)$ and $K_2(s, t)$ are covariances (symmetric, nonnegative definite complex-valued functions defined for $s, t \in T$), show that the following are also covariances:

    (i) $K_1(s, t) + K_2(s, t)$
    (ii) Re $K_1(s, t)$
    (iii) $K_1(s, t)K_2(s, t)$.

Is Im $K_1(s, t)$ a covariance?

(b) Show that a pointwise limit of a sequence of a sequence of covariances is a covariance.

8. Prove the converse of property 1.2.2(f): If $a_1 X_1 + \cdots + a_n X_n$ is Gaussian for all real $a_1, \ldots, a_n$, show that $X_1, \ldots, X_n$ are jointly Gaussian.

9. Consider the random walk $\{X_n(t)\}$ of 1.2.9, and let $0 \le t_1 < \cdots < t_k$. Assume that each t_j is an integral multiple of the time interval Δt. If $(\Delta x)^2/\Delta t \to \sigma^2 > 0$ as $\Delta t \to 0$, show that $(X_n(t_1), \ldots, X_n(t_k))$ converges in distribution as $n \to \infty$ to $(B(t_1), \ldots, B(t_k))$.

10. (*The Compound Poisson Process*) Let $\{M(t), t \ge 0\}$ be a stochastic process such that each $M(t)$ takes values on the nonnegative integers. Let $Y_1, Y_2, \ldots$ be iid (independent, identically distributed) random variables, and assume that $(Y_1, Y_2, \ldots)$ and $(M(t), t \ge 0)$ are independent. Define

$$X(t) = \sum_{j=1}^{M(t)} Y_j, \qquad t \ge 0$$

(take $X(t) = 0$ if $M(t) = 0$). Intuitively, we may think of $M(t)$ as the number of customers who have arrived on or before time t. If customer j makes a purchase costing Y_j dollars ($Y_j < 0$ corresponds to a refund), then $X(t)$ is the total amount of money received up to time t. If $\{M(t), t \geq 0\}$ is a Poisson process, $\{X(t), t \geq 0\}$ is called a *compound* Poisson process.

(a) If $M(t)$ and the Y_i have finite mean, show that

$$E[X(t)] = E[M(t)]E(Y_1).$$

(b) If $M(t)$ and the Y_i are in L^2, show that

$$\text{Var } [X(t)] = E[M(t)] \text{ Var } Y_1 + [E(Y_1)]^2 \text{ Var } M(t).$$

(c) Compute the covariance function of the compound Poisson process.

11. (*The Nonuniform Poisson Process*) Let $\{N(t), t \geq 0\}$ be a stochastic process such that each $N(t)$ is nonnegative integer valued, and:

1. $\{N(t), t \geq 0\}$ has independent increments;
2. $P\{N(t + h) - N(t) = 1\} = \lambda(t)h + o(h)$ as $h \downarrow 0$;
   $P\{N(t + h) - N(t) = 0\} = 1 - \lambda(t)h + o(h)$ as $h \downarrow 0$;
   $P\{N(t + h) - N(t) < 0\} = 0$ for $h > 0$;
3. $P\{N(0) = 0\} = 1$;

where λ is a nonnegative continuous function on $[0, \infty)$.

(a) If $p_n(t) = P\{N(t) = n\}$, assumed to be a differentiable function of t, show that

$$\frac{dp_n}{dt} = \lambda(t)[p_{n-1}(t) - p_n(t)], \qquad n \geq 1,$$

$$\frac{dp_0}{dt} = -\lambda(t)p_0(t).$$

If p_{n-1} is known, the differential equation for p_n is first order linear, and hence may be solved explicitly. The result is

$$p_n(t) = e^{-L(t)}[L(t)]^n/n!$$

where $L(t) = \int_0^t \lambda(x) \, dx$. Thus $N(t)$ has the Poisson distribution with parameter $L(t)$; we may interpret $\lambda(t)$ as the arrival rate at time t; if $\lambda(t)$ is constant, we obtain the ordinary Poisson process.

(b) Show that if $h > 0$, $N(t + h) - N(t)$ has the Poisson distribution with parameter $\int_t^{t+h} \lambda(x) \, dx$.

(c) Find the covariance function of the process.

(d) Assume $\lambda > 0$ everywhere. Show that there is a strictly increasing continuous function f on some interval $[0, a)$, $0 < a \leq \infty$, such that

$\{N(f(t)), \ 0 \leq t < a\}$ is a Poisson process with parameter $\lambda = 1$. Thus by changing the time scale we may change $\{N(t)\}$ into a Poisson process with uniform arrival rate.

(e)   Use the Kolmogorov extension theorem to show that there actually exists a process satisfying (1), (2), and (3).

12.   (*Shot Noise*)   Let $h\colon R \to R$, with $h(t) = 0$ for $t < 0$; assume that h is bounded and has at most countably many discontinuities. Define

$$X(t) = \sum_{n=1}^{\infty} h(t - A_n)$$

where the A_n are the arrival times of a nonuniform Poisson process $\{N(t)\}$ (Problem 11) with associated function λ.

(The careful reader may observe that "arrival times" have not been defined for the nonuniform Poisson process. One approach is to construct the process from basic random variables T_1, T_2, ..., as in 1.1.5, and let $A_n = T_1 + \cdots + T_n$. However, the T_n are no longer independent, and the details become quite tedious. Another approach is to show that a process exists having the specified finite-dimensional distributions, with the property that the sample functions are step functions with jumps of height 1 at times $A_1, A_2, \ldots$. This approach involves the concept of separability which will not be discussed until Chapter 4. Let us instead take the easy way out; assume that $\lambda > 0$ everywhere, and apply Problem 11(d). If $Y_1, Y_2, \ldots$ are the arrival times for the Poisson process $\{N(f(t))\}$, set $A_n = f(Y_n)$. (Note that since $h(t)$ vanishes for $t < 0$, the sum defining $X(t)$ is a.e. finite.) Physically, we can regard $h(t - A_n)$ as the current at time t due to the arrival at time A_n of an electron at the anode of a vacuum tube.)

(a)   Show that the joint characteristic function of $X(t_1)$, ..., $X(t_n)$, $0 \leq t_1 < \cdots < t_n$, is

$$\exp\left[\int_0^{\infty} \lambda(x)\left(\exp\left[i\sum_{j=1}^{n} u_j h(t_j - x)\right] - 1\right) dx\right]$$

(the upper limit of integration can be replaced by t_n since $h(t) = 0$, $t < 0$).

(b)   Show that $\{X(t), t \geq 0\}$ is an L^2 process.

(c)   By (b), we may find the mean and covariance function of the process by differentiating the joint characteristic function [as in RAP, 8.1.5(e), p. 325]. Carry out the computation to show that

$$E[X(t)] = \int_0^{\infty} \lambda(x) h(t - x) \, dx,$$

$$\text{Cov}\,[X(t_1), X(t_2)] = \int_0^{\infty} \lambda(x) h(t_1 - x) h(t_2 - x) \, dx.$$

(d) If $\lambda(t) = k\lambda_0(t)$ and we denote the shot noise process by $\{X_k(t)\}$, show that as $k \to \infty$, $Y_k(t) = k^{-1/2}[X_k(t) - EK_k(t)]$ tends to a Gaussian process $\{Y(t)\}$ with 0 mean and covariance function

$$K(s,\, t) = \int_0^\infty \lambda_0(x)h(s - x)h(t - x)\, dx,$$

in the sense that if $0 \leq t_1 < \cdots < t_n$, $(Y_k(t_1), \ldots, Y_k(t_n))$ converges in distribution to $(Y(t_1), \ldots, Y(t_n))$.

13. (*Alternative Proof of Bochner's Theorem*) Let K be the covariance function of a stationary L^2 process; assume K continuous at 0, and hence continuous everywhere by 1.2.5(d).

(a) Define $G_T(x) = T^{-1} \int_0^T \int_0^T e^{-iux} \overline{e^{-ivx}} K(u - v)\, du\, dv$; note that $G_T \geq 0$ since the approximating Riemann sums are ≥ 0 by nonnegative definiteness of K. Show that

$$G_T(x) = \int_R e^{-itx} g_T(t)\, dt,$$

where

$$g_T(t) = \left(1 - \frac{|t|}{T}\right) K(t) I_{[-T,\, T]}(t).$$

(b) Show that

$$\frac{1}{2\pi} \int_{-M}^M e^{iux}\left(1 - \frac{|x|}{M}\right) G_T(x)\, dx = \frac{1}{2\pi} \int_R f_M(u - t) g_T(t)\, dt,$$

where

$$f_M(x) = \frac{\sin^2 \tfrac{1}{2}Mx}{M(\tfrac{1}{2}x)^2}.$$

(c) Show that $(2\pi)^{-1} \int_R f_M(u - t) g_T(t)\, dt \to g_T(u)$ as $M \to \infty$.

(d) Show that K is the characteristic function of a finite measure μ, proving Bochner's theorem.

1.3   Second Order Calculus

In this section, $\{X(t),\, t \in T\}$ will be an L^2 process, with T an interval of reals. We shall try to develop a theory in which it is possible to talk about continuity, differentiation, and integration of the process. Since knowledge of the covariance function does not directly reveal any properties of the sample functions $(X(t, \omega), t \in T)$, we develop these concepts in the L^2 sense.

1.3.1 Lemma

If $Y_n \xrightarrow{L^2} Y$ and $Z_m \xrightarrow{L^2} Z$ (with all random variables in L^2), then $E(Y_n \bar{Z}_m) \to E(Y\bar{Z})$ as $n, m \to \infty$.

PROOF This is simply the statement that the inner product is continuous in both variables (see RAP, 3.2.3, p. 117). ∥

The next result indicates how the existence of an L^2 limit can be inferred from the existence of limits of sequences of complex numbers.

1.3.2 Theorem

Let $\{Y(s), s \in T\}$ be an L^2 process, and let $s_0 \in T$. There exists $Y \in L^2$ such that $Y(s) \xrightarrow{L^2} Y$ as $s \to s_0$ iff there is a complex number L such that for all sequences $s_n \to s_0$, $s_m' \to s_0$, we have

$$E[Y(s_n)\overline{Y(s_m')}] \to L \qquad \text{as} \quad n, m \to \infty.$$

PROOF If $Y(s) \xrightarrow{L^2} Y$, then by 1.3.1, $E[Y(s_n)\overline{Y(s_m')}] \to E(|Y|^2) = L$. Conversely, suppose there is an L with the indicated properties. Pick any sequence $s_n \to s_0$. Then

$$E[|Y(s_n) - Y(s_m)|^2] = E[(Y(s_n) - Y(s_m))(\overline{Y(s_n) - Y(s_m)})] \to 0$$

by hypothesis, so by completeness of L^2, $Y(s_n)$ converges in L^2 to a limit Y. If we take another sequence $t_n \to s_0$, then

$$\|Y(t_n) - Y\| \le \|Y(t_n) - Y(s_n)\| + \|Y(s_n) - Y\|$$

and

$$E[|Y(t_n) - Y(s_n)|^2] = E[Y(t_n)\overline{Y(t_n)}] - E[Y(t_n)\overline{Y(s_n)}]$$
$$- E[Y(s_n)\overline{Y(t_n)}] + E[Y(s_n)\overline{Y(s_n)}]$$
$$\to L - L - L + L = 0.$$

Thus $Y(t_n) \xrightarrow{L^2} Y$, and the result follows. ∥

As usual, K will denote the covariance function of the L^2 process $\{X(t), t \in T\}$, and m the mean value function, defined by $m(t) = E[X(t)], t \in T$.

1.3.3   Definition

The process is said to be L^2-*continuous* at the point $t \in T$ iff $X(t + h) \xrightarrow{L^2} X(t)$ as $h \to 0$, L^2-*differentiable* at $t \in T$ iff $(X(t + h) - X(t))/h$ converges to a limit $X'(t)$ as $h \to 0$.

We have the following criterion for L^2-continuity in terms of the covariance function.

1.3.4   Theorem

Assume m continuous on T. Then the process is L^2-continuous at r iff K is continuous at (r, r).

PROOF   Since $\{X(t),\ t \in T\}$ is L^2 continuous iff $\{X(t) - m(t),\ t \in T\}$ is L^2-continuous, and $\{X(t) - m(t),\ t \in T\}$ also has covariance function K, we may assume that $m \equiv 0$. If the process is L^2-continuous at r, then

$$X(r + h) \xrightarrow{L^2} X(r),\ X(r + h') \xrightarrow{L^2} X(r) \qquad \text{as}\quad h, h' \to 0,$$

hence by 1.3.1, $K(r + h, r + h') \to K(r, r)$. Thus K is continuous at (r, r).

Conversely, if K is continuous at (r, r), then

$$E[\,|X(r + h) - X(r)|^2\,] = E\{[X(r + h) - X(r)][\overline{X(r + h) - X(r)}]\}$$
$$= K(r + h, r + h) - K(r, r + h) - \overline{K(r, r + h)} + K(r, r) \to 0. \quad \|$$

1.3.5   Corollary

If K is continuous at (t, t) for all $t \in T$, then K is continuous at (s, t) for all $s, t \in T$.

PROOF   Again assume $m \equiv 0$. By 1.3.4, $X(s + h) \xrightarrow{L^2} X(s)$ and $X(t + h') \xrightarrow{L^2} X(t)$ as $h, h' \to 0$. By 1.3.1, $K(s + h, t + h') \to K(s, t)$. $\quad \|$

There are corresponding results in the stationary case, as follows. (As above, we assume $m \equiv 0$ without loss of generality.)

1.3.6   Theorem

Let $\{X(t),\ t \in T\}$ be a stationary L^2 process with covariance $K = K(t)$, $t \in I = \{u - v : u, v \in T\}$.

(a)   If the process is L^2-continuous at a particular point s, then K is continuous at the origin.

(b)   If K is continuous at the origin, it is continuous everywhere, and the process is L^2-continuous for all t.

PROOF   (a)   We have $X(s + t) \overset{L^2}{\to} X(s)$, $X(s) \overset{L^2}{\to} X(s)$ as $t \to 0$, hence by 1.3.1, $K(t) \to K(0)$.

(b)   Since $E[\,|X(t + h) - X(t)|^2] = K(0) - K(h) - \overline{K(h)} + K(0) \to 0$ as $h \to 0$, the process is L^2-continuous for all t. Thus

$$X(s + t + h) \overset{L^2}{\to} X(s + t) \qquad \text{as} \quad h \to 0 \qquad (\text{and } X(s) \overset{L^2}{\to} X(s)),$$

so by 1.3.1, $K(t + h) \to K(t)$. $\|$

We now relate L^2-differentiability and differentiability of the covariance function in the stationary case.

1.3.7   Theorem

Let $\{X(t), t \in T\}$ be a stationary L^2 process with covariance function $K = K(t)$. If the process is L^2-differentiable at all points $t \in T$, then K is twice differentiable for all t, and $\{X'(t), t \in T\}$ is a stationary L^2 process with covariance function $-K''(t)$.

PROOF   Since

$$X(s + t) \overset{L^2}{\to} X(s + t), \qquad \frac{X(s + h) - X(s)}{h} \overset{L^2}{\to} X'(s),$$

it follows from 1.3.1 that

$$\frac{K(t - h) - K(t)}{h} \to E[X(s + t)\overline{X'(s)}].$$

Thus K is differentiable for all t and

$$-K'(t) = E[X(s + t)\overline{X'(s)}]. \tag{1}$$

But

$$\frac{X(s + t + h') - X(s + t)}{h'} \overset{L^2}{\to} X'(s + t), \qquad X'(s) \overset{L^2}{\to} X'(s),$$

hence by 1.3.1 and (1),

$$\frac{-K'(t + h') + K'(t)}{h'} \to E[X'(s + t)\overline{X'(s)}].$$

Therefore $K''(t)$ exists for all t and equals $-E[X'(s + t)\overline{X'(s)}]$. $\|$

For extensions of this result, see Problems 5 and 6.
We now consider integration in the L^2 sense.

1.3.8   Definition of the L^2-integral

Let $\{X(t), a \le t \le b\}$ be an L^2 process $(a, b$ finite$)$ with covariance function K and mean value function m, and let g be a complex-valued function on $[a, b]$. We define $\int_a^b g(t)X(t)\, dt$ as follows:

Let $\Delta: a = t_0 < t_1 < \cdots < t_n = b$ be a partition of $[a, b]$, with $|\Delta| = \max_{1 \le i \le n} |t_i - t_{i-1}|$. Define

$$I(\Delta) = \sum_{k=1}^{n} g(t_k)X(t_k)(t_k - t_{k-1}).$$

If $I(\Delta)$ converges in L^2 to some random variable I as $|\Delta| \to 0$, then we say that $g(t)X(t)$ is L^2-integrable on $[a, b]$ and we write

$$I = \int_a^b g(t)X(t)\, dt.$$

We have the following basic sufficient condition for L^2-integrability.

1.3.9   Theorem

If m and g are continuous on $[a, b]$ and K is continuous on $[a, b] \times [a, b]$, then $g(t)X(t)$ is L^2-integrable on $[a, b]$.

PROOF   We may assume $m \equiv 0$. Let $\Delta: a = s_0 < s_1 < \cdots < s_m = b$, $\Delta':$ $a = t_0 < t_1 < \cdots < t_n = b$. Then

$$I(\Delta)\overline{I(\Delta')} = \sum_{j=1}^{m} \sum_{k=1}^{n} g(s_j)\overline{g(t_k)}X(s_j)\overline{X(t_k)}(s_j - s_{j-1})(t_k - t_{k-1});$$

hence

$$E[I(\Delta)\overline{I(\Delta')}] = \sum_{j,k} g(s_j)\overline{g(t_k)}K(s_j, t_k)(s_j - s_{j-1})(t_k - t_{k-1}),$$

an approximating sum to a Riemann integral. By 1.3.2, $I(\Delta)$ converges in L^2 to a limit I as $|\Delta| \to 0$. $\|$

Note that the hypothesis on g can be weakened to continuity a.e. (Lebesgue measure).

Theorem 1.3.9 is a special case of the following result. If f is a continuous map on $[a, b]$ into a Banach space, then the Riemann integral $\int_a^b f(t)\, dt$ exists. (In 1.3.9, $f(t) = g(t)X(t)$ defines a continuous map of $[a, b]$ into L^2.) This can be proved by imitating one of the standard proofs of the existence of the Riemann integral of a continuous real-valued function. (Use the proof involving Riemann sums rather than one involving upper and lower sums since the latter uses the order properties of the reals.)

The following properties of the integral will often be used.

1.3.10 Theorem

If $m \equiv 0$, g and h are continuous on $[a, b]$, and K is continuous on $[a, b] \times [a, b]$, then

$$E\left[\int_a^b g(s)X(s)\,ds \overline{\int_a^b h(t)X(t)\,dt}\right] = \int_a^b \int_a^b g(s)\overline{h(t)}K(s, t)\,ds\,dt.$$

Also

$$E\left[\int_a^b g(s)X(s)\,ds\right] = E\left[\int_a^b h(t)X(t)\,dt\right] = 0.$$

PROOF   Let

$$I(\Delta) = \sum_{j=1}^m g(s_j)X(s_j)(s_j - s_{j-1}),$$

$$J(\Delta') = \sum_{k=1}^n h(t_k)X(t_k)(t_k - t_{k-1}),$$

$$I = \int_a^b g(s)X(s)\,ds, \qquad J = \int_a^b h(t)X(t)\,dt.$$

By 1.3.9, $I(\Delta) \xrightarrow{L^2} I$, $J(\Delta') \xrightarrow{L^2} J$. By 1.3.1, $E[I(\Delta)\overline{J(\Delta')}] \to E[I\overline{J}]$. But as in 1.3.9,

$$E[I(\Delta)\overline{J(\Delta')}] \to \int_a^b \int_a^b g(s)\overline{h(t)}K(s, t)\,ds\,dt,$$

proving the first statement.

Now $I(\Delta) \xrightarrow{L^2} I$, $1 \xrightarrow{L^2} 1$, hence by 1.3.1, $E[I(\Delta)] \to E(I)$. But $E[I(\Delta)] \equiv 0$, so $E(I) = 0$, and similarly $E(J) = 0$.   ‖

1.3.11 Theorem

If $m \equiv 0$, h is continuous on $[a, b]$, and K is continuous on $[a, b] \times [a, b]$, then

$$E\left[X(s)\overline{\int_a^b h(t)X(t)\,dt}\right] = \int_a^b K(s, t)\overline{h(t)}\,dt.$$

PROOF   Let

$$J(\Delta') = \sum_{k=1}^n h(t_k)X(t_k)(t_k - t_{k-1}), \qquad J = \int_a^b h(t)X(t)\,dt.$$

Then $J(\Delta') \xrightarrow{L^2} J$. But as in 1.3.10, $E[X(s)\overline{J(\Delta')}] \to \int_a^b K(s, t)\overline{h(t)}\,dt$, and the result follows.   ‖

Problems

1. (a) Let $g_k(t)$, $-\infty < t < \infty$, $k = 1, 2, \ldots$, be arbitrary complex-valued functions. Define

$$K(s, t) = \sum_{k=1}^{\infty} g_k(s)\overline{g_k(t)}.$$

Assume that series converges absolutely for all s, t. Show that K is the covariance function of some L^2 process.

 (b) If $K(t) = \int_{-\infty}^{\infty} f(\lambda)e^{it\lambda} \, d\lambda$, where f is a nonnegative Lebesgue integrable function on the real line, show that K is the covariance function of some stationary L^2 process. If f is even, the process can be taken to be real valued.

2. Give an example of a stationary L^2 process that is not L^2-continuous.

3. Show that Brownian motion is L^2-continuous but not L^2-differentiable.

4. If an L^2 process with zero mean and covariance $K(s, t)$ is L^2-differentiable at t_0, show that $\partial^2 K/\partial s \, \partial t$ exists at (t_0, t_0).

5. If $\{X(t)\}$ is stationary and L^2-differentiable at a point, show that it is L^2-differentiable everywhere.

6. If $\{X(t)\}$ is stationary with covariance $K(t)$, show that if $X(t)$ has at least max (n, m) L^2-derivatives, then

$$E[X^{(n)}(s + t)\overline{X^{(m)}(s)}] = (-1)^m K^{(n+m)}(t).$$

7. (a) Let $\{X(t), t \in R\}$ be a stationary L^2 process with a spectral density, that is, assume that the covariance function can be expressed as

$$K(t) = \int_{-\infty}^{\infty} e^{it\lambda}f(\lambda) \, d\lambda$$

for some nonnegative Lebesgue integrable function f (processes of this type will be treated in detail in Chapter 2). Assume f is bounded, and that $E[X(t)] = 0$. If g is a continuous function from R to C that vanishes off $[a, b]$, show that

$$\left\| \int_a^b g(t)X(t) \, dt \right\|_2 \leq [2\pi(\sup |f|)]^{1/2}\|g\|_2$$

where $\|g\|_2^2 = \int_a^b |g(t)|^2 \, dt$. Conclude that the mapping $g \to \int_{-\infty}^{\infty} g(t)X(t) \, dt$ has a unique extension to a bounded linear operator from $L^2(-\infty, \infty)$ to $L^2\{X(t), -\infty < t < \infty\}$.

 (b) Show that 1.3.10 holds for $g, h \in L^2(-\infty, \infty)$.

8. (a)   If $\{X(t)\}$ is L^2-continuous on $[a, b]$, show that in the L^2 sense,

$$\frac{d}{dt} \int_a^t X(s)\, ds = X(t), \qquad a \le t \le b.$$

(b)   If $\{X(t)\}$ is L^2-differentiable on $[a, b]$, show that

$$\int_a^t X'(s)\, ds = X(t) - X(a), \qquad a \le t \le b.$$

1.4   Karhunen–Loève Expansion

Let $\{X(t), \ a \le t \le b\}$, a, b finite, be an L^2 process with zero mean and continuous covariance K. We are going to investigate the possibility of making an orthogonal expansion of $X(t)$:

$$X(t) = \sum_{k=1}^{\infty} Z_k e_k(t), \qquad a \le t \le b,$$

where the series converges in L^2. We require that the Z_k be orthogonal $(E(Z_j \bar{Z}_k) = 0, \ j \ne k)$ random variables in L^2 with zero mean. Furthermore, the functions e_k are required to be orthonormal, that is,

$$\int_a^b e_j(t)\overline{e_k(t)}\, dt = \begin{cases} 0, & j \ne k \\ 1, & j = k. \end{cases}$$

Now if $\sum_{j=1}^n Z_j e_j(s) \overset{L^2}{\to} X(s)$ and $\sum_{k=1}^n Z_k e_k(t) \overset{L^2}{\to} X(t)$, then by 1.3.1,

$$E\left[\sum_{j,\,k=1}^n Z_j \bar{Z}_k e_j(s)\overline{e_k(t)} \right] \to K(s, t),$$

that is,

$$K(s, t) = \sum_{k=1}^{\infty} \lambda_k e_k(s)\overline{e_k(t)}$$

where $\lambda_k = E(|Z_k|^2)$. If term by term integration is permitted, we have

$$\int_a^b K(s, t)e_n(t)\, dt = \lambda_n e_n(s), \qquad a \le s \le b. \tag{1}$$

Thus if an expansion of the above type is possible, it appears that the function e_k must be eigenfunctions of the integral operator associated with the covariance function of the given process, and the variances λ_k of the random variables Z_k must be the eigenvalues of the operator.

Note that if $\lambda_n \neq 0$, then e_n is continuous. (Divide (1) by λ_n and use the dominated convergence theorem.)

Before showing that such an expansion is possible, we must state a few results from Hilbert space theory.

Let K be a continuous covariance, that is, a continuous, symmetric, nonnegative definite function, defined on $[a, b] \times [a, b]$. Let A be the integral operator on $L^2[a, b]$ (the class of all Borel measurable $x: [a, b] \to C$ such that $\int_a^b |x(t)|^2 \, dt < \infty$) associated with K, defined by

$$(Ax)(s) = \int_a^b K(s, t)x(t) \, dt, \qquad a \le s \le b, \quad x \in L^2[a, b].$$

The eigenfunctions of A span $L^2[a, b]$; in other words, the smallest closed subspace of $L^2[a, b]$ containing the eigenfunctions is $L^2[a, b]$ itself. The operator A has at most countably many eigenvalues, all real, with 0 as the only possible limit point. The nonzero eigenvalues are greater than 0 by nonnegative definiteness. The eigenspace corresponding to the eigenvalue $\lambda_n > 0$ is finite dimensional.

Let $\{e_n, n = 1, 2, \ldots\}$ be an orthonormal basis for the space spanned by the eigenvectors corresponding to the *nonzero* eigenvalues. If the basis is taken so that e_n is an eigenvector corresponding to the eigenvalue λ_n, then *Mercer's theorem* states that

$$K(s, t) = \sum_{n=1}^{\infty} \lambda_n e_n(s)\overline{e_n(t)}, \qquad s, t \in [a, b],$$

where the series converges absolutely and also converges uniformly in both variables.

Proofs of the results quoted above may be found in Riesz and Sz.-Nagy (1955). A discussion specifically adapted for our present purposes is in the appendix of Ash (1965), with some minor changes necessary to convert from real to complex scalars.

We may now state the expansion theorem.

1.4.1   Karhunen–Loève Theorem

Let $\{X(t), a \le t \le b\}$, a, b finite, be an L^2 process with zero mean and continuous covariance K. Let $\{e_n, n = 1, 2, \ldots\}$ be an orthonormal basis for the space spanned by the eigenfunctions of the nonzero eigenvalues of the integral operator associated with K, with e_n taken as an eigenvector corresponding to the eigenvalue λ_n. Then

$$X(t) = \sum_{n=1}^{\infty} Z_n e_n(t), \qquad a \le t \le b,$$

where $Z_n = \int_a^b X(t)\overline{e_n(t)}\, dt$, and the Z_n are orthogonal random variables with $E(Z_n) = 0$, $E[\,|Z_n|^2] = \lambda_n$. The series converges in L^2 to $X(t)$, uniformly in t; in other words,

$$E\left[\left\|X(t) - \sum_{k=1}^{n} Z_k e_k(t)\right\|^2\right] \to 0 \qquad \text{as} \quad n \to \infty,$$

uniformly for $t \in [a, b]$.

PROOF By 1.3.9, $\int_a^b X(t)\overline{e_n(t)}\, dt$ defines a random variable Z_n in L^2. By 1.3.10, $E(Z_n) = 0$ and

$$E(Z_j \bar{Z}_k) = \int_a^b \overline{e_j(s)} \int_a^b K(s, t) e_k(t)\, dt\, ds$$

$$= \lambda_k \int_a^b \overline{e_j(s)} e_k(s)\, ds$$

$$= \begin{cases} 0 & \text{if } j \neq k \\ \lambda_k & \text{if } j = k. \end{cases}$$

Let $S_n(t) = \sum_{k=1}^{n} Z_k e_k(t)$. Then

$$E[\,|S_n(t) - X(t)|^2] = E[\,|S_n(t)|^2] - 2\,\mathrm{Re}\,E[X(t)\overline{S_n(t)}]$$

$$+ E[\,|X(t)|^2]$$

$$= \sum_{k=1}^{n} \lambda_k\,|e_k(t)|^2 - 2\,\mathrm{Re}\,\sum_{k=1}^{n} E[X(t)\bar{Z}_k]\overline{e_k(t)}$$

$$+ K(t, t).$$

By 1.3.11, $E[X(t)\bar{Z}_k] = \int_a^b K(t, u)e_k(u)\, du = \lambda_k e_k(t)$. Thus

$$E[\,|S_n(t) - X(t)|^2] = K(t, t) - \sum_{k=1}^{n} \lambda_k\,|e_k(t)|^2 \to 0 \qquad \text{as} \quad n \to \infty,$$

uniformly for $t \in [a, b]$, by Mercer's theorem. $\|$

The Karhunen–Loève expansion assumes a special form when the process is Gaussian. The development is based on the following result.

1.4.2 Theorem

For each $n = 1, 2, \ldots$, let $I_1^n, I_2^n, \ldots, I_r^n$ be complex jointly Gaussian random variables. Assume that $I_j^n \xrightarrow{L^2} I_j$ as $n \to \infty$, $j = 1, 2, \ldots, r$. Then $I_1, \ldots, I_r$ are jointly Gaussian.

PROOF   Since L^2 convergence of complex random variables is equivalent to the L^2 convergence of real and imaginary parts, we may assume all random variables real. The joint characteristic function of $I_1^n, \ldots, I_r^n$ is

$$h_n(u_1, \ldots, u_r) = E\left[\exp\left(i\sum_{j=1}^{r} u_j I_j^n\right)\right]$$

$$= \exp\left[i\sum_{j=1}^{r} u_j b_j^n\right] \exp\left[-\frac{1}{2}\sum_{j,m=1}^{r} u_j \sigma_{jm}^n u_m\right]$$

where $b_j^n = E(I_j^n)$, $\sigma_{jm}^n = \text{Cov}(I_j^n, I_m^n)$. By 1.3.1, $b_j^n \to b_j = E(I_j)$, $\sigma_{jm}^n \to \sigma_{jm} = \text{Cov}(I_j, I_m)$. Thus

(1) $h_n(u_1, \ldots, u_r) \to \exp[iu^t b] \exp[-\frac{1}{2} u^t K u]$, $K = [\sigma_{jm}]$.

But $u_1 I_1^n + \cdots + u_r I_r^n$ converges to $u_1 I_1 + \cdots + u_r I_r$ in L^2, hence in probability, hence in distribution. Therefore

$E[\exp(it(u_1 I_1^n + \cdots + u_r I_r^n))]$

$\to E[\exp(it(u_1 I_1 + \cdots + u_r I_r))]$ for all t,

in particular for $t = 1$. Thus

(2) $h_n(u_1, \ldots, u_r) \to h(u_1, \ldots, u_r)$, the joint characteristic function of $I_1, \ldots, I_r$.

By (1) and (2), $I_1, \ldots, I_r$ are jointly Gaussian.   ‖

1.4.3   Theorem

In the Karhunen–Loève expansion 1.4.1 for a Gaussian process, the random variables Z_k form a Gaussian sequence, that is, for each r, $Z_1, \ldots, Z_r$ are jointly Gaussian. If the random variables of the process are real, the Z_k are independent.

PROOF   Let $I_j(\Delta) = \sum_{m=1}^{n} X(t_m)\overline{e_j(t_m)}(t_m - t_{m-1})$, $j = 1, \ldots, r$, be an approximating sum to $Z_j = \int_a^b X(t)e_j(t)\,dt$. By 1.2.2(d), $I_1(\Delta), \ldots, I_r(\Delta)$ are jointly Gaussian. But $I_j(\Delta) \to Z_j$ as $|\Delta| \to 0, j = 1, \ldots, r$, hence by 1.4.2, $Z_1, \ldots, Z_r$ are jointly Gaussian. In the real case, the Z_k are independent by 1.2.2(g). (Complex jointly Gaussian random variables may be uncorrelated but not independent; see Section 1.2, Problem 6.)   ‖

Thus in the real-valued Gaussian case, the Karhunen–Loève expansion is a series of independent random variables. Since the series converges in L^2, hence in distribution, it follows from Problem 9 of RAP (p. 335) that *for each fixed t*, the series converges with probability 1. Thus (see Problem 1) there is a set N of probability 0 such that for every $\omega \notin N$, $\sum_{n=1}^{\infty} Z_n(\omega)e_n(t)$ converges to $X(t, \omega)$ for *almost* every t (Lebesgue measure). Note that if N_t is the set of all $\omega \in \Omega$ such that the series does not converge to $X(t, \omega)$, then the set on which (pointwise) convergence fails for at least one $t \in [a, b]$ is $\bigcup_{a \le t \le b} N_t$, an *uncountable* union of sets of probability 0, and therefore not necessarily of probability 0. Thus we cannot conclude that for ω outside a set of probability 0, the series converges to $X(t, \omega)$ for all $t \in [a, b]$.

1.4.4 Example

Let $K(s, t) = \min (s, t)$, $s, t \in [0, 1]$. (If in addition we take the process to be Gaussian, we obtain Brownian motion restricted to $[0, 1]$.)

To find the eigenvalues of the integral operator associated with the covariance function, we must solve the integral equation

$$\int_0^1 \min (s, t)e(t) \, dt = \lambda e(s), \qquad 0 \le s \le 1,$$

that is,

(1) $\int_0^s te(t) \, dt + s \int_s^1 e(t) \, dt = \lambda e(s), 0 \le s \le 1$.

If $\lambda \ne 0$, then e is continuous, and we may differentiate with respect to s to obtain

(2) $\int_s^1 e(t) \, dt = \lambda e'(s)$.

Differentiate again to obtain

(3) $-e(s) = \lambda e''(s)$.

If $\lambda = 0$, the above development yields $e(s) = 0$ a.e., so 0 is not an eigenvalue.

The solution of the differential equation (3) is

(4) $e(s) = A \sin \dfrac{s}{\sqrt{\lambda}} + B \cos \dfrac{s}{\sqrt{\lambda}}$.

Set $s = 0$ in (1) to obtain $e(0) = 0$; thus $B = 0$ in (4). Set $s = 1$ in (2) to obtain $e'(1) = 0$. Thus $\cos 1/\sqrt{\lambda} = 0$, or $1/\sqrt{\lambda} = (2n - 1)\pi/2$, $n = 1, 2, \ldots$. Thus the eigenvalues are $\lambda_n = 4/(2n - 1)^2\pi^2$, and the orthonormalized eigenfunctions are $e_n(t) = \sqrt{2} \sin (2n - 1)\frac{1}{2}\pi t$, $n = 1, 2, \ldots$.

Finally, if we define $Z_n^* = Z_n/\sqrt{\lambda_n}$, where the Z_n are as given in the Karhunen–Loève expansion, we obtain

$$X(t) = \sqrt{2} \sum_{n=1}^{\infty} Z_n^* \frac{\sin (n - \frac{1}{2})\pi t}{(n - \frac{1}{2})\pi}$$

where the Z_n^* are orthogonal random variables with mean 0 and variance 1.

In the Gaussian case, the Z_n^* are independent, hence for each t the series converges a.e. In fact it can be shown (see Problem 5) that there is a set N of probability 0 such that if $\omega \notin N$,

$$\sqrt{2} \sum_{k=1}^{2^n} \frac{Z_k^*(\omega)}{(k - \frac{1}{2})\pi} \sin \left(k - \frac{1}{2} \right) \pi t$$

converges as $n \to \infty$, say to $Y(t, \omega)$, *uniformly* for $t \in [0, 1]$. Therefore if $\omega \notin N$, $Y(t, \omega)$ is continuous in t, and if we define $Y(t, \omega) = 0$ for all t if $\omega \notin N$, then $Y(t, \omega)$ is continuous in t for all ω.

Now for each t, $X(t, \omega) = Y(t, \omega)$ for almost every ω, and in this sense, $\{X(t), 0 \leq t \leq 1\}$ is equivalent to a process $\{Y(t), 0 \leq t \leq 1\}$ with continuous sample functions. In particular, the two processes have the same finite-dimensional distributions, and hence the same covariance function.

Problems

1. In the Karhunen–Loève expansion of a real-valued Gaussian process, show that there is a set N of probability 0 such that for $\omega \notin N$, the series converges to $X(t, \omega)$ for almost every t (Lebesgue measure).

2. Use the calculations in 1.4.4 to obtain the Karhunen–Loève expansion on the interval $[0, a]$ of the L^2 process with covariance $\sigma^2 \min (s, t)$.

3. If λ_n, $n = 1, 2, \ldots$ are the nonzero eigenvalues of the integral equation $\int_a^b K(s, t)e(t)\, dt = \lambda e(s)$, $a \leq s \leq b$, where K is the continuous covariance function of an L^2 process, show that

$$\sum_{n=1}^{\infty} \lambda_n = \int_a^b K(t, t)\, dt$$

$$= K(0)(b - a) \qquad \text{in the stationary case}$$

(each λ_n is repeated as many times as its multiplicity).

4. (a) Find the Karhunen–Loève expansion on the interval $[0, 1]$ of an L^2 process with covariance function $K(s, t) = st$.

   (b) Show that zero is an eigenvalue of the integral operator associated with the covariance of part (a); find (explicitly) an infinite collection of linearly independent functions in the associated eigenspace. Indicate also a procedure for constructing a basis for the eigenspace.

5. Let Z_1^*, Z_2^*, ... be independent random variables with mean 0 and variance 1, and define

$$S_n(t) = \sum_{k=2^n+1}^{2^{n+1}} \frac{\sin (k - \frac{1}{2})\pi t}{(k - \frac{1}{2})\pi} Z_k^*,$$

$$T_n = \sup \{S_n(t) : 0 \leq t \leq 1\}.$$

(a) Show that

$$\pi^2 \, |E(T_n)|^2 \le \sum_{k=2^n+1}^{2^{n+1}} \frac{1}{(k-\frac{1}{2})^2}$$

$$+ 2 \sum_{p=1}^{2^{n+1}-2^n-1} \left[\sum_{h=2^n+1}^{2^{n+1}} \frac{1}{(h-\frac{1}{2})^2(h+p-\frac{1}{2})^2} \right]^{1/2}.$$

(b) Show that $\sum_{n=0}^{\infty} E(T_n) < \infty$, so that $\sum_{n=0}^{\infty} T_n < \infty$ a.e. Thus by the Weierstrass M-test, for almost every ω, $\sum_{n=0}^{\infty} S_n(t, \omega)$ converges uniformly in t.

1.5 Estimation Problems

Let $\{X_t, \, t \in T\}$ be an L^2 process, and let $S = L^2\{X_t, \, t \in T\}$ be the closed subspace spanned by the X_t, that is, S consists of all L^2 limits of finite linear combinations $\sum_{k=1}^{n} c_k X_{t_k}$, $t_1, \ldots, t_n \in T$, $n = 1, 2, \ldots$. We may regard S as the space of random variables obtainable by a linear operation on the X_t. Note that L^2 derivatives and integrals of $\{X_t, \, t \in T\}$, if they exist, belong to S.

Now the map $X = (X_t, \, t \in T)$ given by $X(\omega) = (X_t(\omega), \, t \in T)$ is measurable: $(\Omega, \mathscr{F}) \to (R^T, \mathscr{B}(R)^T)$ (or $(C^T, \mathscr{B}(C)^T)$ in the complex case); see 1.1, Problem 1. If $\mathscr{F}$ is replaced by the σ-field induced by X, namely, $\mathscr{F}(X) = \{X^{-1}(A) : A \in \mathscr{B}(R)^T\}$, X is still measurable. Also, if $Z : \Omega \to R$, then by RAP, 6.4.2(c), p. 251, Z is measurable relative to $\mathscr{F}(X)$ and $\mathscr{B}(R)$ iff $Z = g(X)$ for some $g : (R^T, \mathscr{B}(R)^T) \to (R, \mathscr{B}(R))$. (Replace R by C in the complex case.) We shall call such a Z a *Borel measurable function of* X. The class of Borel measurable functions of X that belong to L^2 (in other words, $L^2(\Omega, \mathscr{F}(X), P)$) will be denoted by S_0.

We shall consider two estimation problems. Let Y be a fixed random variable in L^2.

1. Find the element $\hat{Y}$ of S closest to Y, that is, $\|\hat{Y} - Y\| = \inf\{\|W - Y\| : W \in S\}$. Thus $\hat{Y}$ is the best estimate of Y based on a *linear* operation on the X_t.

2. Find the element Y^* of S_0 closest to Y. Thus Y^* is the best estimate of Y based on an *arbitrary* Borel measurable operation on the X_t.

In Hilbert space terms, $\hat{Y}$ is the *projection* of Y on S, characterized (if random variables that agree a.e. are identified) as the element of S such that

$$Y - \hat{Y} \perp S$$

(see Figure 1.3).

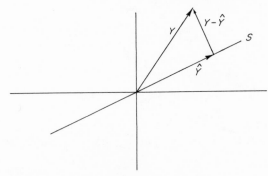

FIGURE 1.3

Equivalently, $Y - \hat{Y}$ is orthogonal to all X_t, or

$$\langle Y, X_t \rangle = \langle \hat{Y}, X_t \rangle \qquad \text{for all} \quad t \in T. \tag{1}$$

For a discussion of projections in Hilbert space, see RAP (3.2, p. 116 ff).

1.5.1   Example

Let $X(n) = Z(n) + W(n)$, where n ranges over the integers; we interpret Z as a "signal" and W as unwanted "noise." Assume that $\{Z(n)\}$ and $\{W(n)\}$ are stationary with 0 mean and covariance functions K_Z and K_W. Assume also that $Z(n)$ and $W(m)$ are uncorrelated for all n, m. We estimate a random variable Y on the basis of a linear operation on $X(r)$, $X(r - 1)$, ..., $X(r - M)$; thus in this case $T = \{r, r - 1, \ldots, r - M\}$ and S consists of all linear combinations $\sum_{j=0}^{M} c_j X(r - j)$.

For example, if $Y = Z(r + \alpha)$, $\alpha > 0$, we have the *prediction* problem; if $Y = Z(r)$ we have the *smoothing* problem.

Now since $\hat{Y} \in S$ we may write $\hat{Y} = \sum_{j=0}^{M} h_j X(r - j)$ for some constants h_j; by (1), $\hat{Y} = \hat{Y}(r)$ is characterized by $\langle Y, X(r - i) \rangle = \langle \hat{Y}, X(r - i) \rangle$, $i = 0, 1, \ldots, M$, that is,

$$E[Y\overline{X(r - i)}] = \sum_{j=0}^{M} K_X(i - j)h_j, \qquad i = 0, 1, \ldots, M.$$

If $Y = Z(r + \alpha)$, then

$$E[Y\overline{X(r - i)}] = E[Z(r + \alpha)[\overline{Z(r - i)} + \overline{W(r - i)}]] = K_Z(\alpha + i);$$

also $K_X = K_Z + K_W$. Thus the equations characterizing $\hat{Y}$ become

$$\begin{bmatrix} K_X(0) & K_X(-1) & \cdots & K_X(-M) \\ K_X(1) & K_X(0) & \cdots & K_X(-M + 1) \\ \vdots & & & \\ K_X(M) & K_X(M - 1) & \cdots & K_X(0) \end{bmatrix} \begin{bmatrix} h_0 \\ h_1 \\ \vdots \\ h_M \end{bmatrix} = \begin{bmatrix} K_Z(\alpha) \\ K_Z(\alpha + 1) \\ \vdots \\ K_Z(\alpha + M) \end{bmatrix}.$$

If K_X is singular (so that $X(r), \ldots, X(r - M)$ are linearly dependent; see 1.2.2(c)), there will be infinitely many solutions to this equation, but all solutions will correspond (a.e.) to the same $\hat{Y}$ since $\hat{Y}$ is unique by the projection theorem. Because of the linear dependence, the same element of S can be represented by different linear combinations.

We now consider the second problem. We claim that

$$Y^* = E(Y \mid X) = E(Y \mid X_t, t \in T). \qquad (2)$$

(A comment about notation is in order here. By definition, $E(Y \mid X)$ is the conditional expectation of Y, given the smallest σ-field $\mathscr{F}(X)$ of subsets of Ω making X measurable relative to $\mathscr{F}(X)$ and $\mathscr{B}(R)^T$. By $E(Y \mid X_t, t \in T)$ we mean the conditional expectation of Y, given the smallest σ-field $\mathscr{G}$ of subsets of Ω making each X_t measurable relative to $\mathscr{G}$ and $\mathscr{B}(R)$. But by RAP (6.4.2(b), p. 251), and its accompanying remark, the two σ-fields, hence the two conditional expectations, are equal.)

To prove (2), note that Y^* is the projection of Y on S_0, hence $\langle Y, Z \rangle = \langle Y^*, Z \rangle$ for all $Z \in S_0$. Set $Z = I_A$, $A \in \mathscr{F}(X)$; then

$$\langle Y, Z \rangle = \int_A Y \, dP, \qquad \langle Y^*, Z \rangle = \int_A Y^* \, dP.$$

But $Y^* \in S_0$, so that Y^* is $\mathscr{F}(X)$-measurable; since A is an arbitrary set in $\mathscr{F}(X)$, the result follows.

Note that since $S \subset S_0$, it follows that $\| Y - Y^* \| \leq \| Y - \hat{Y} \|$. The notation $\hat{E}(Y \mid X_t, t \in T)$ is sometimes used for $\hat{Y}$.

There is an importance case in which $\hat{Y} = Y^*$, so that the best linear estimate coincides with the best overall estimate.

1.5.2 Theorem

If $\{Y\} \cup \{X_t, t \in T\}$ is a Gaussian process, with all random variables having 0 mean, then $\hat{Y} = Y^*$.

PROOF   Since $\hat{Y} \in S$, $\hat{Y}$ is an L^2 limit of some sequence of finite linear combinations of the X_t, say

$$Y_n = \sum_{j=1}^{r_n} c_{nj} X(t_{nj}) \overset{L^2}{\to} \hat{Y}.$$

For any $t_1, \ldots, t_m$, $Y - Y_n, X_{t_1}, \ldots, X_{t_m}$ are jointly Gaussian by 1.2.2(d), hence by 1.4.2, $Y - \hat{Y}, X_{t_1}, \ldots, X_{t_m}$ are jointly Gaussian. Thus $\{Y - \hat{Y}\} \cup \{X_t, t \in T\}$ is a Gaussian process.

Fix $t_1, \ldots, t_n$. Since $Y - \hat{Y}$ is orthogonal to all X_t, the covariance matrix of $Y - \hat{Y}, X_{t_1}, \ldots, X_{t_n}$ looks like

$$K = \begin{bmatrix} a & 0 & 0 & \cdots & 0 \\ 0 & & & & \\ 0 & & B & & \\ \vdots & & & & \\ 0 & & & & \end{bmatrix}.$$

If $a = \text{Var}\,(Y - \hat{Y}) = 0$, then $Y = \hat{Y}$ a.e., hence $Y \in S \subset S_0$. But then $Y = Y^*$ a.e., and we are finished. Thus assume $a > 0$. If B is nonsingular, then K^{-1} has the same form as K. It follows from the form of the joint density of $Y - \hat{Y}, X_{t_1}, \ldots, X_{t_n}$ (see 1.2.2(c)) that $Y - \hat{Y}$ and $(X_{t_1}, \ldots, X_{t_n})$ are independent. If B is singular, the same argument shows that $Y - \hat{Y}$ and $(X_{s_1}, \ldots, X_{s_r})$ are independent, where $\{X_{s_1}, \ldots, X_{s_r}\}$ is a maximal linearly independent subset of $\{X_{t_1}, \ldots, X_{t_n}\}$. Since $(X_{t_1}, \ldots, X_{t_n})$ can be expressed as a linear function of $(X_{s_1}, \ldots, X_{s_r})$, $Y - \hat{Y}$ and $(X_{t_1}, \ldots, X_{t_n})$ are independent in this case also. Since $t_1, \ldots, t_n$ are arbitrary, $Y - \hat{Y}$ and X are independent; hence $E(Y - \hat{Y} \mid X) = E(Y - \hat{Y}) = 0$. Now $\hat{Y} \in S \subset S_0$, so that $\hat{Y}$ is $\mathscr{F}(X)$-measurable; consequently $E(\hat{Y} \mid X) = \hat{Y}$. Therefore $E(Y \mid X) = \hat{Y}$. ‖

Problems

1. Let $\{Z(n)\}$ be a sequence of independent random variables, with $P\{Z(n) = 2\} = P\{Z(n) = -2\} = \frac{1}{2}$ for all n. Let $\{W(n)\}$ be a sequence of independent random variables, with $P\{W(n) = 1\} = P\{W(n) = -1\} = \frac{1}{2}$ for all n. Assume that $Z(n)$ and $W(m)$ are uncorrelated for all n, m. Consider the smoothing problem $(Y = Z(r))$ of 1.5.1, with $M = 1$.

(a) Evaluate $\hat{Y}$ and Y^*.
(b) Evaluate $\sigma^2 = \|Y - \hat{Y}\|^2$. (Note that $\sigma^2 = \|Y\|^2 - \|\hat{Y}\|^2$ by the Pythagorean theorem.)
(c) Show that $\sigma^{*2} = \|Y - Y^*\|^2$ is 0. Thus the best linear estimate is not optimal.

2. (*The Kalman Filter*) Consider the following model for a random process:

$$X(k + 1) = \Phi(k)X(k) + U(k), \qquad k = 0, 1, \ldots,$$

where $X(k)$ and $U(k)$ are n-dimensional random column vectors and $\Phi(k)$ is a known n by n matrix. If $X(k)$ represents the state of a system at time k, then $X(k + 1)$ is a known linear transformation of $X(k)$ plus a random noise $U(k)$. We assume the $U(k)$ have 0 mean and are orthogonal:

$E[U(j)U^*(k)] = Q(k)\delta_{jk}$ where $*$ denotes conjugate transpose, δ_{jk} is the Kronecker delta, and $Q(k)$ is an n by n nonnegative definite matrix.

Suppose we cannot observe the process $\{X(k)\}$ directly, but instead we observe a related measurement process $\{V(k)\}$:

$$V(k) = M(k)X(k) + W(k)$$

where $V(k)$ and $W(k)$ are m-dimensional random column vectors and $M(k)$ is a known m by n matrix. Thus $V(k)$, the observable at time k, is a known linear transformation of $X(k)$ plus a random noise $W(k)$. We also assume the $W(k)$ have 0 mean and are orthogonal: $E[W(j)W^*(k)] = R(k)\delta_{jk}$.

We want the best least squares estimate of $X(k)$ based on previous observations $V(0), \ldots, V(k-1)$; thus let $\hat{X}(k)$ be the n-vector whose ith component $\hat{X}_i(k)$ is the projection of $X_i(k)$ on the subspace of $L^2(\Omega, \mathscr{F}, P)$ spanned by the components of $V(0), \ldots, V(k-1)$.

We start with a random vector $X(0)$ and an initial estimate $\hat{X}(0) = EX(0)$; the initial error covariance matrix is

$$P(0) = E([X(0) - \hat{X}(0)][X(0) - \hat{X}(0)]^*).$$

We assume that the noise processes and $X(0)$ are mutually orthogonal, that is, for all $j, k = 0, 1, \ldots, E[U(j)W^*(k)], E[X(0)U^*(j)], E[X(0)W^*(k)]$ are zero matrices.

We observe $V(0), \ldots, V(k-1)$ and wish to compute the estimate $\hat{X}(k)$ of $X(k)$ and the error covariance matrix

$$P(k) = E([X(k) - \hat{X}(k)][X(k) - \hat{X}(k)]^*);$$

we shall find recursion formulas for $\hat{X}(k+1)$ and $P(k+1)$ in terms of $\hat{X}(k)$, $P(k)$ and the new observation $V(k)$.

To summarize, the ingredients are:

(1) system states $X(k)$: $X(k+1) = \Phi(k)X(k) + U(k)$,
(2) measurements $V(k)$: $V(k) = M(k)X(k) + W(k)$,
(3) moments: $EU(k) = 0$, $EW(k) = 0$,
     $E[U(j)U^*(k)] = Q(k)\delta_{jk}$, $E[W(j)W^*(k)] = R(k)\delta_{jk}$;
     $X(0)$, $\{U(k)\}$, and $\{W(k)\}$ are mutually orthogonal.

(a) Let S_k be the subspace of $L^2(\Omega, \mathscr{F}, P)$ spanned by the components of $V(0), \ldots, V(k)$. Show that $\hat{X}(k)$ is the projection of $X(k)$ on $S_{k-1}^n = S_{k-1} \times \cdots \times S_{k-1}$ (n factors) in $[L^2(\Omega, \mathscr{F}, P)]^n$, the Hilbert space of random vectors $U = (U_1, \ldots, U_n)$, $U_i \in L^2(\Omega, \mathscr{F}, P)$, with inner product $\langle U, V \rangle = \sum_{i=1}^{n} \langle U_i, V_i \rangle_{L^2(\Omega, \mathscr{F}, P)}$.

(b)   If the components of $X(0)$, $U(k)$, $k \geq 1$, and $W(k)$, $k \geq 1$ form a Gaussian process (so that $X(0)$, $\{U(k)\}$, and $\{W(k)\}$ are independent), it follows from 1.5.2 that

$$\hat{X}(k) = E[X(k) \,|\, S_{k-1}] \qquad (= E[X(k) \,|\, V(0), \ldots, V(k-1))$$

the conditional expectation of $X(k)$ given the components of $V(0)$, ..., $V(k-1)$. (The conditional expectation of a random vector is taken componentwise.) Thus

$$\begin{aligned}
\hat{X}(k+1) &= E[X(k+1) \,|\, S_k] \\
&= E[\Phi(k)X(k) + U(k) \,|\, S_k] \\
&= E[\Phi(k)\hat{X}(k) \,|\, S_k] + E[\Phi(k)(X(k) - \hat{X}(k)) \,|\, S_k] \\
&\quad + E[U(k) \,|\, S_k].
\end{aligned}$$

We analyze these three summands.

(i)   Show that $E[\Phi(k)\hat{X}(k) \,|\, S_k] = \Phi(k)\hat{X}(k)$.

(ii)   Show that the second summand can be written as

$$\Phi(k)E[X(k) - \hat{X}(k) \,|\, V(k) - E\{V(k) \,|\, V(0), \ldots, V(k-1)\}].$$

(iii)   Show that

$$\begin{aligned}
V(k) &- E[V(k) \,|\, V(0), \ldots, V(k-1)] \\
&= M(k)(X(k) - \hat{X}(k)) + W(k) = V(k) - M(k)\hat{X}(k).
\end{aligned}$$

(iv)   Let X and Y be random column vectors, not necessarily of the same dimension; assume (X, Y) is Gaussian, and that $E[YY^*]$ is strictly positive definite and hence nonsingular. Show that

$$E(X \,|\, Y) = E(XY^*)[E(YY^*)]^{-1}Y.$$

(v)   Show that, assuming $R(k)$ is strictly positive definite, the second summand is $A(k)[V(k) - M(k)\hat{X}(k)]$ where

$$A(k) = \Phi(k)P(k)M^*(k)[M(k)P(k)M^*(k) + R(k)]^{-1}.$$

(vi)   Show that $E[U(k) \,|\, S_k] = 0$.

It follows that

$$\hat{X}(k+1) = \Phi(k)\hat{X}(k) + A(k)[V(k) - M(k)\hat{X}(k)].$$

Thus the new estimate $\hat{X}(k+1)$ is a known linear transformation of the old estimate $\hat{X}(k)$ plus an updating term that takes into account the new observation $V(k)$. The linear transformation $\Phi(k)$ is the same as the one governing the deterministic aspect of the dynamics of the process $\{X(t)\}$ (see (1)).

(c) Show that

$$P(k + 1) = \Phi_0(k)P(k)\Phi_0^*(k) + Q(k) + A(k)R(k)A^*(k)$$

where $\Phi_0(k) = \Phi(k) - A(k)M(k)$; therefore, starting with $\hat{X}(0)$ and $P(0)$, the estimates $X(k)$ and error covariances $P(k)$ may be computed recursively.

(d) Let $\hat{X}(k \mid j)$ denote the best least squares estimate of $X(k)$ based on $V(0), \ldots, V(j)$. Show that for $k \geq 2$,

$$\hat{X}(j + k \mid j) = \Phi(j + k - 1)\Phi(j + k - 2) \cdots \Phi(j + 1)\hat{X}(j + 1 \mid j).$$

COMMENT The Kalman filter is perhaps a more realistic model for estimation problems than the sophisticated schemes to be studied in Chapter 2; stationarity is not assumed, and only a finite number of observations are necessary to construct any particular estimate.

1.6 Notes

References on L^2 processes will be given at the end of Chapter 2.

Chapter 2

Spectral Theory and Prediction

2.1 Introduction; L^2 Stochastic Integrals

In this chapter we discuss the prediction problem for L^2 processes. The basic approach is via the spectral decomposition theorem, which establishes an isometric isomorphism between $L^2\{X(t), t \in T\}$ and a certain L^2 space of functions. To get at this idea, we must first develop in detail an L^2 integration theory.

Consider the process $X(t) = \sum_{j=1}^k Z_j e^{i\lambda_j t}$, where the Z_j are orthogonal random variables in L^2 with 0 mean, and $\lambda_1, \dots, \lambda_k$ are real numbers. Then $\{X(t), t \in T\}$ is a stationary L^2 process with covariance function

$$K(t) = \sum_{j=1}^k E |Z_j|^2 e^{i\lambda_j t}.$$

Throughout this chapter, we take T to be either R or the integers. In the latter case, we may assume that all λ_j belong to $[-\pi, \pi)$; this will not change the values of $X(t)$.

We shall see that processes of this type are basic in the sense that any stationary L^2 process with 0 mean and a continuous covariance function can be expressed as a limit of such processes. If we write $X(t) = \sum_{j=1}^k e^{i\lambda_j t} \Delta Z(\lambda_j)$, with $\Delta Z(\lambda_j) = Z_j$, the notation suggests a Riemann–Stieltjes sum approximating an integral of the form $\int e^{i\lambda t} dZ(\lambda)$. Thus the investigation that is suggested involves the definition and properties of integrals of the form $\int f(\lambda) dZ(\lambda)$ for an appropriate class of functions f. If $\{Z(\lambda), \lambda \in T\}$ is a stochastic process, then for each ω in the probability space we have a sample function $(Z(\lambda, \omega), \lambda \in T)$; it is natural to consider integrals of the form $\int_T f(\lambda) Z(d\lambda, \omega)$. The problem is that the sample functions need not be of bounded variation on any interval $[a, b]$. This is typical of L^2 theory; the covariance function does not yield information about sample function behavior. Thus we shall define integrals in the L^2 sense.

2.1.1 Definitions and Comments

Let $(S, \mathscr{S})$ be a measurable space, and $\mathscr{S}_0$ a field whose minimal σ-field is $\mathscr{S}$. Let m be a measure on $\mathscr{S}_0$ (see RAP, 1.2.3, p. 6), and let H be a Hilbert space with inner product $\langle \cdot, \cdot \rangle$. (If H is an L^2 space of functions, we identify functions that agree almost everywhere.)

An *elementary measure with orthogonal values* (abbreviated *elementary mov*) is a function $Z: \mathscr{S}_0 \to H$ such that for all $E_1, E_2 \in \mathscr{S}_0$,

(a) $\langle Z(E_1), Z(E_2) \rangle = m(E_1 \cap E_2)$; in particular, $m(E_1) = \|Z(E_1)\|^2$.
(b) $Z(E_1 \cup E_2) = Z(E_1) + Z(E_2)$ if $E_1 \cap E_2 = \varnothing$.

We call m the *measure associated with* Z. Note that (a) forces m to be nonnegative and finite valued on $\mathscr{S}_0$; the addition of (b) implies that m must be finitely additive (but not necessarily countably additive) on $\mathscr{S}_0$. However, we have countable additivity by virtue of the hypothesis that m is a measure on $\mathscr{S}_0$; thus by the Carathéodory extension theorem, m has a unique extension to $\mathscr{S}$. The extension will be denoted by m also.

Note that $Z(\varnothing) = 0$; this follows from (a) by setting $E_1 = \varnothing$. Also, by (a), if E_1 and E_2 are disjoint, then $Z(E_1) \perp Z(E_2)$.

Now let $f = \sum_{k=1}^{n} a_k I_{E_k}$, with the E_k disjoint sets in $\mathscr{S}_0$, and the $a_k \in C$; call such a function *simple $\mathscr{S}_0$-measurable*. Define

$$\int_S f \, dZ = \sum_{k=1}^{n} a_k Z(E_k) \in H.$$

By (b), the integral is well defined; the proof is the same as the corresponding argument for the integral with respect to an ordinary measure (see RAP, 1.5.3, p. 36).

If $f = \sum_{j=1}^{n} a_j I_{E_j}$, $g = \sum_{k=1}^{n} b_k I_{E_k}$ are simple $\mathscr{S}_0$-measurable functions (nothing is lost by using the same n and E's in both cases since $\mathscr{S}_0$ is a field), then

$$\left\langle \int_S f \, dZ, \int_S g \, dZ \right\rangle = \sum_{j,k=1}^{n} a_j \bar{b}_k \langle Z(E_j), Z(E_k) \rangle$$

$$= \sum_{k=1}^{n} a_k \bar{b}_k m(E_k) \qquad \text{by (a)}$$

$$= \int_S f \bar{g} \, dm$$

$$= \langle f, g \rangle_{L^2(S, \mathscr{S}, m)}. \tag{1}$$

Thus the linear map $\Phi(f) = \int_S f\,dZ$, from the subspace L of $L^2(S, \mathcal{S}, m)$ consisting of simple $\mathcal{S}_0$-measurable functions, to the Hilbert space H, preserves inner products. But L is dense in $L^2(S, \mathcal{S}, m)$ (see RAP, 1.3.11, p. 20; 2.4.13, p. 88), so Φ has a unique extension (also denoted by Φ) to an isometric isomorphism of $L^2(S, \mathcal{S}, m)$ and a closed subspace of H. This allows us to extend the integral; if $f \in L^2(S, \mathcal{S}, m)$, define

$$\int_S f\,dZ = \Phi(f).$$

Since Φ is an isometric isomorphism, the following properties are immediate.

2.1.2   Properties of the Integral

If $f, f_n, g \in L^2(S, \mathcal{S}, m)$, $a, b \in C$, then

(a) $\int_S f\,dZ \in H$;

(b) $\langle \int_S f\,dZ, \int_S g\,dZ \rangle = \int_S f\bar{g}\,dm$;

(c) $\| \int_S f\,dZ \|^2 = \int_S |f|^2\,dm$;

(d) $\int_S (af + bg)\,dZ = a \int_S f\,dZ + b \int_S g\,dZ$;

(e) $f_n \to f$ in $L^2(S, \mathcal{S}, m)$ iff $\int_S f_n\,dZ \to \int_S f\,dZ$ in H.

If $f_1, f_2, \dots$ are simple $\mathcal{S}_0$-measurable functions converging to f in $L^2(S, \mathcal{S}, m)$, and

$$f_n = \sum_{k=1}^{m_n} a_k^{(n)} I_{E_k^{(n)}},$$

the $E_k^{(n)}$, $1 \leq k \leq m_n$, disjoint sets in $\mathcal{S}_0$, then by (e),

$$\int_S f\,dZ = \lim_{n \to \infty} \int_S f_n\,dZ = \lim_{n \to \infty} \sum_{k=1}^{m_n} a_k^{(n)} Z(E_k^{(n)}).$$

If $H = L^2(\Omega, \mathcal{F}, P)$, the $Z(E_k^{(n)})$ are orthogonal random variables, so that the integral is approximated by sums of the type discussed at the beginning of the section.

We may use the integral to extend Z from $\mathcal{S}_0$ to $\mathcal{S}$; if $E \in \mathcal{S}$, we define

$$Z_1(E) = \Phi(I_E) = \int_S I_E\,dZ.$$

Since m is a finite measure, $I_E \in L^2(S, \mathcal{S}, m)$, so $\int_S I_E\,dZ$ is defined; furthermore, $Z_1 = Z$ on $\mathcal{S}_0$ by definition of the integral. From now on we drop the subscript and denote the extension of Z by Z also.

The extension still satisfies properties (a) and (b) of 2.1.1; in fact, more is true.

2.1.3   Theorem

The function $Z: \mathscr{S} \to H$ has the following properties. If $E_1, E_2, \ldots \in \mathscr{S}$,

(a) $\langle Z(E_1), Z(E_2)\rangle = m(E_1 \cap E_2)$.

(b) $Z(E_1 \cup E_2) = Z(E_1) + Z(E_2)$ if $E_1 \cap E_2 = \varnothing$.

(c)   If $E_1, E_2, \ldots$ are disjoint,

$$Z\left(\bigcup_{n=1}^{\infty} E_n\right) = \sum_{n=1}^{\infty} Z(E_n)$$

in the sense that

$$\left\| Z\left(\bigcup_{n=1}^{\infty} E_n\right) - \sum_{k=1}^{n} Z(E_k)\right\| \to 0 \qquad \text{as } n \to \infty.$$

PROOF

(a) $\langle Z(E_1), Z(E_2)\rangle = \langle \Phi(I_{E_1}), \Phi(I_{E_2})\rangle = \int_S I_{E_1} I_{E_2} \, dm$

$$= m(E_1 \cap E_2).$$

(b) $Z(E_1 \cup E_2) = \Phi(I_{E_1 \cup E_2}) = \Phi(I_{E_1} + I_{E_2})$

$$= \Phi(I_{E_1}) + \Phi(I_{E_2}) = Z(E_1) + Z(E_2)$$

(c)   Let $E = \bigcup_{n=1}^{\infty} E_n; f_n = I_E - \sum_{k=1}^{n} I_{E_k} = I_{E - \cup_{1 \le k \le n} E_k}$; then

$$\left\| Z(E) - \sum_{k=1}^{n} Z(E_k)\right\|^2 = \left\| \int_S f_n \, dZ\right\|^2$$

$$= \int_S |f_n|^2 \, dm \qquad \text{by 2.1.2(c)}$$

$$\to 0 \qquad \text{by the dominated convergence theorem.} \quad \|$$

2.1.4   Definition

A function Z from a σ-field $\mathscr{S}$ to a Hilbert space, satisfying, for some measure m on $\mathscr{S}$, properties (a), (b), and (c) of 2.1.3 is called a *measure with orthogonal values* (abbreviated mov); m is called the *measure associated with Z*.

A mov differs from an elementary mov in that a mov is defined on a σ-field and is countably additive, while an elementary mov need only be defined on a field. For further properties and an alternative approach, see Problems 3 and 4.

We now develop some additional facts about integration with respect to a mov. For the remainder of the chapter, $L^2(m)$ will abbreviate $L^2(S, \mathscr{S}, m)$.

2.1.5   Theorem

Let Z be a mov on $(S, \mathscr{S})$ with values in H and associated measure m. Let $g \in L^2(S, \mathscr{S}, m)$, and define $Z_1: \mathscr{S} \to H$, $m_1: \mathscr{S} \to R$ by

$$Z_1(E) = \int_E g \, dZ \left(= \int_S g I_E \, dZ \right)$$

$$m_1(E) = \int_E |g|^2 \, dm.$$

Then:

(a) Z_1 is a mov with associated measure m_1.

(b)   If $f \in L^2(m_1)$, then $fg \in L^2(m)$, and

$$\int_S f \, dZ_1 = \int_S fg \, dZ.$$

(c)   If $|g| > 0$ a.e. $[m]$, then

$$Z(E) = \int_E g^{-1} \, dZ_1, \qquad E \in \mathscr{S}.$$

PROOF   (a)   Since Z is finitely additive, so is Z_1; if $E_1, E_2 \in \mathscr{S}$,

$$\langle Z_1(E_1), Z_1(E_2) \rangle = \left\langle \int_S g I_{E_1} \, dZ, \int_S g I_{E_2} \, dZ \right\rangle$$

$$= \int_S |g|^2 I_{E_1 \cap E_2} \, dm$$

$$= m_1(E_1 \cap E_2).$$

The proof that Z_1 is countably additive is done exactly as in 2.1.3(c), with Z replaced by Z_1 and f_n by gf_n; this proves (a).

(b)   If f is simple, (b) holds by definition of Z_1. In general, let $\{f_n\}$ be a sequence of simple functions converging to f in $L^2(m_1)$; then $\int_S f_n \, dZ_1 = \int_S f_n g \, dZ$ for all n. As $n \to \infty$, $\int_S f_n \, dZ_1 \to \int_S f \, dZ_1$ by 2.1.2(e), and

$$\left\| \int_S (f_n g - fg) \, dZ \right\|^2 = \int_S |f_n - f|^2 |g|^2 \, dm = \int_S |f_n - f|^2 \, dm_1 \to 0$$

(see RAP, p. 69, Problem 4; in particular, this shows that $(f_n - f)g \in L^2(m)$). Therefore $\int_S f \, dZ_1 = \int_S fg \, dZ$.

(c)   Note that $|g| > 0$ a.e. $[m_1]$ by definition of m_1. Set $g^{-1} = 0$ on $\{g = 0\}$; since $|g| > 0$ a.e. $[m]$, we have $g^{-1} g = 1$ a.e. $[m]$. Thus

$$\int_S |g^{-1} I_E|^2 \, dm_1 = \int_S |g|^{-2} I_E |g|^2 \, dm = m(E) < \infty.$$

Thus $g^{-1}I_E \in L^2(m_1)$, and part (b) applies to give

$$\int_S g^{-1}I_E \, dZ_1 = \int_S g^{-1}I_E g \, dZ = Z(E). \quad \|$$

Our main interest will be in integrating with respect to a mov that takes values in $H = L^2(\Omega, \mathscr{F}, P)$, where $(\Omega, \mathscr{F}, P)$ is a probability space. In other words, our mov's will have L^2 random variables as their values; a mov of this type will be called a *stochastic mov*. Stochastic mov's are generated by processes with orthogonal increments in the following way.

2.1.6 The Basic Construction

Let $S = [a, b]$ be a bounded interval of reals and $\{Z(\lambda), \lambda \in S\}$ an L^2 process with *orthogonal increments*; that is, if $\lambda_1 < \lambda_2 \leq \lambda_3 < \lambda_4$, then

$$E[(Z(\lambda_2) - Z(\lambda_1))(\overline{Z(\lambda_4) - Z(\lambda_3)})] = 0.$$

Assume also that the process is L^2 *right-continuous*, that is, $Z(\lambda) \xrightarrow{L^2} Z(\lambda_0)$ as λ approaches λ_0 from above. If $E = (x, y]$ is a subinterval of S, we define $Z(E) = Z(y) - Z(x)$: Z may be extended to a finitely additive set function on the field $\mathscr{S}_0$ of finite disjoint unions of right-semiclosed intervals of S. In fact, Z is an elementary mov on $\mathscr{S}_0$. To see this, define $F(x) = \|Z(x) - Z(a)\|^2$, $x \in S$. Since $\{Z(t)\}$ is L^2 right-continuous, F is right-continuous and since Z has orthogonal increments, we have, for $x < y$,

$$\|Z(y) - Z(a)\|^2 = \|Z(y) - Z(x)\|^2 + \|Z(x) - Z(a)\|^2$$
$$\geq \|Z(x) - Z(a)\|^2.$$

Thus F is increasing; we conclude that F is a distribution function. If m is the corresponding Lebesgue–Stieltjes measure, the above argument shows that

$$m(x, y] = F(y) - F(x) = \|Z(x, y]\|^2, \qquad x < y,$$

hence by orthogonality, $m(E) = \|Z(E)\|^2$ for all $E \in \mathscr{S}_0$. But if $E_1, E_2 \in \mathscr{S}_0$, then

$$\langle Z(E_1), Z(E_2) \rangle = \langle Z(E_1 \cap E_2) + Z(E_1 - E_2), Z(E_1 \cap E_2)$$
$$+ Z(E_2 - E_1) \rangle$$
$$= \|Z(E_1 \cap E_2)\|^2$$
$$= m(E_1 \cap E_2).$$

Thus (a) and (b) of 2.1.1 are satisfied, so that Z is an elementary mov with associated measure m.

There is no difficulty in constructing L^2 right-continuous processes with orthogonal increments. For example, if F is an arbitrary distribution function on S, construct an L^2 process $Z(\lambda)$, $\lambda \in S$, with covariance function $K(\lambda_1, \lambda_2) = \min (F(\lambda_1), F(\lambda_2))$. (Since $\min (s, t)$ is a realiable covariance and F is increasing, K is realizable.) As in the discussion of Brownian motion in 1.2.9, it follows that $\{Z(\lambda)\}$ has orthogonal increments; also, $E[\, |Z(\lambda_2) - Z(\lambda_1)|^2] = F(\lambda_2) - F(\lambda_1)$, proving L^2 right-continuity.

If S is an unbounded interval, the given procedure will not work in general since the distribution function F may be unbounded, and thus the measure m will be infinite. What we do in this case is approximate S by bounded intervals. For example, if $\{Z(\lambda), \lambda \in R\}$ is an L^2 right-continuous process with orthogonal increments, there is a corresponding mov Z_{ab} on each bounded interval $[a, b]$, and if $a \leq a' < b' \leq b$, then Z_{ab} restricted to subsets of $[a', b']$ coincides with $Z_{a'b'}$, and similarly for the associated measures m_{ab}. Define

$$\int_R f \, dZ = L^2 \lim_{\substack{a \to -\infty \\ b \to \infty}} \int_{[a, b]} f \, dZ_{ab}$$

provided the limit exists. If m is the unique measure on $\mathscr{B}(R)$ that coincides with m_{ab} on $\mathscr{B}[a, b]$ for all $a < b$, the limit will always exist for $f \in L^2(m)$; see Problem 5. Furthermore, if $f \in L^2(m)$, then $f I_{[-n, n]}$ converges to f in $L^2(m)$ as $n \to \infty$, and it follows that properties (a)–(d) of 2.1.2 are satisfied (with $S = R$, $\mathscr{S} = \mathscr{B}(R)$). Since property (e) is implied by (c), all properties are satisfied. Thus, as before, we have an isometric isomorphism of $L^2(m)$ and a closed subspace of $H = L^2(\Omega, \mathscr{F}, P)$.

Problems

1. Let μ be a finite measure on $(S, \mathscr{S})$, and let $H = L^2(S, \mathscr{S}, \mu)$. Define $Z: \mathscr{S} \to H$ by $Z(E) = I_E$, $E \in \mathscr{S}$. Show that Z is a mov with associated measure μ, and $\int_S f \, dZ = f$ for all $f \in L^2(S, \mathscr{S}, \mu)$.

2. Let $\{e_1, e_2, \ldots\}$ be an orthogonal set in a Hilbert space H, and assume that $\sum_{k=1}^{\infty} \|e_k\|^2 < \infty$. Let $(S, \mathscr{S})$ be a measurable space, and let $x_1, x_2, \ldots \in S$, with each $\{x_i\} \in \mathscr{S}$. Define $Z: \mathscr{S} \to H$ by

$$Z(E) = \sum_{k=1}^{\infty} e_k I_E(x_k), \qquad E \in \mathscr{S}.$$

Show that Z is a mov with associated measure given by

$$m(E) = \sum_{k=1}^{\infty} \|e_k\|^2 I_E(x_k);$$

thus m assigns mass $\|e_k\|^2$ to the point x_k. Show also that

$$\int_S f \, dZ = \sum_{k=1}^{\infty} f(x_k)e_k \qquad \text{for all} \quad f \in L^2(S, \mathscr{S}, m).$$

3.   Let Z be a mov on $(S, \mathscr{S})$ with associated measure m. If $E_1, E_2 \in \mathscr{S}$, establish the following relations:

(a) $\|Z(E_1) - Z(E_2)\|^2 = m(E_1 \triangle E_2) = m(E_1) + m(E_2)$
$\qquad\qquad\qquad\qquad\qquad\qquad\qquad\quad - 2m(E_1 \cap E_2).$
(b)   If $E_2 \subset E_1$, then $\|Z(E_1) - Z(E_2)\|^2 = m(E_1) - m(E_2).$
(c)   If $E_2 \subset E_1$, then $\|Z(E_2)\| \le \|Z(E_1)\|.$
(d)   If $E_2 \subset E_1$ and $Z(E_1) = 0$, then $Z(E_2) = 0.$
(e) $Z(E_1 - E_2) = Z(E_1) - Z(E_1 \cap E_2).$
(f)   If $E_2 \subset E_1$, then $Z(E_1 - E_2) = Z(E_1) - Z(E_2).$
(g) $Z(E_1 \cup E_2) = Z(E_1) + Z(E_2) - Z(E_1 \cap E_2).$
(h) $Z(E_1 \triangle E_2) = Z(E_1) + Z(E_2) - 2Z(E_1 \cap E_2).$

4.   Let Z be a function from a σ-field $\mathscr{S}$ to a Hilbert space H. Show that if Z satisfies conditions (b) and (c) of 2.1.3, then Z is a mov iff whenever E_1 and E_2 are disjoint sets in $\mathscr{S}$, then $Z(E_1) \perp Z(E_2).$

5.   In 2.1.6, show that $\int_R f \, dZ$ exists for all $f \in L^2(m).$

6.   Let Z be a mov on $(S, \mathscr{S})$ with values in H and associated measure m, and let ψ be an isometric isomorphism of H and the Hilbert space H'. Show that $Z'(E) = \psi(Z(E))$, $E \in \mathscr{S}$, defines a mov with values in H' and associated measure m.

2.2   Decomposition of Stationary Processes

Let $\{X(t), t \in T\}$ be a stationary L^2 process, with T the set of integers (the *discrete parameter* case) or the set of reals (the *continuous parameter* case). In the discrete parameter case, let $S = [-\pi, \pi]$, $\mathscr{S} = \mathscr{B}[-\pi, \pi]$, and in the continuous parameter case, let $S = R$, $\mathscr{S} = \mathscr{B}(R)$. If the covariance function K is continuous (this is automatically satisfied in the discrete parameter case), by 1.2.6–1.2.8 there is a unique finite measure μ on $\mathscr{S}$ such that

$$K(t) = \int_S e^{itu} \, d\mu(u), \qquad t \in T.$$

We call μ the *spectral measure* of the process; if F is the distribution function corresponding to μ, adjusted so that $F(-\infty) = 0$, F is called the *spectral distribution function*. If μ is absolutely continuous with respect to Lebesgue measure, with density f, f is called the *spectral density* of the process. The *spectrum* of the process is the set of points $s \in S$ such that $\mu(U) > 0$ for every open set $U \subset S$ containing s. The intuitive content of these concepts will be discussed following the proof of the spectral decomposition theorem. This result shows that $\{X(t)\}$ can be represented as the Fourier transform of a measure with orthogonal values.

2.2.1   Spectral Decomposition Theorem

Let $\{X(t), \, t \in T\}$ be a stationary L^2 process with 0 mean, continuous covariance function K, and spectral measure μ. Then there is a unique mov Z defined on $\mathscr{S}$ with values in $H = L^2\{X(t), \, t \in T\}$ such that

$$ X(t) = \int_S e^{it\lambda} \, dZ(\lambda), \qquad t \in T; $$

furthermore, the measure associated with Z is μ. This formula is referred to as the *spectral representation* of the process, and Z is called the *spectral mov.*

PROOF   If $t \in T$, let e_t be the complex exponential function, that is, $e_t(\lambda) = e^{it\lambda}$. Define $\Phi(e_t) = X(t)$, so that Φ maps a subset of $L^2(\mu)$ to a subset of H; observe that Φ preserves inner products:

$$ \langle e_s, e_t \rangle = \int_S e^{i(s-t)\lambda} \, d\mu(\lambda) = K(s - t) = E[X(s)\overline{X(t)}]. $$

We may extend Φ by linearity to an inner-product-preserving map of the space M of finite linear combinations of complex exponentials; note that Φ is well defined on M, for if $\sum_{i=1}^n a_i e_{t_i} = 0$ a.e. $[\mu]$, then

$$ \left\| \sum_{i=1}^n a_i \Phi(e_{t_i}) \right\| = \left\| \sum_{i=1}^n a_i e_{t_i} \right\| = 0, $$

hence $\sum_{i=1}^n a_i \Phi(e_{t_i}) = 0$ a.e. $[P]$. Now since μ is finite, M is dense in $L^2(\mu)$ (in the discrete parameter case, this follows from RAP, p. 127, Problem 9; in the continuous parameter case, use the technique of RAP, 8.1.3, p. 323, along with RAP, p. 188, Problem 3). Therefore Φ extends to an isometric isomorphism of $L^2(\mu)$ and a closed subspace of H. In fact Φ maps onto H, since the $X(t), \, t \in T$, span H.

Define $Z(E) = \Phi(I_E), \, E \in \mathscr{S}$; by Problems 1 and 6, Section 2.1, Z is a mov on $\mathscr{S}$ with values in H and associated measure μ. We claim that

$$ \Phi(f) = \int_S f \, dZ \qquad \text{for all} \quad f \in L^2(\mu). $$

If f is a simple function, this is true by definition of Z; in general, it holds by 2.1.2(e) and the continuity of Φ. Set $f = e_t$ to obtain

$$X(t) = \int_S e^{it\lambda} \, dZ(\lambda).$$

To prove uniqueness, assume $\int_S e^{it\lambda} \, dZ_1(\lambda) = \int_S e^{it\lambda} \, dZ_2(\lambda)$, $t \in T$, where Z_1 and Z_2 are mov's on $\mathscr{S}$ with values in H and associated measures μ_1 and μ_2. As above, the space M is dense in $L^2(\mu_1 + \mu_2)$, so that if $E \in \mathscr{S}$, there is a sequence of finite linear combinations of complex exponentials converging to I_E in both $L^2(\mu_1)$ and $L^2(\mu_2)$; by 2.1.2(e), $Z_1(E) = Z_2(E)$. $\parallel$

2.2.2 Comments

(a) If $\{X(t), t \in R\}$ has a spectral representation, that is, there is a mov Z such that $X(t) = \int_R e^{it\lambda} \, dZ(\lambda)$ for all $t \in R$, then (assuming $E[X(t)] \equiv 0$) $\{X(t)\}$ must have a continuous covariance function. For if m is the measure associated with Z,

$$K(t) = E[X(s + t)\overline{X(s)}]$$

$$= E\left[\int_R e^{i(s+t)\lambda} \, dZ(\lambda) \int_R e^{-is\lambda} \, dZ(\lambda)\right]$$

$$= \int_R e^{it\lambda} \, dm(\lambda) \qquad \text{by 2.1.2(b),}$$

which is continuous by the dominated convergence theorem.

(b) The spectral mov Z can be obtained from a process with orthogonal increments. Define $Z(0) \equiv 0$, and

$$Z(\lambda) = \begin{cases} Z(0, \lambda], & \lambda > 0 \\ -Z(\lambda, 0], & \lambda < 0. \end{cases}$$

Then $\{Z(\lambda)\}$ has orthogonal increments, and Z is obtained from this process as discussed in 2.1.6. In the continuous parameter case, λ ranges over R, and in the discrete parameter case, over $[-\pi, \pi]$.

(c) The random variable Y belongs to $H = L^2\{X(t), t \in T\}$ if and only if $Y = \int_S f \, dZ$ for some $f \in L^2(\mu)$.

If $f \in L^2(\mu)$, then $\int_S f \, dZ \in H$ by definition of Z. Conversely, if $Y \in H$, then Y is an L^2-limit of finite linear combinations Y_n of the $X(t)$. If $f_n = \Phi^{-1}(Y_n)$, then $f_n \in L^2(\mu)$ and, since Φ is an isometric isomorphism, f_n

converges to a limit f in $L^2(\mu)$. By continuity of Φ, $Y_n \to \Phi(f)$, hence $Y = \Phi(f) = \int_S f \, dZ$.

(d) The spectral decomposition theorem implies that for each t, $X(t)$ can be approximated in the L^2 sense by a finite sum of the form $X_n(t) = \sum_{j=1}^{n} e^{it\lambda_j} \Delta Z(\lambda_j)$, where the $\Delta Z(\lambda_j)$ are orthogonal random variables in H (see the discussion after 2.1.2). We may write $X_n(t)$ as $\int_S e^{it\lambda} \, dW(\lambda)$ where W is the mov on $\mathscr{S}$ determined by specifying that $W\{\lambda_j\} = \Delta Z(\lambda_j)$, $1 \le j \le n$, $W\{\lambda_1, \ldots, \lambda_n\}^c = 0$. Thus we approximate $X(t)$ by a sum of complex exponentials with frequencies λ_j and complex amplitudes $\Delta Z(\lambda_j)$.

If we imagine $X_n(t)$ to be the complex representation of the voltage across a resistor (so that the physical voltage is the real part of $X_n(t)$), we expect that the average power dissipated in the resistor is proportional to

$$E[\,|X_n(t)|^2\,] = \sum_{j=1}^{n} E[\,|\Delta Z(\lambda_j)|^2\,]$$

by orthogonality. Now we may construct $X_n(t)$ so that $\Delta Z(\lambda_j)$ is of the form $Z(\lambda_j, \lambda_{j+1}]$; since the measure associated with Z is the spectral measure μ,

$$
\begin{aligned}
E[\,|\Delta Z(\lambda_j)|^2\,] &= \mu(\lambda_j, \lambda_{j+1}] \\
&= F(\lambda_{j+1}) - F(\lambda_j) \\
&= \int_{\lambda_j}^{\lambda_{j+1}} f(\lambda) \, d\lambda
\end{aligned}
$$

if there is a spectral density f.

Thus the increment in the spectral distribution function (or the integral of the spectral density) between two points λ and λ' can be thought of as the contribution to the "average power" of the process $\{X(t)\}$ due to complex exponentials whose frequencies are between λ and λ'. This interpretation is often exploited in the engineering literature (see, for example, Papoulis, 1967).

We now apply the spectral theorem to prove a version of the weak law of large numbers for stationary processes.

2.2.3 Theorem

Under the hypothesis of the spectral decomposition theorem 2.2.1, we have, in the discrete parameter case,

(a) $\dfrac{1}{n} \sum\limits_{k=0}^{n-1} X(k) \overset{L^2}{\to} Z\{0\}$ as $n \to \infty$;

and

(b) $\dfrac{1}{n} \displaystyle\sum_{k=0}^{n-1} K(k) \to \mu\{0\}$ as $n \to \infty$.

In the continuous parameter case,

(c) $\dfrac{1}{T} \displaystyle\int_0^T X(t)\, dt \overset{L^2}{\to} Z\{0\}$ as $T \to \infty$;

and

(d) $\dfrac{1}{T} \displaystyle\int_0^T K(t)\, dt \to \mu\{0\}$ as $T \to \infty$.

PROOF   In the discrete parameter case,

$$\frac{1}{n} \sum_{k=0}^{n-1} X(k) = \int_{-\pi}^{\pi} \frac{1}{n} \sum_{k=0}^{n-1} e^{ik\lambda}\, dZ(\lambda) = \int_{-\pi}^{\pi} g_n(\lambda)\, dZ(\lambda),$$

where

$$g_n(\lambda) = \begin{cases} \dfrac{1}{n} \dfrac{1 - e^{in\lambda}}{1 - e^{i\lambda}} & \text{if } \lambda \neq 0 \\[2mm] 1 & \text{if } \lambda = 0. \end{cases}$$

Now

$$|g_n(\lambda)| = \left| \frac{\sin (n\lambda/2)}{n \sin (\lambda/2)} \right|,$$

and since $|\sin x| \geq 2|x|/\pi$, $|x| \leq \pi/2$ (draw a picture), we have

$$|g_n(\lambda)| \leq \frac{\pi}{2} \left| \frac{\sin (n\lambda/2)}{n\lambda/2} \right| \leq \frac{\pi}{2}, \qquad -\pi \leq \lambda \leq \pi.$$

By the dominated convergence theorem, $g_n \to I_{\{0\}}$ in $L^2(\mu)$, hence

$$\int_{-\pi}^{\pi} g_n(\lambda)\, dZ(\lambda) \to \int_{-\pi}^{\pi} I_{\{0\}}(\lambda)\, dZ(\lambda) = Z\{0\},$$

proving (a). To prove (b), note that $K(k) = \int_{-\pi}^{\pi} e^{ik\lambda}\, d\mu(\lambda)$, hence

$$\frac{1}{n} \sum_{k=0}^{n-1} K(k) = \int_{-\pi}^{\pi} g_n(\lambda)\, d\mu(\lambda) \to \int_{-\pi}^{\pi} I_{\{0\}}(\lambda)\, d\mu(\lambda) = \mu\{0\}.$$

To prove (c), note that if $\Delta t_k = T/n$, $t_k = k\, \Delta t_k$, $k = 1, \ldots, n$, then

$$\sum_{k=1}^{n} \exp (it_k \lambda)\, \Delta t_k \to \int_0^T e^{it\lambda}\, dt \qquad \text{pointwise for each } \lambda \in R;$$

since the left-hand side is bounded by T, we also have convergence in $L^2(\mu)$. Since $\int_R \exp{(it_k \lambda)}\, dZ(\lambda) = X(t_k)$, 2.1.2(e) yields

$$\sum_{k=1}^{n} X(t_k)\, \Delta t_k \to \int_R \left[\int_0^T e^{it\lambda}\, dt \right] dZ(\lambda),$$

in other words,

$$\frac{1}{T} \int_0^T X(t)\, dt = \int_R g_T(\lambda)\, dZ(\lambda)$$

where

$$g_T(\lambda) = \begin{cases} \dfrac{e^{iT\lambda} - 1}{iT\lambda} & \text{if } \lambda \neq 0 \\[2mm] 1 & \text{if } \lambda = 0. \end{cases}$$

The proof is now completed as in (a).

Finally, to prove (d), write

$$\frac{1}{T} \int_0^T K(t)\, dt = \frac{1}{T} \int_0^T \int_R e^{it\lambda}\, d\mu(\lambda)\, dt$$

$$= \int_R g_T(\lambda)\, d\mu(\lambda) \qquad \text{by Fubini's theorem.}$$

The proof is completed as in (b). $\parallel$

If $\mu\{0\} = 0$ in 2.2.3, then $E[\,|Z\{0\}|^2] = \mu\{0\} = 0$, so that $Z\{0\} = 0$ a.e. Thus in this case the arithmetic average of the process converges in L^2 to $E[X(t)] = 0$. Conditions 2.2.3(b) and (d) yield simple conditions under which this will happen. In the discrete parameter case, $K(n) \to 0$ as $n \to \infty$ is sufficient; in particular, this is implied by $\sum_n |K(n)| < \infty$. The analogous conditions in the continuous parameter case are $K(t) \to 0$ as $t \to \infty$ and $\int_0^\infty |K(t)|\, dt < \infty$.

The spectral theorem is a "frequency domain" decomposition in the sense that the process $\{X(t)\}$ is approximated by sums of complex exponentials with various frequencies. The Wold decomposition, to be discussed next, is a "time domain" results; here, $\{X(t)\}$ is represented as the sum of two mutually orthogonal L^2 processes $\{X_1(t)\}$ and $\{X_2(t)\}$. The process $\{X_2(t)\}$ is deterministic in the sense that if $X_2(s)$ is known for $s \leq t$, perfect prediction of $X_2(t + \tau)$ is possible for $\tau > 0$. The process $\{X_1(t)\}$ is purely nondeterministic in the sense that in the long run, no useful prediction is possible. In other words, if the present time is t and we wish to predict $X_1(t + \tau)$ for very large τ, we can do just as well without any knowledge of the past history $X_1(s)$, $s \leq t$, as we can with this information.

First, we indicate the terminology to be used.

2.2.4   Definitions and Comments

Let $\{X(t), t \in T\}$ be a stationary L^2 process; define

$$H(X) = L^2\{X(t), t \in T\} \tag{1}$$

$$H_t(X) = L^2\{X(s), s \leq t\}, \qquad t \in T \tag{2}$$

$$H_{-\infty}(X) = \bigcap_{t \in T} H_t(X). \tag{3}$$

Note that $H_t(X) \downarrow H_{-\infty}(X)$ as $t \downarrow -\infty$, and $H_t(X) \uparrow H(X)$ as $t \uparrow \infty$.

If $Y \in L^2(\Omega, \mathscr{F}, P)$, $P_t Y$ will denote the projection of Y on $H_t(X)$; thus in the terminology of Section 1.5,

$$P_t Y = \hat{E}(Y \mid X(s), s \leq t).$$

Define the *shift operator* T_t on $H(X)$ by the requirement that

$$T_t\left[\sum_{k=1}^{n} a_k X(t_k) \right] = \sum_{k=1}^{n} a_k X(t_k + t);$$

by stationarity, T_t is a linear, inner-product-preserving map on the space L of finite linear combinations of the $X(t)$. Thus T_t extends to an isometric isomorphism of $H(X)$ with itself.

If $H_{-\infty}(X) = H(X)$, or equivalently, $H_t(X)$ is the same for all t, we call $\{X(t), t \in T\}$ a *deterministic* process; the process is called *nondeterministic* if $H_{-\infty}(X) \subsetneqq H(X)$, *purely nondeterministic* if $H_{-\infty}(X) = \{0\}$. Thus the only process that is both deterministic and purely nondeterministic is the zero process.

Prediction always means linear prediction; thus to predict $X(t + \tau)$ given $X(s), s \leq t$, we look for the element of the space $H_t(X)$ that is closest in the L^2 sense to $X(t + \tau)$, namely, $P_t X(t + \tau)$.

It follows from the definition of the shift operator that

$$T_t H_s(X) = H_{s+t}(X), \qquad s, t \in T; \tag{4}$$

we also have

$$T_r P_s X(t) = P_{r+s} X(r + t), \qquad r, s, t \in T. \tag{5}$$

To see this, note that $P_s X(t) \in H_s(X)$ and $X(t) - P_s X(t) \perp H_s(X)$; therefore, $T_r P_s X(t) \in H_{r+s}(X)$ and since T_r preserves inner products, $T_r[X(t) - P_s X(t)] \perp T_r H_s(X)$, in other words, $X(r + t) - T_r P_s X(t) \perp H_{r+s}(X)$. Thus $T_r P_s X(t) = P_{r+s} X(r + t)$.

By (5), the best predictor of $X(r + t)$ given $X(u), u \leq r + s$, can be obtained by shifting the best predictor of $X(t)$ given $X(u), u \leq s$, by r. Note also that since the shift operator is norm-preserving, (5) yields

$$E[\, |X(t) - P_s X(t)|^2] = E[\, |X(r + t) - P_{r+s} X(r + t)|^2],$$

$$r, s, t \in T. \tag{6}$$

Thus the *prediction error*, given by

$$\delta(\tau) = E[\,|\,X(t + \tau) - P_t X(t + \tau)|^2], \qquad t, \tau \in T, \tag{7}$$

is in fact independent of t. Since $H_s(X) \subset H_t(X)$ for $s \le t$, $\delta(\tau)$ increases with τ, and $\delta(\tau) = 0$ for $\tau \le 0$.

We may characterize deterministic and purely nondeterministic processes in terms of the prediction error.

2.2.5   Theorem

(a)   If $\{X(t),\, t \in T\}$ is deterministic, then $\delta(\tau) = 0$ for all τ.

(b)   If $\delta(\tau) = 0$ for some $\tau > 0$, the processes is deterministic.

(c)   The process is purely nondeterministic iff $\delta(\tau) \to K(0)$ as $\tau \to \infty$.

PROOF   (a)   By definition of deterministic, $X(t + \tau) \in H_t(X)$, hence $P_t X(t + \tau) = X(t + \tau)$.

(b)   We have $\delta(\tau') = 0$ for $\tau' \le \tau$, hence

$$X(t + \tau') = P_t X(t + \tau') \in H_t(X)$$

for $\tau' \le \tau$ and all t. Thus $H_s(X) \subset H_t(X)$ for all $s \le t + \tau$, and it follows that $H_t(X)$ is the same for all t.

(c)   Let $\delta(\tau) \to K(0)$. Since $H_{-\infty}(X) \subset H_s(X)$ for all s, we have

$$\|X(t) - P_s X(t)\| \le \|X(t) - P_{-\infty} X(t)\|,$$

and since

$$\|X(t)\|^2 = \|P_s X(t)\|^2 + \|X(t) - P_s X(t)\|^2 \tag{1}$$

(with a similar equation if P_s is replaced by $P_{-\infty}$), we have

$$\|P_{-\infty} X(t)\| \le \|P_s X(t)\|.$$

But $\|X(t)\|^2 = K(0)$ for all t, and $\|X(t) - P_s X(t)\|^2 = \delta(t - s) \to K(0)$ as $s \to -\infty$; therefore by (1), $\|P_s X(t)\| \to 0$, so that $P_{-\infty} X(t) = 0$. Thus $X(t) \perp H_{-\infty}(X)$ for all t, so $H_{-\infty}(X) = \{0\}$.

Conversely, assume $H_{-\infty}(X) = \{0\}$. Let $W = L^2 \lim_{\tau \to \infty} P_{t-\tau} X(t)$ (the L^2 limit exists; see Problem 4). Then $W \in H_{-\infty}(X) = \{0\}$, so that

$$\lim_{\tau \to \infty} \delta(\tau) = \lim_{\tau \to \infty} \|X(t + \tau) - P_t X(t + \tau)\|^2$$

$$= \lim_{\tau \to \infty} \|X(t) - P_{t-\tau} X(t)\|^2$$

$$= \|X(t)\|^2 = K(0). \quad \|$$

In the purely nondeterministic case, knowledge of the past history of the process does not help us predict $X(t + \tau)$ for very large τ. For if we set $\hat{X}(t) = m = E[X(t)]$, then $\|X(t) - \hat{X}(t)\|^2 = K(0)$, so that $\hat{X}(t)$ is, asymptotically, an optimal predictor.

We now establish the desired decomposition.

2.2.6 Wold Decomposition Theorem

A stationary L^2 process $\{X(t), t \in T\}$ can be represented uniquely as

$$X(t) = X_1(t) + X_2(t)$$

where $\{X_1(t)\}$ is purely nondeterministic, $\{X_2(t)\}$ is deterministic, $\{X_1(t)\}$ and $\{X_2(t)\}$ are mutually orthogonal ($X_1(s) \perp X_2(t)$ for all s, t), and, for each t, $X_1(t)$ and $X_2(t)$ belong to $H_t(X)$.

PROOF   Define $X_2(t) = P_{-\infty} X(t)$, $X_1(t) = X(t) - X_2(t)$; then

$$X_2(t) \in H_{-\infty}(X) \subset H_t(X),$$

and hence $X_1(t) \in H_t(X)$. Since $X_2(t) \in H_{-\infty}(X)$ and $X_1(s) \perp H_{-\infty}(X)$, $\{X_1(t)\}$ and $\{X_2(t)\}$ are mutually orthogonal.

Since $X_1(t) \in H_t(X)$ for all t, we have $H_{-\infty}(X_1) \subset H_{-\infty}(X)$; but $X_1(t) \perp H_{-\infty}(X)$ for all t, so $H_{-\infty}(X_1) \perp H_{-\infty}(X)$. Therefore $H_{-\infty}(X_1) = \{0\}$, proving $\{X_1(t)\}$ purely nondeterministic.

To prove that $\{X_2(t)\}$ is deterministic, note first that $X_2(t) \in H_{-\infty}(X)$ for all t, hence $H(X_2) \subset H_{-\infty}(X)$. Also, $X(t) = X_1(t) + X_2(t)$, so that $H_t(X) \subset H_t(X_1) \oplus H_t(X_2)$ (the orthogonal direct sum of two closed subspaces is closed, by continuity of the projection operators); the reverse inclusion also holds because $X_1(t)$, $X_2(t) \in H_t(X)$. Therefore, $H_{-\infty}(X) \subset H_t(X_1) \oplus H_t(X_2)$ for all t. But $H_{-\infty}(X) \perp X_1(t)$, so that $H_{-\infty}(X) \subset H_t(X_2)$. We conclude that $H(X_2) = H_t(X_2)$ ($= H_{-\infty}(X)$) for all t, as desired.

Finally, we prove uniqueness. If $X(t) = X_1(t) + X_2(t)$, where $\{X_1(t)\}$ and $\{X_2(t)\}$ satisfy the given requirements, the argument of the preceding paragraph shows that $H_t(X) = H_t(X_1) \oplus H_t(X_2) = H_t(X_1) \oplus H(X_2)$ because $\{X_2(t)\}$ is deterministic. It follows that $H_{-\infty}(X) = H_{-\infty}(X_1) \oplus H(X_2)$ (note that if $Z \in H_{-\infty}(X)$ and $Z \perp H(X_2)$, then $Z \in H_t(X_1)$ for each t). Since $\{X_1(t)\}$ is purely nondeterministic, $H_{-\infty}(X) = H(X_2)$. Thus we have $X(t) = X_1(t) + X_2(t)$, $X_2(t) \in H(X_2) = H_{-\infty}(X)$, $X_1(t) \perp H(X_2) = H_{-\infty}(X)$; consequently, $X_2(t) = P_{-\infty} X(t)$, and the proof is complete.   ∥

The basic advantage of the Wold decomposition is that the deterministic component is perfectly predictable, and an explicit formula may be obtained for predicting the purely nondeterministic component. We do this now for the discrete parameter case; the continuous parameter case will be considered beginning in Section 2.6.

2.2.7   Theorem

Let $\{X(n), \; n = 0, \; \pm 1, \; \ldots\}$ be a discrete parameter L^2 process, with $X(n) \not\equiv 0$. The process is purely nondeterministic iff it can be represented as

$$X(n) = \sum_{j=0}^{\alpha} a_j W(n - j), \qquad n = 0, \; \pm 1, \ldots,$$

where the $W(n)$ are mutually orthonormal random variables, $H_n(W) = H_n(X)$ for all n, and $\sum_{j=0}^{\infty} |a_j|^2 < \infty$.

In this case, the optimal predictor of $X(n)$ is given by

$$P_m X(n) = \sum_{j=n-m}^{\infty} a_j W(n - j)$$

with prediction error

$$\delta(n - m) = \sum_{j=0}^{n-m-1} |a_j|^2;$$

this holds for all n and m if we set $a_j = 0$ for $j < 0$.

PROOF   Assume $\{X(n)\}$ purely nondeterministic. Write $H_n(X) = H_{n-1}(X) \oplus D_n$; since $H_{n-1}(X) \cup \{aX(n) : a \in C\}$ spans $H_n(X)$, dim $D_n \le 1$. If dim $D_n = 0$, then $H_n(X) = H_{n-1}(X)$, so that $H_n(X) = H(X)$ for all n, and thus $\{X(n)\}$ is deterministic, a contradiction. Therefore dim $D_n = 1$ for all n. Moreover, if $m < n$, we have $D_m \subset H_{n-1}(X) \perp D_n$, so the D_n are mutually orthogonal.

Choose $W(n) \in D_n$ so that $\|W(n)\| = 1$; the $W(n)$ are then mutually orthonormal, so that if $k \ge 0$, $\{W(m), \; n - k \le m \le n\}$ is an orthonormal basis for $D_{n-k} \oplus \cdots \oplus D_n$. Since $H_n(X) = D_{n-k} \oplus \cdots \oplus D_n \oplus H_{n-k-1}(X)$,

$$X(n) = \sum_{j=0}^{k} a_j W(n - j) + P_{n-k-1} X(n)$$

where $a_j = \langle X(n), W(n - j) \rangle = \langle X(0), W(-j) \rangle$; note that $\sum_{j=0}^{\infty} |a_j|^2 < \infty$ by Bessel's inequality. Now as $k \to \infty$, $P_{n-k-1} X(n) \to P_{-\infty} X(n) = 0$ since $\{X(n)\}$ is purely nondeterministic (see Problem 4), hence $X(n) = \sum_{j=0}^{\infty} a_j W(n - j)$. This representation shows that $H_n(X) \subset H_n(W)$, and since $W(k) \in D_k \subset H_k(X)$, we have $H_n(W) \subset H_n(X)$.

Now if $X(n)$ has a representation of the given form, then by orthogonality, the projection of $W(n - j)$ on $H_m(X)$ is 0 for $n - j > m$, and since $W(n - j) \in H_{n-j}(X)$, the projection is $W(n - j)$ for $n - j \leq m$. Therefore

$$P_m X(n) = \sum_{j=n-m}^{\infty} a_j W(n - j).$$

The prediction error is

$$\delta(k) = \|X(n) - P_{n-k}X(n)\|^2 = \left\| \sum_{j=0}^{k-1} a_j W(n - j) \right\|^2$$

$$= \sum_{j=0}^{k-1} |a_j|^2 \to \|X(n)\|^2 = K(0)$$

as $k \to \infty$. By 2.2.5(c), $\{X(n)\}$ must be purely nondeterministic. ‖

If we drop the requirement that $H_n(W) = H_n(X)$ for all n, we may prove a result analogous to 2.2.7.

2.2.8   Theorem

The process $\{X(n)\}$ is purely nondeterministic iff it can be represented as

$$X(n) = \sum_{j=0}^{\infty} a_j W(n - j), \qquad n = 0, \pm 1, \ldots,$$

where the $W(n)$ are mutually orthonormal random variables and $\sum_{j=0}^{\infty} |a_j|^2 < \infty$.

PROOF   Let $\{X(n)\}$ be purely nondeterministic. If $X(n) \not\equiv 0$, the desired representation follows from 2.2.7, and if $X(n) \equiv 0$, the result is immediate. Conversely, assume $X(n) = \sum_{j=0}^{\infty} a_j W(n - j)$; then the $W(n)$ form an orthonormal basis for $H(X)$. Since $X(n)$ is a linear combination of the random variables $W(m)$, $m \leq n$, we have $H_n(X) \subset H_n(W)$. Thus for all n we have $H_{-\infty}(X) \subset H_{n-1}(W)$. But $W(n) \perp H_{n-1}(W)$, so that $W(n) \perp H_{-\infty}(X)$ for all n. Since $\{W(n)\}$ is an orthonormal basis for $H(X)$, we must have $H_{-\infty}(X) = \{0\}$, proving $\{X(n)\}$ purely nondeterministic. ‖

Note that if $X(n) = \sum_{j=0}^{\infty} a_j W(n - j)$, $n = 0, \pm 1, \ldots$, where the $W(n)$ are mutually orthonormal and $\sum_{j=0}^{\infty} |a_j|^2 < \infty$, it does not follow that $H_n(X) = H_n(W)$ for all n, even if $X(n) \not\equiv 0$. All that 2.2.7 implies is that $X(n)$ has a representation $\sum_{j=0}^{\infty} \hat{a}_j \hat{W}(n - j)$ in terms of a (possibly) *different* orthonormal set $\{\hat{W}(n)\}$, where $H_n(X) = H_n(\hat{W})$ for all n.

Problems

1. (*A Sampling Theorem*) Let $\{X(t), t \in R\}$ have spectral representation $\int_R e^{it\lambda} dZ(\lambda)$ and spectral measure μ. If the process is *band-limited*, that is, $\mu[R - (-2\pi W, 2\pi W)] = 0$ for some $W > 0$, show that for each t,

$$X(t) = \sum_{n=-\infty}^{\infty} X\left(\frac{n}{2W}\right) \frac{\sin (2\pi Wt - n\pi)}{2\pi Wt - n\pi}$$

where the series converges in $L^2(\Omega, \mathscr{F}, P)$. (*Hint:* Expand $e^{it\lambda}$, $-2\pi W \leq \lambda \leq 2\pi W$, in a Fourier series.)

2. Let $X(t) = \cos (2\pi Wt + \theta)$, where $W > 0$ and θ is uniformly distributed between 0 and 2π; $\{X(t)\}$ is stationary with covariance function $K(t) = \frac{1}{2} \cos 2\pi Wt$, so that $\mu\{-2\pi W\} = \mu\{2\pi W\} = \frac{1}{4}$, $\mu\{-2\pi W, 2\pi W\}^c = 0$. Show that the sampling theorem of Problem 1 fails for $\{X(t)\}$. (The trouble is that μ assigns positive measure to the endpoints of $(-2\pi W, 2\pi W)$.)

3. If n is an integer, let $X_n = e^{in\theta}$, where θ is uniformly distributed between 0 and 2π. Show that $\{X_n\}$ is purely nondeterministic, but that perfect *nonlinear* prediction is possible; specifically, for each n there is a Borel measurable $f_n: C \times C \to C$ such that if $m < n$, $X_n = f_n(X_{m-1}, X_m)$.

4. If $\{X(t), t \in T\}$ is a stationary L^2 process and $Y \in L^2\{X(t), t \in T\}$, show that $P_s Y \overset{L^2}{\to} P_{-\infty} Y$ as $s \to -\infty$.

5. Let $X(t) = X_1(t) + X_2(t)$ be the Wold decomposition of $\{X(t)\}$, and assume that $\{X(t)\}$ has 0 mean and a continuous covariance function with spectral measure μ_X.

 (a) If T_t is the shift operator on $H(X)$ (see 2.2.4), show that $T_t X_i(r) = X_i(r + t)$, $i = 1, 2$.
 (b) If $\varphi \in L^2(\mu_X)$, show that

$$T_t \int_S \varphi(\lambda) dZ_X(\lambda) = \int_S e^{it\lambda} \varphi(\lambda) dZ_X(\lambda).$$

 (c) Show that there exist $\varphi_i \in L^2(\mu_X)$, $i = 1, 2$, such that

$$X_i(t) = \int_S e^{it\lambda} \varphi_i(\lambda) dZ_X(\lambda) \qquad \text{for all } t.$$

6. In the Wold decomposition of a process with 0 mean and a continuous covariance function, show that the spectral measure of $\{X(t)\}$ is the sum of the spectral measures of $\{X_1(t)\}$ and $\{X_2(t)\}$.

2.3 Examples of Discrete Parameter Processes

When we begin to discuss the solution to the prediction problem in the next section, it will be useful to have a collection of examples; in this section we examine the properties of several stationary discrete parameter L^2 processes. All processes are assumed to have 0 mean; the interval $[-\pi, \pi]$ will be denoted by S, and $\mathscr{B}[-\pi, \pi]$ by $\mathscr{S}$.

2.3.1 White Noise

Let $\{W(n)\}$ be a sequence of orthonormal random variables, that is, $E[W(m)\overline{W(n)}] = 0$ for $n \neq m$, $E[\,|W(n)|^2] = 1$. The covariance function of $\{W(n)\}$ is given by

$$K_W(n) = \begin{cases} 1 & \text{if} \quad n = 0 \\ 0 & \text{if} \quad n \neq 0. \end{cases}$$

Now if the process has a spectral density $f(\lambda)$, $-\pi \leq \lambda \leq \pi$, then

$$K_W(n) = \int_S e^{in\lambda} f(\lambda)\, d\lambda, \tag{1}$$

so that the $K_W(n)$ are the Fourier coefficients of f. We therefore expect that

$$f(\lambda) = \frac{1}{2\pi} \sum_{n=-\infty}^{\infty} K_W(n) e^{-in\lambda} = \frac{1}{2\pi}, \qquad -\pi \leq \lambda \leq \pi,$$

and it may be verified that this choice of f does indeed satisfy (1). Since the spectral density is constant, all frequencies contribute equally to the average power; hence the term "white noise."

2.3.2 Cyclic Model

Let $X(n) = \sum_{j=1}^{N} \exp(in\lambda_j) Z_j$, where $-\pi \leq \lambda_j \leq \pi$, $1 \leq j \leq N$, the Z_j are mutually orthogonal random variables, and $E[\,|Z_j|^2] = \sigma_j^2 > 0$. The covariance function is given by

$$K(n) = \sum_{j=1}^{N} e^{in\lambda_j} \sigma_j^2,$$

so that the spectral measure μ is concentrated on $\{\lambda_1, \ldots, \lambda_N\}$, with $\mu\{\lambda_j\} = \sigma_j^2$.

2.3.3   Generalized Cyclic Model

This is a natural limiting case of 2.3.2. Let $X(n) = \int_S e^{in\lambda} \, dZ(\lambda)$, where Z is a mov defined on $\mathscr{S}$ with values in $L^2(\Omega, \mathscr{F}, P)$ and associated measure μ. By 2.1.2(b), the covariance function is given by

$$K(n) = \int_S e^{in\lambda} \, d\mu(\lambda),$$

so that μ is the spectral measure of the process. By the spectral decomposition theorem, every stationary discrete parameter L^2 process can be represented as a generalized cyclic model.

2.3.4   Linear Operations

If $X(n) = \sum_{j=-\infty}^{\infty} a_j Y(n-j)$, $n = 0, \pm 1, \ldots$, where the series is assumed to converge in L^2, and $\{Y(n)\}$ has spectral representation

$$Y(n) = \int_S e^{in\lambda} \, dZ_Y(\lambda),$$

let us try to find the spectral properties of $\{X(n)\}$ in terms of those of $\{Y(n)\}$. We have

$$\sum_{j=-N}^{N} a_j Y(n-j) = \int_S e^{in\lambda} \sum_{j=-N}^{N} a_j e^{-ij\lambda} \, dZ_Y(\lambda).$$

If the series $\sum_{j=-\infty}^{\infty} a_j e^{-ij\lambda}$ converges in $L^2(\mu_Y)$, we may let $N \to \infty$ to obtain, by 2.1.2(e),

$$X(n) = \int_S e^{in\lambda} \varphi(\lambda) \, dZ_Y(\lambda) \tag{1}$$

where

$$\varphi(\lambda) = \sum_{j=-\infty}^{\infty} a_j e^{-ij\lambda}.$$

If the stationary process $\{Y(n)\}$ is given, and a process $\{X(n)\}$ is defined by (1), where $\varphi \in L^2(\mu_Y)$ (not necessarily given by $\varphi(\lambda) = \sum_{j=-\infty}^{\infty} a_j e^{-ij\lambda}$), we say that $\{X(n)\}$ is obtained from $\{Y(n)\}$ by a *linear operation* or a *linear filter* with *transfer function* φ.

The physical significance of φ may be seen as follows. If we define

$$Z(E) = \int_E \varphi(\lambda) \, dZ_Y(\lambda), \qquad E \in \mathscr{S},$$

then by 2.1.5, Z is a mov on $\mathscr{S}$ and

$$\int_S e^{in\lambda} \, dZ(\lambda) = \int_S e^{in\lambda} \varphi(\lambda) \, dZ_Y(\lambda) = X(n).$$

Thus by the uniqueness part of the spectral decomposition theorem, $Z = Z_X$, the spectral mov of $\{X(n)\}$. We may therefore regard $X(n)$ as being approximated (see 2.2.2(d)) by a finite sum of the form $\sum_{k=1}^n e^{in\lambda_k} \varphi(\lambda_k) \, \Delta Z_Y(\lambda_k)$. In effect the complex exponential $\Delta Z_Y(\lambda_k) e^{in\lambda_k}$, which is a component of the "input" process $\{Y(n)\}$, is multiplied by $\varphi(\lambda_k)$ to obtain the corresponding component of the "output" process $\{X(n)\}$; note that multiplication by $\varphi(\lambda_k)$ changes the amplitude and phase of the "input signal," but not the frequency λ_k.

From (1) we see that $\{X(n)\}$ is stationary, with covariance function

$$K_X(n) = E[X(n+m)\overline{X(m)}] = \int_S e^{in\lambda} \, |\varphi(\lambda)|^2 \, d\mu_Y(\lambda) \qquad (2)$$

by 2.1.2(b). Therefore the spectral measure of $\{X(n)\}$ is given by

$$\mu_X(E) = \int_E |\varphi(\lambda)|^2 \, d\mu_Y(\lambda). \qquad (3)$$

In particular, if $\{Y(n)\}$ has a spectral density f_Y, then $\{X(n)\}$ has a spectral density f_X, where

$$f_X(\lambda) = |\varphi(\lambda)|^2 f_Y(\lambda). \qquad (4)$$

Now let us try to reverse the above process. Suppose that we are given a finite measure μ on $\mathscr{S}$, and a stationary L^2 process $\{X(n)\}$ whose spectral measure is given by

$$\mu_X(E) = \int_E |\varphi(\lambda)|^2 \, d\mu, \qquad E \in \mathscr{S},$$

for some function $\varphi \in L^2(\mu)$. We wish to find a stationary process $\{Y(n)\}$, with spectral measure μ, such that (1) will be satisfied.

Assume $|\varphi(\lambda)| > 0$ a.e. $[\mu]$ (note that $|\varphi(\lambda)| > 0$ a.e. $[\mu_X]$ by definition of μ_X). Then, as in the proof of 2.1.5(c),

$$\mu(E) = \int_E |\varphi(\lambda)|^{-2} \, d\mu_X(\lambda), \qquad (5)$$

so that $\varphi^{-1} \in L^2(\mu_X)$. Since we want $\{Y(n)\}$ to have spectral measure μ, it is natural to define

$$Y(n) = \int_S e^{in\lambda} [\varphi(\lambda)]^{-1} \, dZ_X(\lambda). \qquad (6)$$

Then $\{Y(n)\}$ is stationary, with covariance function

$$K_Y(n) = \int_S e^{in\lambda} |\varphi(\lambda)|^{-2} d\mu_X(\lambda)$$

$$= \int_S e^{in\lambda} d\mu(\lambda) \qquad \text{by (5)},$$

so that $\{Y(n)\}$ does indeed have spectral measure μ. The spectral mov of $\{Y(n)\}$ is given by

$$Z_Y(E) = \int_E [\varphi(\lambda)]^{-1} dZ_X(\lambda), \qquad E \in \mathscr{S} \tag{7}$$

(see the discussion after (1)). Thus by 2.1.5(b),

$$\int_S e^{in\lambda} \varphi(\lambda) \, dZ_Y(\lambda) = \int_S e^{in\lambda} \, dZ_X(\lambda) = X(n),$$

proving (1).

Finally, if $\varphi(\lambda) = \sum_{j=-\infty}^{\infty} a_j e^{-ij\lambda}$, where the series converges in $L^2(\mu)$, then

$$X(n) = \int_S e^{in\lambda} \, dZ_X(\lambda)$$

$$= \int_S e^{in\lambda} \varphi(\lambda) \, dZ_Y(\lambda) \qquad \text{by (7) and 2.1.5}$$

$$= \sum_{j=-\infty}^{\infty} a_j \int_S e^{i(n-j)\lambda} \, dZ_Y(\lambda)$$

$$= \sum_{j=-\infty}^{\infty} a_j Y(n-j).$$

To summarize: If $\{Y(n)\}$ is given and $\{X(n)\}$ is defined by (1), we assume only that $\varphi \in L^2(\mu_Y)$; the spectral properties of $\{X(n)\}$ are then given by (2)–(4). If $X(n) = \sum_{j=-\infty}^{\infty} a_j Y(n-j)$, with convergence in $L^2(\Omega, \mathscr{F}, P)$, then $\varphi(\lambda) = \sum_{j=-\infty}^{\infty} a_j e^{-ij\lambda}$, where the series is assumed to converge in $L^2(\mu_Y)$.

If $\{X(n)\}$ is given and we wish to find $\{Y(n)\}$ satisfying (1), it is assumed that we have a finite measure μ on $\mathscr{S}$ such that $\mu_X \ll \mu$, with density $|\varphi|^2 > 0$ a.e. $[\mu]$. A process $\{Y(n)\}$ can then be constructed so that $\mu_Y = \mu$ and (1) is satisfied. If $\varphi(\lambda) = \sum_{j=-\infty}^{\infty} a_j e^{-ij\lambda}$, where the series converges in $L^2(\mu)$, then $X(n) = \sum_{j=-\infty}^{\infty} a_j Y(n-j)$.

We now apply these results to some specific cases.

2.3.5 Moving Average

Let $X(n) = \sum_{j=-\infty}^{\infty} a_j W(n-j)$, where $\{W(n)\}$ is white noise; assume $\sum_{j=-\infty}^{\infty} |a_j|^2 < \infty$, so that the series converges in L^2. (If $a_j = 0$ for $j < 0$, we call $\{X(n)\}$ a *one-sided moving average*.) By 2.3.1, $\{W(n)\}$ has spectral density $f(\lambda) = 1/2\pi$, $-\pi \le \lambda \le \pi$, hence $\{X(n)\}$ has a spectral density given by

$$f_X(\lambda) = |\varphi(\lambda)|^2/2\pi$$

where

$$\varphi(\lambda) = \sum_{j=-\infty}^{\infty} a_j e^{-ij\lambda}.$$

The spectral measure of $\{W(n)\}$ is $(2\pi)^{-1}$ times Lebesgue measure, so the series converges in $L^2(\mu_W)$. The covariance function is given by

$$\begin{aligned} K_X(n) &= E[X(m+n)\overline{X(m)}] \\ &= \sum_{j,k} a_j \bar{a}_k E[W(m+n-j)\overline{W(m-k)}] \\ &= \sum_{k=-\infty}^{\infty} a_{n+k}\bar{a}_k. \end{aligned}$$

In fact *any process with a spectral density f such that $f > 0$ a.e. [Lebesgue measure] can be represented as a moving average.*

To prove this, let μ be $(2\pi)^{-1}$ times Lebesgue measure, and choose φ such that $|\varphi|^2 = 2\pi f$. Then $\varphi \in L^2(\mu)$, so that φ has a Fourier expansion $\varphi(\lambda) = \sum_{j=-\infty}^{\infty} a_j e^{-ij\lambda}$, with convergence in $L^2(\mu)$. Thus $X(n) = \sum_{j=-\infty}^{\infty} a_j W(n-j)$, where $\{W(n)\}$ has spectral measure μ; and hence is white noise.

A *moving average of finite extent* is of the form $X(n) = \sum_{j=0}^{N-1} a_j W(n-j)$, where $\{W(n)\}$ is white noise. The spectral density of the process is

$$f_X(\lambda) = \frac{1}{2\pi} \left| \sum_{k=0}^{N-i} a_k e^{-ik\lambda} \right|^2.$$

Conversely, a process with this spectral density can be represented as a moving average with finite extent. To see this, note that $f_X(\lambda) > 0$ a.e. [Lebesgue measure] because a polynomial has only finitely many zeros; thus the above argument applies, with $\varphi(\lambda) = \sum_{k=0}^{N-1} a_k e^{-ik\lambda}$.

An *equal weight moving average of finite extent* is formed by taking $a_j = 1/N$, $0 \le j \le N-1$. The covariance function $\sum_{k=-\infty}^{\infty} a_{n+k}\bar{a}_k$ can then be evaluated explicitly; we obtain the triangular function

$$K_X(n) = \begin{cases} \dfrac{1}{N}\left(1 - \dfrac{|n|}{N}\right) & \text{if } |n| \le N \\ 0 & \text{if } |n| > N. \end{cases}$$

The spectral density is

$$f_X(\lambda) = \frac{1}{2\pi N} \left| \frac{1 - e^{-iN\lambda}}{1 - e^{-i\lambda}} \right|^2 = \frac{1}{2\pi N} \left(\frac{1 - \cos N\lambda}{1 - \cos \lambda} \right) = \frac{1}{2\pi N} \frac{\sin^2 \frac{1}{2}N\lambda}{\sin^2 \frac{1}{2}\lambda}.$$

2.3.6 Autoregressive Scheme

Let b_0, b_1, ... be a sequence of complex numbers such that the power series $B(z) = \sum_{j=0}^{\infty} b_j z^j$ has radius of convergence greater than 1, and $B(z)$ is never 0 when $|z| = 1$. If $\{W(n)\}$ is white noise, an autoregressive scheme with coefficients b_j is a stationary L^2 process $\{Y(n)\}$ that satisfies the stochastic difference equation

$$\sum_{j=0}^{\infty} b_j Y(n-j) = W(n), \qquad n = 0, \pm 1, \ldots,$$

with convergence in $L^2(\Omega, \mathscr{F}, P)$. (In the most common situation, B is a polynomial, and there are no convergence difficulties.)

To prove that such a process can be found, let

$$\varphi(\lambda) = B(e^{-i\lambda}) = \sum_{j=0}^{\infty} b_j e^{-ij\lambda},$$

and let

$$\mu(E) = \int_E |\varphi(\lambda)|^{-2} \, d\mu_W(\lambda) = \int_E |\varphi(\lambda)|^{-2} \frac{d\lambda}{2\pi}, \qquad E \in \mathscr{S}.$$

Then $\mu_W(E) = \int_E |\varphi(\lambda)|^2 \, d\mu(\lambda)$, and the series for φ converges uniformly, hence in $L^2(\mu)$. By 2.3.4, there is a process $\{Y(n)\}$ satisfying the stochastic difference equation. Furthermore, the spectral measure of $\{Y(n)\}$ is μ, hence $\{Y(n)\}$ has a spectral density given by

$$f_Y(\lambda) = \frac{1}{2\pi |B(e^{-i\lambda})|^2}.$$

Note that if $b_0 \neq 0$, we may solve the difference equation for $Y(n)$; we obtain

$$Y(n) = -b_0^{-1} \sum_{j=1}^{\infty} b_j Y(n-j) + b_0^{-1} W(n).$$

In statistical terminology, $Y(n)$ is a "linear regression" on the "past" $Y(n-1)$, $Y(n-2)$, ..., plus an "innovation" $b_0^{-1} W(n)$ at time n. This is the origin of the term "autoregressive scheme."

All L^2 processes satisfying the stochastic difference equation have the same spectral density. For if $\{Y(n)\}$ is such a process, then the series for $\varphi(\lambda) = B(e^{-i\lambda})$ converges uniformly, hence in $L^2(\mu_Y)$. By Eq. (3) of 2.3.4,

$$\mu_W(E) = \int_E |B(e^{-i\lambda})|^2 \, d\mu_Y(\lambda),$$

hence

$$\mu_Y(\lambda) = \int_E \frac{d\lambda}{2\pi \,|B(e^{-i\lambda})|^2},$$

as desired.

Conversely, if $\{Y(n)\}$ has spectral density $(2\pi)^{-1} |B(e^{-i\lambda})|^{-2}$, then $\sum_{j=0}^{\infty} b_j Y(n - j)$ is white noise, by Eq. (4) of 2.3.4.

Note that since $f_Y(\lambda) > 0$ for all $\lambda \in S$, an autoregressive scheme can be represented as a moving average.

2.3.7   Rational Spectral Densities

Let $\{X(n)\}$ have spectral density

$$f(\lambda) = \frac{1}{2\pi} \left| \frac{A(e^{-i\lambda})}{B(e^{-i\lambda})} \right|^2$$

where A and B are polynomials and B has no zeros in $\{z : |z| = 1\}$.

If $A(z) = \sum_{j=0}^{M} a_j z^k$, $B(z) = \sum_{k=0}^{N} b_k z^k$, we know from 2.3.4 that $\sum_{k=0}^{N} b_k X(n - k)$ has spectral density

$$|B(e^{-i\lambda})|^2 f(\lambda) = (2\pi)^{-1} |A(e^{-i\lambda})|^2.$$

We claim that there is a white noise process $\{W(n)\}$ such that

$$X_0(n) = \sum_{k=0}^{N} b_k X(n - k) = \sum_{j=0}^{M} a_j W(n - j).$$

To see this, let μ be $(2\pi)^{-1}$ times Lebesgue measure; the spectral measure μ_0 of $\{X_0(n)\}$ is absolutely continuous with respect to μ with density $|A(e^{-i\lambda})|^2 > 0$ a.e. $[\mu]$. (Note that $A(z)$ has only finitely many zeros.) The result follows from 2.3.4.

Conversely, if $\sum_{k=0}^{N} b_k X(n - k) = \sum_{j=0}^{M} a_j W(n - j)$, where $\{W(n)\}$ is white noise, and $B(z)$ has no zeros for $|z| = 1$, then $\{X(n)\}$ has spectral density $(2\pi)^{-1} |A(e^{-i\lambda})/B(e^{-i\lambda})|^2$. For by 2.3.4, $\{\sum_{j=0}^{M} a_j W(n - j)\}$ has spectral density $(2\pi)^{-1} |A(e^{-i\lambda})|^2$, and $\{\sum_{k=0}^{N} b_k X(n - k)\}$ has spectral measure

$$\mu_0(E) = \int_E |B(e^{-i\lambda})|^2 \, d\mu_X(\lambda),$$

so that

$$\int_E |B(e^{-i\lambda})|^2 \, d\mu_X(\lambda) = \int_E |A(e^{-i\lambda})|^2 \frac{d\lambda}{2\pi}, \qquad E \in \mathcal{S}.$$

Since $|B(e^{-i\lambda})| > 0$ for all $\lambda \in S$,

$$\mu_X(E) = \int_E \frac{1}{2\pi} \left| \frac{A(e^{-i\lambda})}{B(e^{-i\lambda})} \right|^2 d\lambda, \qquad E \in \mathscr{S},$$

as desired.

If $\sum_{k=0}^N b_k X(n-k) = \sum_{j=0}^M a_j W(n-j)$, where $\{W(n)\}$ is white noise, $\{X(n)\}$ is sometimes called a *mixed scheme*; if $M = 0$, we have an autoregressive scheme, and if $N = 0$, we have a moving average.

A process with a rational spectral density can be represented as a moving average, since $f(\lambda) > 0$ a.e. [Lebesgue measure] (see 2.3.5).

Problems

1. Let $\{X(n)\}$ be an autoregressive scheme, with $\sum_{j=0}^\infty b_j X(n-j) = W(n)$.

   (a) If $m < n$, show that $E[X(m)\overline{W(n)}]$ is given by the contour integral

   $$\frac{1}{2\pi i} \int_{|z|=1} \frac{z^{n-m-1}}{B(z)} \, dz$$

where the path of integration is counterclockwise.

   (b) Show that if $B(z)$ has no zeros for $|z| \leq 1$, then $X(m) \perp W(n)$ whenever $m < n$.

2. If $\{Y(n)\}$ is an autoregressive scheme and $B(z)$ has no zeros for $|z| \leq 1$, show that $\{Y(n)\}$ is a one-sided moving average.

3. Consider the *first-order autoregressive scheme* $Y(n) - \alpha Y(n-1) = W(n), 0 < |\alpha| < 1$. Express $\{Y(n)\}$ as a one-sided moving average, and find the covariance function and spectral density.

4. Let $\{X(n)\}$ be a mixed scheme, with

$$X_0(n) = \sum_{k=0}^N b_k X(n-k) = \sum_{j=0}^M a_j W(n-j).$$

If $B(z)$ has no zeros for $|z| \leq 1$, show that $X(m) \perp W(n)$ whenever $m < n$; show also that $\{X(n)\}$ can be represented as a one-sided moving average.

5. If $\{X_1(n)\}$ is obtained from $\{Y(n)\}$ by a linear filter with transfer function φ_1, and $\{X_2(n)\}$ is obtained from $\{X_1(n)\}$ by a linear filter with transfer function φ_2, show that $\{X_2(n)\}$ can be obtained from $\{Y(n)\}$ by a linear filter with transfer function $\varphi(\lambda) = \varphi_1(\lambda)\varphi_2(\lambda)$.

6. If $X(n) = \frac{1}{3}[Y(n) + Y(n-1) + Y(n-2)]$, find the transfer function of the associated linear filter. Show that this operation has the effect of attenuating the frequency components of $\{Y(n)\}$ near $\pm 2\pi/3$.

2.4 Discrete Parameter Prediction: Special Cases

The proof of the spectral decomposition theorem 2.2.1 shows that there is an isometric isomorphism between the space $H = L^2\{X(t), \ t \in T\}$ and $\hat{H} = L^2(S, \mathscr{S}, \mu)$, where μ is the spectral measure of the process $\{X(t)\}$. In the isomorphism, $X(t)$ corresponds to the function $e_t(\lambda) = e^{it\lambda}$, $\lambda \in S$. We may therefore translate the prediction problem into a problem in approximation of functions. If we wish to predict $X(t + \tau)$ given $X(s), s \leq t$, we look for the projection $P_t X(t + \tau)$ of $X(t + \tau)$ on the space $H_t(X)$ spanned by the $X(s)$, $s \leq t$. Because of the isometric isomorphism, this is equivalent to finding the projection $\hat{P}_t e_{t+\tau}$ on the space $\hat{H}_t(X)$ spanned by the functions $e_s, s \leq t$. This represents a solution in principle to the prediction problem, but it remains to find the predictor explicitly. In certain cases this is feasible; we consider some examples.

2.4.1 Example

Let $C(z) = \sum_{j=0}^{\infty} c_j z^j$ have radius of convergence $r > 1$, and assume that $C(z)$ has no zeros for $|z| \leq 1$. Let $\{X(n)\}$ be a process with spectral density

$$f(\lambda) = |C(e^{-i\lambda})|^2/2\pi;$$

by 2.3.5, $\{X(n)\}$ is a one-sided moving average: $X(n) = \sum_{j=0}^{\infty} c_j W(n - j)$, where $\{W(n)\}$ is white noise. Now if τ is a positive integer, we wish to project $e_{n+\tau}$ on the space spanned by the $e_m, m \leq n$. We claim that the projection is given by

$$\hat{P}_n e^{i(n+\tau)\lambda} = e^{i(n+\tau)\lambda} \frac{C_\tau(e^{-i\lambda})}{C(e^{-i\lambda})}, \qquad \tau \geq 0, \tag{1}$$

where $C_\tau(z) = \sum_{j=\tau}^{\infty} c_j z^j$.

First observe that $1/C(z)$ is analytic on an open set $U \supset \{z : |z| \leq 1\}$, hence $1/C(e^{-i\lambda})$ can be expressed as a power series in $e^{-i\lambda}$. Also, $C_\tau(e^{-i\lambda})$ can be expressed in terms of $e^{-ik\lambda}$, $k \geq \tau$. It follows that $e^{i(n+\tau)\lambda} C_\tau(e^{-i\lambda})/C(e^{-i\lambda}) \in \hat{H}_n(X)$. Now

$$\int_S \left[e^{i(n+\tau)\lambda} - \frac{e^{i(n+\tau)\lambda} C_\tau(e^{-i\lambda})}{C(e^{-i\lambda})} \right] e^{-ir\lambda} \, d\mu(\lambda)$$

$$= \frac{1}{2\pi} \int_S e^{i(n+\tau-r)\lambda} \left[1 - \frac{C_\tau(e^{-i\lambda})}{C(e^{-i\lambda})} \right] |C(e^{-i\lambda})|^2 \, d\lambda$$

$$= \frac{1}{2\pi} \int_S e^{i(n+\tau-r)\lambda} \sum_{j=0}^{\tau-1} c_j e^{-ij\lambda} \sum_{k=0}^{\infty} \bar{c}_k e^{ik\lambda} \, d\lambda.$$

Since $e^{i\tau\lambda} \sum_{j=0}^{\tau-1} c_j e^{-ij\lambda}$ can be expressed in terms of exponentials $e^{im\lambda}$ with $m \geq 1$, it follows that the integral is 0 for $r \leq n$, and hence

$$e^{i(n+\tau)\lambda} - e^{i(n+\tau)\lambda} \frac{C_\tau(e^{-i\lambda})}{C(e^{-i\lambda})} \perp \hat{H}_n(X),$$

as desired.

Now let us translate (1) back to the space H. If $f \in L^2(\mu)$, the element of H corresponding to f in the isomorphism is $\int_S f \, dZ$, where Z is the spectral mov of $\{X(n)\}$. Thus (1) becomes

$$P_n X(n + \tau) = \int_S e^{in\lambda} e^{i\tau\lambda} \frac{C_\tau(e^{-i\lambda})}{C(e^{-i\lambda})} \, dZ(\lambda).$$

In other words the optimal predictor is a linear filter with transfer function

$$\varphi(\lambda) = \frac{e^{i\tau\lambda} C_\tau(e^{-i\lambda})}{C(e^{-i\lambda})}.$$

(A natural question at this point is why not use $\varphi(\lambda) = e^{i\tau\lambda}$, for then $\int_S e^{i(n+\tau)\lambda} \, dZ(\lambda) = X(n + \tau)$, giving perfect prediction. The trouble is that $e^{i(n+\tau)\lambda}$ does not belong to the space $\hat{H}_n(X)$ if $\tau > 0$.)

2.4.2 Example

Let $\{X(n)\}$ have a rational spectral density

$$f(\lambda) = \frac{1}{2\pi} \left| \frac{A(e^{-i\lambda})}{B(e^{-i\lambda})} \right|^2,$$

where $A(z)$ and $B(z)$ have no zeros for $|z| \leq 1$. We may write

$$C(z) = \frac{A(z)}{B(z)} = \sum_{j=0}^{\infty} c_j z^j,$$

with radius of convergence $r > 1$. We now are in the situation described in the previous example, and the optimal predictor is given by formula (1) of 2.4.1.

Let us carry out a numerical example. Suppose that $\{X(n)\}$ has covariance function $K(n) = a^n$, $n \geq 0$, where $|a| < 1$. (If $n < 0$, $K(n) = \overline{K(-n)} = (\bar{a})^{-n}$.) If there is a spectral density f, then

$$K(n) = \int_{-\pi}^{\pi} e^{in\lambda} f(\lambda) \, d\lambda,$$

and as in 2.3.1 we expect that $f(\lambda)$ is given by a Fourier series:

$$
\begin{aligned}
f(\lambda) &= \frac{1}{2\pi} \sum_{n=-\infty}^{\infty} K(n)e^{-in\lambda} \\
&= \frac{1}{2\pi} \left[\sum_{n=0}^{\infty} a^n e^{-in\lambda} + \sum_{n=-\infty}^{-1} \bar{a}^{|n|} e^{-in\lambda} \right] \\
&= \frac{1}{2\pi} \left[\sum_{n=0}^{\infty} (ae^{-i\lambda})^n + \sum_{n=0}^{\infty} (\bar{a}e^{i\lambda})^n - 1 \right] \\
&= \frac{1}{2\pi} \left[\frac{1}{1 - ae^{-i\lambda}} + \frac{1}{1 - \bar{a}e^{i\lambda}} - 1 \right] \\
&= \frac{1}{2\pi} \frac{1 - |a|^2}{|1 - ae^{-i\lambda}|^2}.
\end{aligned}
$$

Thus we have $A(z) = (1 - |a|^2)^{1/2}$, $B(z) = 1 - az$; therefore

$$
C(z) = \frac{A(z)}{B(z)} = (1 - |a|^2)^{1/2} \sum_{j=0}^{\infty} a^j z^j.
$$

Now

$$
C_\tau(e^{-i\lambda}) = (1 - |a|^2)^{1/2} \sum_{j=\tau}^{\infty} a^j e^{-ij\lambda} = a^\tau e^{-i\tau\lambda} C(e^{-i\lambda}),
$$

and hence

$$
\varphi(\lambda) = a^\tau.
$$

Thus

$$
P_n X(n + \tau) = a^\tau X(n),
$$

so that optimal prediction τ seconds ahead is simply multiplication by a^τ.

Problems

1. If $\{X(n)\}$ has spectral density $f(\lambda) = (1 - |a|^2)/2\pi \, |1 - ae^{-i\lambda}|^2$ as in 2.4.2, then by the discussion of 2.3.6, $(1 - |a|^2)^{-1}(X(n) - aX(n - 1))$ is white noise $W(n)$, so that $\{X(n)\}$ is a first order autoregressive scheme. Furthermore, $X(m) \perp W(n)$ for $m < n$, by Problem 1 of Section 2.3. Use these results to give an alternative proof that optimal prediction τ seconds ahead is multiplication by a^τ.

2. Consider the *second order autoregressive scheme*

$$X(n) - (\alpha_1 + \alpha_2)X(n-1) + \alpha_1\alpha_2 X(n-2) = W(n),$$

where $\{W(n)\}$ is white noise and $|\alpha_1| < 1$, $|\alpha_2| < 1$, $\alpha_1 \neq \alpha_2$.

(a)   Find the spectral density of $\{X(n)\}$.
(b)   Express the predictor $P_n X(n+\tau)$ in terms of $X(n)$ and $X(n-1)$.
(c)   Find the prediction error for predicting one step ahead ($\tau = 1$).

3. Let $\varphi_\tau(\lambda)$ be the projection of $e^{i\tau\lambda}$ on the subspace of $L^2(\mu_X)$ spanned by the functions $e^{im\lambda}$, $m \leq 0$. Show that φ_τ is the transfer function of the optimal predictor of $X(n+\tau)$, that is,

$$P_n X(n+\tau) = \int_{-\pi}^{\pi} e^{in\lambda}\varphi_\tau(\lambda)\, dZ_X(\lambda).$$

4. If the covariance function of $\{X(n)\}$ is a sum of exponentials, then, as in 2.4.2, the spectral density can be expressed as

$$f(\lambda) = \frac{R(e^{-i\lambda})}{S(e^{-i\lambda})} = Ae^{-in\lambda}\frac{\prod_j (e^{-i\lambda} - \alpha_j)}{\prod_k (e^{-i\lambda} - \beta_k)}$$

where R and S are relatively prime polynomials, the α_j and β_k are the nonzero roots of R and S, A is constant, and n is an integer; since f is integrable, none of the β_k has modulus 1.

Let f be a nonnegative Lebesgue integrable function expressible in the above form. We are going to show that f is a rational spectral density as defined in 2.3.7.

(a)   Show that z is a zero of $R(z)/S(z)$ iff $1/\bar{z}$ is a zero of $R(z)/S(z)$, and similarly for poles.

(b)   Show that $f(\lambda)$ can be written in the form

$$A_0 e^{-im\lambda}\frac{|\prod_{|\alpha_j|>1} (e^{-i\lambda} - \alpha_j)|^2}{|\prod_{|\beta_k|>1} (e^{-i\lambda} - \beta_k)|^2}\prod_{|\alpha_j|=1} (e^{-i\lambda} - \alpha_j)$$

(Actually, we can choose the roots α_j and β_k inside the unit circle rather than outside; the above choice is made so that the solution to the prediction problem given in 2.4.2 will be applicable.)

(c)   Show that if $|\alpha_j| = 1$, then α_j is a root of even multiplicity. Conclude that $f(\lambda)$ can be expressed as a rational spectral density as defined in 2.3.7.

COMMENT   In 2.4.2, the case in which some $|\alpha_j| = 1$ is excluded. In this case, $\int_{-\pi}^{\pi} \ln f(\lambda)\, d\lambda = -\infty$, so that, as we show in 2.5.4, $\{X(n)\}$ is deterministic and hence (at least in principle) perfectly predictable.

2.5   Discrete Parameter Prediction: General Solution

(This section may (and probably should on a first reading) be omitted without loss of continuity.)

Let $\{X(n)\}$ be a stationary discrete parameter L^2 process with 0 mean. By the Wold decomposition theorem, we may write $X(n) = X_1(n) + X_2(n)$, where $\{X_1(n)\}$ is purely nondeterministic and $\{X_2(n)\}$ is deterministic. For the moment, we concentrate on the purely nondeterministic component. By 2.2.7, $\{X_1(n)\}$ is a one-sided moving average; so that by 2.3.5, $\{X_1(n)\}$ has a spectral density $f_1(\lambda)$. In fact we have considerable information about f_1.

2.5.1   Theorem

If $\{X(n)\}$ can be represented as a one-sided moving average, then, assuming $X(n) \not\equiv 0$, the spectral density f of the process satisfies

$$\int_{-\pi}^{\pi} \ln f(\lambda) \, d\lambda > -\infty;$$

in particular, $f(\lambda) > 0$ a.e. [Lebesgue measure]. Conversely, any process having a spectral density satisfying this inequality can be represented as a one-sided moving average.

PROOF   Assume $X(n) = \sum_{j=0}^{\infty} c_j W(n-j)$, where $\{W(n)\}$ is white noise and $\sum_{j=0}^{\infty} |c_j|^2 < \infty$. If $C(z) = \sum_{j=0}^{\infty} c_j z^j$, then $f(\lambda) = (2\pi)^{-1} |C(e^{-i\lambda})|^2$, by 2.3.5. Since $\sum_{j=0}^{\infty} |c_j|^2 < \infty$, $C(z)$ belongs to the Hardy–Lebesgue space H^2, and $\ln f(\lambda) = -\ln 2\pi + 2 \ln |C(e^{-i\lambda})|$, where $|C(e^{i\lambda})|$ is the boundary function of $C(z)$. The result follows from standard results about H^2; for full details see Appendix 1, Theorem A1.5.

Conversely, if $\int_{-\pi}^{\pi} \ln f(\lambda) \, d\lambda > -\infty$, we specify a boundary function by $f_0(e^{i\theta}) = f(-\theta)$, $-\pi \leq \theta \leq \pi$. By Theorem A1.6, we can find $C(z) = \sum_{j=0}^{\infty} c_j z^j$ in H^2 such that $(2\pi)^{-1} |C(e^{-i\lambda})|^2 = f(\lambda)$. By 2.3.5, $\{X(n)\}$ is a moving average, one-sided because $c_j = 0$ for $j < 0$. ‖

The power series expansion of $C(z)$ can be obtained from the spectral density as follows.

2.5.2   Theorem

In the proof of 2.5.1, we have

$$C(z) = (2\pi)^{1/2} \exp \left[\tfrac{1}{2} a_0 + \sum_{k=1}^{\infty} a_k z^k \right], \qquad |z| < 1,$$

where

$$a_k = \frac{1}{2\pi} \int_{-\pi}^{\pi} e^{ik\lambda} \ln f(\lambda) \, d\lambda.$$

PROOF Apply Theorem A1.7. ‖

The next step is to relate the Wold decomposition to the Lebesgue decomposition of the spectral measure.

2.5.3 Theorem

If $X(n) = X_1(n) + X_2(n)$ is the Wold decomposition of $\{X(n)\}$, and $X_1(n) \not\equiv 0$, then the Lebesgue decomposition of the spectral measure μ_X is given by $\mu_X = \mu_{X_1} + \mu_{X_2}$, where μ_{X_i} is the spectral measure of $\{X_i(n)\}$.

PROOF By 2.2.7 and 2.3.5, the spectral density of $\{X_1(n)\}$ is given by $f_1(\lambda) = (2\pi)^{-1} |C(e^{-i\lambda})|^2$, where $C(z) = \sum_{j=0}^{\infty} c_j z^j$, and $\sum_{j=0}^{\infty} |c_j|^2 < \infty$. By 2.2.7, 2.5.1, and 2.3.4, we may write

$$X_1(n) = \int_{-\pi}^{\pi} e^{in\lambda} C(e^{-i\lambda}) \, dZ_W(\lambda)$$

where $\{W(n)\}$ is white noise and $H_n(X_1) = H_n(W)$ for all n. Now if $\varphi \in L^2(\mu_X)$, we claim that

$$\int_{-\pi}^{\pi} \varphi(\lambda) \, dZ_X(\lambda) = \int_{-\pi}^{\pi} \varphi(\lambda) C(e^{-i\lambda}) \, dZ_W(\lambda) + \int_{-\pi}^{\pi} \varphi(\lambda) \, dZ_{X_2}(\lambda). \quad (1)$$

If $\varphi(\lambda) = e^{in\lambda}$, this follows from the Wold decomposition, and the general case is handled by approximating φ by a finite linear combination of complex exponentials. Now for some $\varphi_2 \in L^2(\mu_X)$, we have

$$X_2(n) = \int_{-\pi}^{\pi} e^{in\lambda} \varphi_2(\lambda) \, dZ_X(\lambda)$$

(Problem 5, Section 2.2), and therefore by (1),

$$X_2(n) = \int_{-\pi}^{\pi} e^{in\lambda} \varphi_2(\lambda) C(e^{-i\lambda}) \, dZ_W(\lambda) + \int_{-\pi}^{\pi} e^{in\lambda} \varphi_2(\lambda) \, dZ_{X_2}(\lambda).$$

Thus

$$\int_{-\pi}^{\pi} e^{in\lambda} [1 - \varphi_2(\lambda)] \, dZ_{X_2}(\lambda) = \int_{-\pi}^{\pi} e^{in\lambda} \varphi_2(\lambda) C(e^{-i\lambda}) \, dZ_W(\lambda). \quad (2)$$

But by 2.2.2(c), the left side of (2) belongs to $H(X_2)$, and since $H(W) = H(X_1)$, the right side belongs to $H(X_1)$. Since $H(X_1) \perp H(X_2)$, both sides are 0 for all n. Since $X_1(n) \not\equiv 0$, $|C(e^{-i\lambda})| > 0$ a.e. [Lebesgue measure] by

2.5.1, so that $\varphi_2(\lambda) = 0$ a.e. [Lebesgue measure]; also, $1 - \varphi_2(\lambda) = 0$ a.e. $[\mu_{X_2}]$. If $A = \{\lambda : \varphi_2(\lambda) = 1\}$, then A has Lebesgue measure 0 and A^c has μ_{X_2}-measure 0; consequently, μ_{X_2} is singular with respect to Lebesgue measure. Since μ_{X_1} is absolutely continuous with density f_1, the result follows.   ‖

We may now obtain a criterion for a process to be deterministic.

2.5.4   Theorem

Let μ be the spectral measure of $\{X(n)\}$, and let μ_1 be the absolutely continuous component in the decomposition of μ with respect to Lebesgue measure. If f_1 is the density of μ_1, then $\{X(n)\}$ is deterministic iff $\int_{-\pi}^{\pi} \ln f_1(\lambda)\, d\lambda = -\infty$.

PROOF   If $\{X(n)\}$ is not deterministic, then in the Wold decomposition we have $X_1(n) \not\equiv 0$, so that by 2.5.3, f_1 is the spectral density of $\{X_1(n)\}$. By 2.5.1, $\int_{-\pi}^{\pi} \ln f_1(\lambda)\, d\lambda > -\infty$. Conversely, if $\int_{-\pi}^{\pi} \ln f_1(\lambda)\, d\lambda > -\infty$, let μ_1 be concentrated on A_1 and the singular component μ_2 on $A_2 = A_1^c$. Define mov's Z_1, Z_2 by

$$Z_k(E) = \int_E I_{A_k}\, dZ_X,$$

and processes $\{Y_1(n)\}, \{Y_2(n)\}$ by

$$Y_k(n) = \int_{-\pi}^{\pi} e^{in\lambda}\, dZ_k(\lambda) = \int_{-\pi}^{\pi} e^{in\lambda} I_{A_k}(\lambda)\, dZ_X(\lambda).$$

Since $A_1 \cap A_2 = \varnothing$, $\{Y_1(n)\}$ and $\{Y_2(n)\}$ are mutually orthogonal; the spectral measure of $\{Y_k(n)\}$ is given by

$$\|Z_k(E)\|^2 = \int_{-\pi}^{\pi} |I_E I_{A_k}|^2\, d\mu_X = \mu_X(E \cap A_k) = \mu_k(E).$$

Thus $\{Y_1(n)\}$ has spectral density f_1, so by 2.5.1, $\{Y_1(n)\}$ is a one-sided moving average, and therefore purely nondeterministic by 2.2.8. Now $Y_1(n) + Y_2(n) = X(n)$, so that

$$H_n(X) \subset H_n(Y_1) \oplus H_n(Y_2) \qquad \text{for all} \quad n.$$

Therefore

$$H_{-\infty}(X) \subset H_{-\infty}(Y_1) \oplus H_{-\infty}(Y_2) \qquad \text{(Problem 1)}$$

$$= H_{-\infty}(Y_2) \qquad \text{since } \{Y_1(n)\} \text{ is purely nondeterministic}$$

$$\subset H(Y_2)$$

$$\subsetneqq H(Y_1) \oplus H(Y_2) \qquad \text{since } Y_1(n) \not\equiv 0.$$

Thus $H_{-\infty}(X) \subsetneqq H(X)$, so that $\{X(n)\}$ is not deterministic.   ‖

We now start the analysis of the general prediction problem. Write $X(n) = X_1(n) + X_2(n)$, where $\{X_1(n)\}$ is purely nondeterministic and $\{X_2(n)\}$ is deterministic. By 2.2.7, we may write

$$X_1(n) = \sum_{j=0}^{\infty} c_j W(n - j)$$

where $\{W(n)\}$ is white noise, $H_n(W) = H_n(X_1)$ for all n, and $\sum_{j=0}^{\infty} |c_j|^2 < \infty$. By 2.3.5, $\{X_1(n)\}$ has spectral density $f_1(\lambda) = (2\pi)^{-1} |C(e^{-i\lambda})|^2$, where $C(z) = \sum_{j=0}^{\infty} c_j z^j$. We assume that $\{X(n)\}$ is not deterministic, so that $X_1(n) \not\equiv 0$. By 2.5.1, $|C(e^{-i\lambda})| > 0$ a.e. [Lebesgue measure]; we need one other property of $C(z)$.

2.5.5   Lemma

$C(z)$ has no zeros for $|z| < 1$.

PROOF   If $|z_0| < 1$ and $C(z_0) = 0$, write $C(z) = (z - z_0)B(z)$, where $B(z) = \sum_{j=0}^{\infty} b_j z^j$. Then $c_0 = -z_0 b_0$, and

$$|C(e^{-i\lambda})| = |e^{-i\lambda} - z_0| \left| \sum_{j=0}^{\infty} b_j e^{-ij\lambda} \right|$$

$$= \left| \sum_{j=0}^{\infty} b_j e^{-ij\lambda} \right| |1 - \bar{z}_0 e^{-i\lambda}|$$

since $(e^{-i\lambda} - z_0)/(1 - \bar{z}_0 e^{-i\lambda})$

has absolute value 1

$$= \left| \sum_{j=0}^{\infty} a_j e^{-ij\lambda} \right|$$

where $a_0 = b_0$. For all $d_1, \ldots, d_N \in C$, we have

$$\int_{-\pi}^{\pi} \left| e^{in\lambda} - \sum_{j=1}^{N} d_j e^{i(n-j)\lambda} \right|^2 f_1(\lambda) \, d\lambda$$

$$= \frac{1}{2\pi} \int_{-\pi}^{\pi} \left| \left(1 - \sum_{j=1}^{N} d_j e^{-ij\lambda} \right) \sum_{k=0}^{\infty} a_k e^{-ik\lambda} \right|^2 d\lambda$$

$$\geq |a_0|^2 \qquad \text{by orthogonality.}$$

Now $z_0 \neq 0$, and hence $|a_0|^2 = |b_0|^2 = |c_0|^2/|z_0|^2 > |c_0|^2$. (To see this, recall from 2.2.7 that the one-step prediction error for $\{X_1(n)\}$ is $\delta_1(1) = |c_0|^2$; if $z_0 = 0$, then $c_0 = 0$, so that by 2.2.5(b), $\{X_1(n)\}$ is deterministic, a contradiction.) But if we take the inf over all $d_1, \ldots, d_N$ and all $N \geq 1$, we find that $\delta_1(1) \geq |a_0|^2 > |c_0|^2$, a contradiction. $\|$

To continue with the general prediction problem, we are trying to project $X(n + \tau)$ on the space $H_n(X)$ spanned by the $X(m)$, $m \leq n$. By the proof of 2.2.6, $H_n(X)$ is the orthogonal direct sum of $H_n(X_1)$ and $H_n(X_2)$; thus if $P(Z; B)$ denotes the projection of Z on the subspace B,

$$P_n X(n + \tau) = P[X_1(n + \tau) + X_2(n + \tau); H_n(X_1) \oplus H_n(X_2)]$$

$$= P[X_1(n + \tau); H_n(X_1)] + P[X_2(n + \tau); H_n(X_2)]$$

$$\text{since } \{X_1(n)\} \text{ and } \{X_2(n)\}$$

$$\text{are mutually orthogonal}$$

$$= P[X_1(n + \tau); H_n(X_1)] + X_2(n + \tau)$$

$$\text{since } \{X_2(n)\} \text{ is deterministic}$$

$$= \sum_{j=\tau}^{\infty} c_j W(n + \tau - j) + X_2(n + \tau) \qquad \text{by 2.2.7.}$$

The first term can be expressed as follows.

2.5.6 Theorem

If $C(z) = \sum_{j=0}^{\infty} c_j z^j$, $C_\tau(z) = \sum_{j=\tau}^{\infty} c_j z^j$, then

$$P[X_1(n + \tau); H_n(X_1)] = \int_{-\pi}^{\pi} e^{i(n+\tau)\lambda} \frac{C_\tau(e^{-i\lambda})}{C(e^{-i\lambda})} \, dZ_{X_1}(\lambda).$$

PROOF Let $T_n^{(W)}$ be the shift operator on $H(W)$ (see 2.2.4); then

$$T_n^{(W)} X_1(m) = T_n^{(W)} \sum_{j=0}^{\infty} c_j W(m - j) = \sum_{j=0}^{\infty} c_j W(m + n - j) = X_1(m + n).$$

Thus (remember that $H_n(W) = H_n(X_1)$ for all n, and hence $H(W) = H(X_1)$) $T_n^{(W)}$ coincides with the shift operator $T_n^{(1)}$ on $H(X_1)$. Now since $W(0) \in H_0(W) = H_0(X_1)$, we have, by 2.2.2(c), $W(0) = \int_{-\pi}^{\pi} \varphi(\lambda) \, dZ_{X_1}(\lambda)$ for some $\varphi \in L^2(\mu_{X_1})$; since $W(0)$ is in the space $H_0(X_1)$ spanned by the $X_1(m)$, $m \leq 0$, φ belongs to the space $\hat{H}_0(X_1)$ spanned by the complex exponentials $e^{im\lambda}$, $m \leq 0$. Now

$$W(n) = T_n^{(W)} W(0) = T_n^{(1)} W(0)$$

$$= \int_{-\pi}^{\pi} e^{in\lambda} \varphi(\lambda) \, dZ_{X_1}(\lambda)$$

by Problem 5, Section 2.2.

Thus $\{W(n)\}$ is obtained from $\{X_1(n)\}$ by a linear filter with transfer function φ. But by 2.3.4 and 2.3.5, $\{X_1(n)\}$ is obtained from $\{W(n)\}$ by a linear filter with transfer function $C(e^{-i\lambda})$. Thus (Problem 5, Section 2.3) $\{X_1(n)\}$ can be obtained from itself by a linear filter with transfer function $\varphi(\lambda)C(e^{-i\lambda})$, in other words,

$$X_1(n) = \int_{-\pi}^{\pi} e^{in\lambda}\varphi(\lambda)C(e^{-i\lambda})\,dZ_{X_1}(\lambda).$$

Thus $e^{in\lambda}$ and $e^{in\lambda}\varphi(\lambda)C(e^{-i\lambda})$ both correspond to $X_1(n)$ under the basic isomorphism of the spectral decomposition theorem 2.2.1, and therefore $\varphi(\lambda)C(e^{-i\lambda}) = 1$ a.e. $[\mu_{X_1}]$, in particular, $1/C(e^{-i\lambda}) \in L^2(\mu_{X_1})$. Since $\varphi \in \hat{H}_0(X_1)$, we have $e^{i(n+\tau)\lambda}C_\tau(e^{-i\lambda})/C(e^{-i\lambda}) \in \hat{H}_n(X_1)$. (Note that $C_\tau(e^{-i\lambda}) \in L^2$ [Lebesgue measure], and hence

$$\int_{-\pi}^{\pi} |\varphi(\lambda)C_\tau(e^{-i\lambda})|^2\,d\mu_{X_1}(\lambda)$$

$$= (2\pi)^{-1} \int_{-\pi}^{\pi} |C^{-1}(e^{-i\lambda})C_\tau(e^{-i\lambda})|^2\,|C(e^{-i\lambda})|^2\,d\lambda$$

$$= (2\pi)^{-1} \int_{-\pi}^{\pi} |C_\tau(e^{-i\lambda})|^2\,d\lambda < \infty.$$

Thus $\varphi(\lambda)C_\tau(e^{-i\lambda}) \in L^2(\mu_{X_1})$.) Furthermore, exactly as in 2.4.1,

$$e^{i(n+\tau)\lambda} - e^{i(n+\tau)\lambda}\frac{C_\tau(e^{-i\lambda})}{C(e^{-i\lambda})} \perp \hat{H}_n(X_1),$$

and the result follows. ‖

Theorem 2.5.6 solves the prediction problem for $\{X_1(n)\}$, and in fact the solution is the same for $\{X(n)\}$. For by the discussion before 2.5.6,

$$X(n + \tau) - P_nX(n + \tau) = X_1(n + \tau) - P[X_1(n + \tau); H_n(X_1)],$$

so that the prediction errors are identical: $\delta_X(\tau) = \delta_{X_1}(\tau)$. It follows that the transfer function of an optimal predictor for $\{X_1(n)\}$ must be optimal for $\{X(n)\}$.

2.5.7 Summary

Let $\{X(n)\}$ have spectral measure μ, and let f be the density of the absolutely continuous component of μ in the decomposition with respect to Lebesgue measure. If $\int_{-\pi}^{\pi} \ln f(\lambda)\,d\lambda = -\infty$, then $\{X(n)\}$ is deterministic; if $\int_{-\pi}^{\pi} \ln f(\lambda)\,d\lambda > -\infty$, $\{X(n)\}$ is not deterministic. In the latter case, let

$$C(z) = (2\pi)^{1/2} \exp\left[\tfrac{1}{2}a_0 + \sum_{k=1}^{\infty} a_k z^k\right], \qquad |z| < 1,$$

where

$$a_k = \frac{1}{2\pi} \int_{-\pi}^{\pi} e^{ik\lambda} \ln f(\lambda)\, d\lambda.$$

The optimal predictor is given by

$$P_n X(n + \tau) = \int_{-\pi}^{\pi} e^{in\lambda} \left[e^{i\tau\lambda} \frac{C_\tau(e^{-i\lambda})}{C(e^{-i\lambda})} \right] dZ_X(\lambda).$$

By 2.2.7, the prediction error is given by

$$\delta(\tau) = \sum_{j=0}^{\tau-1} |c_j|^2;$$

in particular, since $c_0 = (2\pi)^{1/2} \exp(\tfrac{1}{2}a_0)$,

$$\delta(1) = 2\pi \exp\left[\frac{1}{2\pi} \int_{-\pi}^{\pi} \ln f(\lambda)\, d\lambda \right].$$

(This is the theorem of Szëgo–Kolmogorov–Krein; for another approach, see Hoffman, 1962, p. 49.)

We conclude this section by discussing the problem of estimating a random variable $Y \in L^2(\Omega, \mathscr{F}, P)$ that is not necessarily of the form $X(n + \tau)$. If we observe $X(m)$, $m \le n$, and then estimate Y, we are looking for the projection $P_n Y$ of Y on $H_n(X)$. We may write $Y = Y_0 + Y_1$, where $Y_0 \in H(X)$, $Y_1 \perp H(X)$; thus $P_n Y = P_n Y_0$. The *estimation error* is defined by

$$\delta(n) = \| Y - P_n Y \|^2$$
$$= \| Y_0 - P_n Y_0 \|^2 + \| Y_1 \|^2 \qquad \text{since} \quad P_n Y_1 = 0.$$

Thus the problem reduces to estimating Y_0. If $\{X(n)\}$ is deterministic, then $Y_0 \in H(X) = H_n(X)$ so the best estimate is simply Y_0. Thus assume $\{X(n)\}$ not deterministic.

If the Wold decomposition of $\{X(n)\}$ is $X(n) = X_1(n) + X_2(n)$, the proof of 2.2.6 shows that $H_n(X) = H_n(X_1) \oplus H_n(X_2) = H_n(X_1) \oplus H_{-\infty}(X)$, and $H(X) = H(X_1) \oplus H_{-\infty}(X)$. Furthermore, by 2.2.7, $\{X_1(n)\}$ can be represented as a one-sided moving average, where the white noise process $\{W(n)\}$ satisfies $H_n(W) = H_n(X_1)$ for all n, and therefore $H(W) = H(X_1)$. Thus the $W(n)$ form an orthonormal basis for $H(X_1)$, so that Y_0 can be written as

$$Y_0 = P[Y_0; H(X_1)] + P_{-\infty} Y_0$$
$$= \sum_{j=-\infty}^{\infty} a_j W(j) + P_{-\infty} Y_0,$$

where $a_j = \langle Y_0, W(j) \rangle = \langle Y, W(j) \rangle = E[Y \overline{W(j)}]$. Since the $W(m)$, $m \le n$, form an orthonormal basis for $H_n(W) = H_n(X_1)$, we have

$$P[Y_0; H_n(X_1)] = \sum_{j=-\infty}^{n} a_j W(j).$$

As in 2.5.6, we write

$$W(j) = \int_{-\pi}^{\pi} \frac{e^{ij\lambda}}{C(e^{-i\lambda})} dZ_{X_1}(\lambda).$$

Hence

$$P[Y_0; H_n(X_1)] = \int_{-\pi}^{\pi} \frac{\sum_{j=-\infty}^{n} a_j e^{ij\lambda}}{C(e^{-i\lambda})} dZ_{X_1}(\lambda).$$

But

$$Y_0 - P_n Y_0 = P[Y_0; H(X_1)] - P[Y_0; H_n(X_1)]$$
$$+ P[Y_0; H(X_2)] - P[Y_0; H_n(X_2)],$$

and since $H(X_2) = H_n(X_2)$ for all n,

$$Y_0 - P_n Y_0 = P[Y_0; H(X_1)] - P[Y_0; H_n(X_1)].$$

As in the discussion preceding 2.5.7, the desired estimate is

$$P_n Y_0 = P_n Y = \int_{-\pi}^{\pi} \frac{\sum_{j=-\infty}^{n} a_j e^{ij\lambda}}{C(e^{-i\lambda})} dZ_X(\lambda).$$

Problem

1. (a) Let $\{A_n\}$ be a decreasing sequence of closed subspaces of a Hilbert space H. If B is a closed subspace and $A_n \perp B$ for all n, show that

$$\left(\bigcap_n A_n \right) \oplus B = \bigcap_n (A_n \oplus B).$$

(b) In the proof of 2.5.4, show that

$$H_{-\infty}(X) \subset H_{-\infty}(Y_1) \oplus H_{-\infty}(Y_2).$$

2.6   Examples of Continuous Parameter Processes

In this section we give some examples of continuous parameter L^2 processes. All processes are assumed to have 0 mean, and throughout the section, S will denote $(-\infty, \infty)$ and $\mathscr{S}$ the Borel sets of S.

2.6.1 Linear Operations

Let $\{Y(t)\}$ be a continuous parameter L^2 process with continuous covariance and spectral representation

$$Y(t) = \int_S e^{it\lambda} \, dZ_Y(\lambda), \qquad t \in S.$$

If $\varphi \in L^2(\mu_Y)$, and we define

$$X(t) = \int_S e^{it\lambda}\varphi(\lambda) \, dZ_Y(\lambda), \qquad t \in S, \tag{1}$$

we say that $\{X(t)\}$ is obtained from $\{Y(t)\}$ by a *linear operation* or *linear filter* with *transfer function* φ; $\{Y(t)\}$ is called the *input* to the filter, $\{X(t)\}$ the *output*.

An important special case is the filter which, in the engineering terminology, has "impulse response" h; here,

$$X(t) = \int_{-\infty}^{\infty} h(s)Y(t-s) \, ds.$$

(See Problem 1 for the definition of the integral under appropriate hypotheses on h and $\{Y(t)\}$.)

The process $\{X(t)\}$ is the continuous parameter analog of the process $X(n) = \sum_{j=-\infty}^{\infty} a_j Y(n-j)$ discussed in 2.3.4.

The general discussion of 2.3.4 may be repeated verbatim to show that $\{X(t)\}$ is a stationary process whose covariance function and spectral measure are given in terms of the spectral properties of $\{Y(t)\}$ by

$$K_X(t) = \int_S e^{it\lambda} \, |\varphi(\lambda)|^2 \, d\mu_Y(\lambda) \tag{2}$$

$$\mu_X(E) = \int_E |\varphi(\lambda)|^2 \, d\mu_Y(\lambda). \tag{3}$$

If $\{Y(t)\}$ has a spectral density, so does $\{X(t)\}$ and

$$f_X(\lambda) = |\varphi(\lambda)|^2 f_Y(\lambda). \tag{4}$$

Conversely, suppose we are given a finite measure μ on $\mathscr{S}$ and a stationary L^2 process $\{X(t)\}$ whose spectral measure μ_X is absolutely continuous with respect to μ with density $|\varphi(\lambda)|^2 > 0$ a.e. $[\mu]$. Again, just as in 2.3.4, we have

$$\mu(E) = \int_E |\varphi(\lambda)|^{-2} \, d\mu_X(\lambda), \tag{5}$$

so that $\varphi^{-1} \in L^2(\mu_X)$. If we define

$$Y(t) = \int_S e^{it\lambda}[\varphi(\lambda)]^{-1} dZ_X(\lambda), \tag{6}$$

then $\{Y(t)\}$ is stationary, with spectral measure μ and spectral mov

$$Z_Y(E) = \int_E [\varphi(\lambda)]^{-1} dZ_X(\lambda). \tag{7}$$

As we indicated in 2.3.4, the input process may be regarded physically as a sum of complex exponentials; an exponential $e^{it\lambda}$ of frequency λ is multiplied by $\varphi(\lambda)$ to obtain the corresponding component of the output process. Thus, for example, if $\varphi(\lambda)$ is constant on an interval A and 0 outside A, the filter is a bandpass filter with "pass band" A; it passes frequencies in A without distortion, and blocks frequencies outside A. As another example, a filter with transfer function $\varphi(\lambda) = e^{-i\tau\lambda}$ yields a "time delay" of τ seconds since

$$X(t) = \int_S e^{it\lambda}e^{-i\tau\lambda} dZ_Y(\lambda) = Y(t - \tau).$$

2.6.2   Differentiation

We now show that L^2-differentiation (see 1.3.3) is a special case of linear filtering. If $X(t)$ is the L^2-derivative $Y'(t)$, then as $h \to 0$, $h^{-1}[Y(t + h) - \overset{L^2}{Y(t)}] \to X(t)$. By 2.2.1 and 2.1.2(e), $h^{-1}[e^{i(t+h)\lambda} - e^{it\lambda}]$ converges in $L^2(\mu_Y)$ as $h \to 0$, necessarily to $d(e^{it\lambda})/dt = i\lambda e^{it\lambda}$. The limit function must belong to $L^2(\mu_Y)$, that is,

$$\int_S \lambda^2 d\mu_Y(\lambda) < \infty. \tag{1}$$

Conversely, if $\{Y(t)\}$ satisfies (1) then by the dominated convergence theorem, $h^{-1}[\exp(i(t + h)\lambda) - \exp(it\lambda)]$ converges to $i\lambda \exp(it\lambda)$ in $L^2(\mu_Y)$, so by 2.1.2(e),

$$\frac{Y(t + h) - Y(t)}{h} \overset{L^2}{\to} \int_S i\lambda e^{it\lambda} dZ_Y(\lambda).$$

Thus $\{Y(t)\}$ is L^2-differentiable iff (1) holds (physically, iff the high frequency components of $\{Y(t)\}$ fall off rapidly enough as $\lambda \to \infty$), and in this case, $\{Y'(t)\}$ is obtained from $\{Y(t)\}$ by a linear filter with transfer function $\varphi(\lambda) = i\lambda$:

$$Y'(t) = \int_S i\lambda e^{it\lambda} dZ_Y(\lambda).$$

By 2.6.1, Eq. (2),

$$K_{Y'}(t) = \int_S e^{it\lambda}\lambda^2 \, d\mu_Y(\lambda) = -K_Y''(t),$$

as obtained in 1.3.7.

A brief induction argument shows that $\{Y(t)\}$ has n L^2-derivatives iff $\int_S \lambda^{2n} \, d\mu_Y(\lambda) < \infty$; the associated linear filter has transfer function $(i\lambda)^n$.

2.6.3   Stochastic Differential Equations

Equations involving derivatives of stochastic processes are called stochastic differential equations; now consider one class of examples. Suppose that $\{Y(t)\}$ is stationary and has n L^2-derivatives; we seek a stationary L^2 process $\{X(t)\}$ with m L^2-derivatives satisfying

$$\sum_{j=0}^{m} b_j X^{(j)}(t) = \sum_{k=0}^{n} a_k Y^{(k)}(t) \tag{1}$$

where the a_k and b_j are given constants.

Let us try to find a solution $\{X(t)\}$ that can be obtained from $\{Y(t)\}$ by a linear operation with transfer function $\varphi(\lambda)$. By 2.6.2,

$$\sum_{j=0}^{m} b_j X^{(j)}(t) = \int_S e^{it\lambda} B(i\lambda)\varphi(\lambda) \, dZ_Y(\lambda)$$

and

$$\sum_{k=0}^{n} a_k Y^{(k)}(t) = \int_S e^{it\lambda} A(i\lambda) \, dZ_Y(\lambda)$$

where

$$A(z) = \sum_{k=0}^{n} a_k z^k, \qquad B(z) = \sum_{j=0}^{m} b_j z^j.$$

Thus if

$$\varphi(\lambda) = A(i\lambda)/B(i\lambda)$$

(assuming $B(z)$ has no zeros on the imaginary axis and $A(i\lambda)/B(i\lambda) \in L^2(\mu_Y)$), and

$$X(t) = \int_S e^{it\lambda}\varphi(\lambda) \, dZ_Y(\lambda),$$

then $\{X(t)\}$ satisfies the stochastic differential equation; furthermore, if $\{Y(t)\}$ has a spectral density, then so does $\{X(t)\}$ and

$$f_X(\lambda) = \left| \frac{A(i\lambda)}{B(i\lambda)} \right|^2 f_Y(\lambda).$$

Thus if $\{Y(t)\}$ had a constant spectral density, we would have a correspondence between rational spectral densities and outputs of linear systems governed by the differential equation (1). The difficulty is that $f_Y(\lambda) \equiv f_0 > 0$ is not Lebesgue integrable, and hence cannot be a spectral density. Yet in many situations it is helpful to imagine a process called "white noise" whose spectral density is constant over the entire real line, and carry through calculations as if the process actually existed. We now try to develop a mathematical foundation for such manipulations.

2.6.4 White Noise

If an L^2 process has a spectral density that is constant over a large interval, we might consider the process as an approximation to our fictitious process with constant spectral density. For example, consider a process $\{W_n(t), -\infty < t < \infty\}$ with spectral density $f_n(\lambda) = f_0$, $|\lambda| \leq n$; $f_n(\lambda) = 0$, $|\lambda| > n$. If $g \in L^2(m)$, where m is Lebesgue measure on $(-\infty, \infty)$, we have defined the integral $\int_{-\infty}^{\infty} g(t)W_n(t)\, dt$ (see Section 1.3, Problem 7). If g, $h \in L^2(m)$, with Fourier transforms G, H (see Appendix 2),

$$E\left[\int_{-\infty}^{\infty} g(t)W_n(t)\, dt\, \overline{\int_{-\infty}^{\infty} h(t)W_n(t)\, dt}\right]$$

$$= \int_{-\infty}^{\infty} G(\lambda)\overline{H(\lambda)} f_n(\lambda)\, d\lambda$$

$$= f_0 \int_{-n}^{n} G(\lambda)\overline{H(\lambda)}\, d\lambda$$

$$\to f_0 \int_{-\infty}^{\infty} G(\lambda)\overline{H(\lambda)}\, d\lambda$$

$$= 2\pi f_0 \int_{-\infty}^{\infty} g(t)\overline{h(t)}\, dt$$

by the Fourier–Plancherel theorem (Appendix 2, Theorem A2.6). Furthermore, the $\{W_n(t)\}$ can be constructed so that $\int_{-\infty}^{\infty} g(t)W_n(t)\, dt$ converges in L^2 to a random variable (see Problem 3). Thus the L^2 properties of integrals involving $W_n(t)$ are approximately what we would expect from a constant spectral density. Therefore we say that a sequence $\{W_n(t), -\infty < t < \infty\}$ of stationary L^2 processes *converges to white noise* iff:

(a) For every $g \in L^2(m)$, $W_n(g) = \int_S g(t)W_n(t)\, dt$ converges in L^2 to a random variable $W(g)$; and

(b) there is an $f_0 > 0$ such that for all g, $h \in L^2(m)$,

$$\lim_{n \to \infty} E[W_n(g)\overline{W_n(h)}] = 2\pi f_0 \int_S g(t)\overline{h(t)}\, dt;$$

hence by 1.3.1,

$$E[W(g)\overline{W(h)}] = 2\pi f_0 \int_S g(t)\overline{h(t)}\, dt.$$

Note that we have defined the phrase "converges to white noise" but not "white noise" itself. The suggestive notation $\int_S g(t)W(t)\, dt$ is often used for $W(g)$, but the integral is not of the type discussed in Section 1.3, and in fact $\{W(t)\}$ does not exist as an L^2 process. The integral should be regarded as a bounded linear operator which maps $g \in L^2(m)$ to the random variable $W(g)$. (Boundedness follows from the observation that W is a strong limit of a sequence of bounded linear operators; see RAP, p. 149, Problem 8.)

We now relate $\int_S g(t)W(t)\, dt$ to the stochastic integrals introduced in 2.1.6. Let $\{Z(\lambda), -\infty < \lambda < \infty\}$ be an L^2 process such that for $\lambda_1 < \lambda_2$,

$$Z(\lambda_2) - Z(\lambda_1) = W(I_{(\lambda_1, \lambda_2]}).$$

(To determine the process completely, we can set $Z(0) \equiv 0$.) If $\lambda_1 < \lambda_2 \le \lambda_3 < \lambda_4$,

$$E[(Z(\lambda_2) - Z(\lambda_1))\overline{(Z(\lambda_4) - Z(\lambda_3))}]$$

$$= E\left[L^2 \lim_{n \to \infty} W_n(I_{(\lambda_1, \lambda_2]})\overline{L^2 \lim_{n \to \infty} W_n(I_{(\lambda_3, \lambda_4]})}\right]$$

$$= \lim_{n \to \infty} E[W_n(I_{(\lambda_1, \lambda_2]})\overline{W_n(I_{(\lambda_3, \lambda_4]})}] \qquad \text{by 1.3.1}$$

$$= 2\pi f_0 \int_S I_{(\lambda_1, \lambda_2]}(t)I_{(\lambda_3, \lambda_4]}(t)\, dt$$

by condition (b) above

$$= 0,$$

and hence $\{Z(\lambda)\}$ has orthogonal increments. Similarly,

$$E[|Z(\lambda_2) - Z(\lambda_1)|^2] = 2\pi f_0(\lambda_2 - \lambda_1), \qquad \lambda_1 < \lambda_2,$$

so that the process is L^2-continuous, and the measure associated with Z is $2\pi f_0 m$. Now if $g = I_{(\lambda_1, \lambda_2]}, \lambda_1 < \lambda_2$, then

$$\int_S g\, dZ = Z(\lambda_2) - Z(\lambda_1) = W(g);$$

since (see 2.1.6) $g \to \int_S g\, dZ$ is an isometric isomorphism of $L^2(2\pi f_0 m)$ and a closed subspace of $L^2(\Omega, \mathscr{F}, P)$, W is a bounded linear operator, and the finite linear combinations of indicators of the above type are dense in $L^2(m)$, it follows that

$$\int_S g\, dZ = W(g) \qquad \text{for every } g \in L^2(m).$$

We now look at white noise from a covariance viewpoint. The covariance function corresponding to the spectral density $f_n(\lambda) = f_0$, $|\lambda| \leq n$; $f_n(\lambda) = 0$, $|\lambda| > n$ is

$$K_n(t) = \int_{-\infty}^{\infty} e^{it\lambda} f_n(\lambda) \, d\lambda = 2f_0 \frac{\sin nt}{t}.$$

It is convenient to normalize by dividing by Var $W_n(t) = K_n(0) = 2nf_0$ to obtain the *correlation function*

$$R_n(t) = \frac{\sin nt}{nt}$$

which approaches 0 for $t \neq 0$. Thus we expect intuitively that for the "white-noise process" $\{W(t)\}$, $W(t_1)$, and $W(t_2)$ are uncorrelated for $t_1 \neq t_2$.

The sequence of covariance functions $K_n(t)$ approximates a "Dirac delta function" $2\pi f_0 \, \delta(t)$, in other words, if g is a continuous complex-valued function on R that vanishes outside a finite interval, then as $n \to \infty$,

$$\int_{-\infty}^{\infty} g(t) K_n(t) \, dt \to 2\pi f_0 g(0).$$

Although no L^2 process satisfies $f(\lambda) \equiv f_0$ or $R(t) = 0$, $t \neq 0$, white noise is a very useful idealization in linear system theory. For example, suppose that $\{W_n(t)\}$ converges to white noise, and let W_n be the input to a linear filter with transfer function φ. The output process is

$$X_n(t) = \int_S e^{it\lambda} \varphi(\lambda) \, dZ_{W_n}(\lambda).$$

If φ_t has Fourier transform $\Phi_t(\lambda) = e^{it\lambda}\varphi(\lambda)$, then

$$W_n(\varphi_t) = \int_S \varphi_t(s) W_n(s) \, ds$$

$$= \int_S \Phi_t(\lambda) \, dZ_{W_n}(\lambda) \qquad \text{(see Problem 2b)}$$

$$= X_n(t).$$

Since $\{W_n(t)\}$ converges to white noise, $X_n(t)$ converges in L^2 to $X(t) = W(\varphi_t)$.

By condition (b) of the definition of convergence to white noise,

$$E[X(s+t)\overline{X(s)}] = 2\pi f_0 \langle \varphi_{s+t}, \varphi_s \rangle_{L^2(m)}$$

$$= f_0 \langle \Phi_{s+t}, \Phi_s \rangle_{L^2(m)}$$

by the Fourier–Plancherel theorem A2.6

$$= \int_S e^{it\lambda} |\varphi(\lambda)|^2 f_0 \, d\lambda \tag{1}$$

so that $\{X(t)\}$ has spectral density $f_0 |\varphi(\lambda)|^2$. This is exactly the result obtained in 2.6.1, Eq. (4), by imagining a process with constant spectral density f_0 applied as the input to the linear filter. In particular, an arbitrary process with spectral density $f(\lambda)$ can always be realized as the output of a linear filter with white-noise input; simply take $\varphi(\lambda) = [f(\lambda)/f_0]^{1/2}$.

In many common engineering situations, the input–output relation of a linear filter is given by

$$X(t) = \int_{-\infty}^{\infty} h(s)Y(t - s) \, ds$$

where h is continuous a.e. $[m]$ and belongs to $L^1(m) \cap L^2(m)$. The transfer function of the filter is $H(-\lambda)$, where H is the Fourier transform of h (see Problem 1). If the input is white noise, the covariance function of the output is, by (1),

$$K(t) = \int_{-\infty}^{\infty} e^{it\lambda} |H(-\lambda)|^2 f_0 \, d\lambda$$

$$= f_0 \int_{-\infty}^{\infty} e^{-it\lambda} H(\lambda)\overline{H(\lambda)} \, d\lambda$$

$$= 2\pi f_0 \int_{-\infty}^{\infty} h(s + t)\overline{h(s)} \, ds$$

by Theorems A2.1(b) and A2.6.

This is often obtained in the engineering literature by an argument of the following form:

$$E[X(s + t)\overline{X(s)}]$$

$$= E\left[\int_{-\infty}^{\infty} \int_{-\infty}^{\infty} h(u)\overline{h(v)}W(s + t - u)\overline{W(s - v)} \, du \, dv\right]$$

$$= \int_{-\infty}^{\infty} \int_{-\infty}^{\infty} h(u)\overline{h(v)}E[W(s + t - u)\overline{W(s - v)}] \, du \, dv$$

$$= \int_{-\infty}^{\infty} \int_{-\infty}^{\infty} h(u)\overline{h(v)}2\pi f_0 \delta(t - u + v) \, du \, dv$$

$$= 2\pi f_0 \int_{-\infty}^{\infty} h(u)\overline{h(u - t)} \, du$$

$$= 2\pi f_0 \int_{-\infty}^{\infty} h(s + t)\overline{h(s)} \, ds.$$

Problems

1. Let $\{Y(t)\}$ be a continuous parameter L^2 process with bounded spectral density, and let h be a complex-valued function on $(-\infty, \infty)$, continuous a.e. $[m]$ where m is Lebesgue measure, with $h \in L^1(m) \cap L^2(m)$. Define

$$\int_{-\infty}^{\infty} h(s)Y(t-s)\,ds = L^2 \lim_{T \to \infty} \int_{-T}^{T} h(s)Y(t-s)\,ds$$

where the integral on $[-T, T]$ is defined in 1.3.8.

Show that the L^2 limit exists and defines a process $\{X(t)\}$ obtainable by a linear operation on $\{Y(t)\}$; show also that the transfer function of the filter is $H(-\lambda)$, where H is the Fourier transform of h.

2. Let $\{X(t), -\infty < t < \infty\}$ be a stationary L^2 process with bounded spectral density f. Let $g \in L^2(m)$ where m is Lebesgue measure, and let G be the Fourier transform of g. Let

$$X(g) = \int_{-\infty}^{\infty} G(\lambda)\,dZ_X(\lambda).$$

(a) Show that X is a continuous linear mapping from $L^2(m)$ to $L^2\{X(t), -\infty < t < \infty\}$.

(b) If g is continuous with compact support, show that

$$X(g) = \int_{-\infty}^{\infty} g(t)X(t)\,dt$$

where the integral is defined in 1.3.8. Thus by Section 1.3, Problem 7,

$$\int_{-\infty}^{\infty} g(t)X(t)\,dt = \int_{-\infty}^{\infty} G(\lambda)\,dZ_X(\lambda) \qquad \text{for all} \quad g \in L^2(m).$$

3. Let $\{f_n(\lambda)\}$ be a uniformly bounded sequence of spectral densities such that $f_n(\lambda) \to f_0 > 0$ for all λ. Show that there exist L^2 processes $\{W_n(t), -\infty < t < \infty\}$ with spectral densities f_n such that $\{W_n(t)\}$ converges to white noise in the sense of 2.6.4.

4. Let $B_1(t)$ and $B_2(t)$, $t \geq 0$, be Brownian motion processes (see 1.2.9) with the same σ^2, and assume that $B_1(s)$ and $B_2(t)$ are independent for all s, t. Define an *extended Brownian motion* by

$$B(t) = \begin{cases} B_1(t), & t \geq 0 \\ B_2(-t), & t < 0. \end{cases}$$

(a) Show that the covariance function of $\{B(t)\}$ is

$$K(s, t) = \tfrac{1}{2}\sigma^2(|s| + |t| - |s-t|)$$

and therefore

$$E[\,|\,B(s) - B(t)\,|^2] = \sigma^2\,|\,s - t\,|.$$

(b)   Define $W_n(t) = n[B(t + 1/n) - B(t)]$; show that $\{W_n(t)\}$ is stationary, and find its covariance function and spectral density.

(c)   Show that $\{W_n(t)\}$ converges to white noise (with $f_0 = \sigma^2/2\pi$), so that in a sense, the derivative of Brownian motion is white noise.

(d)   Call white noise *Gaussian* iff $W(g)$ is a Gaussian random variable for all $g \in L^2(m)$. Let $\{Z(\lambda),\ -\infty < \lambda < \infty\}$ be defined as in 2.6.4, with $Z(0) = 0$. Show that W is Gaussian iff $\{Z(\lambda),\ -\infty < \lambda < \infty\}$ is extended Brownian motion.

5.   Let W' be any norm-preserving linear map of $L^2(m)$ into $L^2(\Omega, \mathscr{F}, P)$, and let $W = (2\pi f_0)^{1/2} W'$, $f_0 > 0$. Show that W is white noise in the sense that there are L^2 processes $\{W_n(t),\ -\infty < t < \infty\}$ converging to white noise, with $W_n(g) \to W(g)$ for each $g \in L^2(m)$.

(b)   Let $\{\hat{W}(t)\}$ be white noise. Show that there is another white noise $\{W(t)\}$ such that for all $h \in L^2(m)$, $\hat{W}(H) = W(h)$, where H is the Fourier transform of h.

6.   Consider the stochastic differential equation of 2.6.3; assume as in that section that $A(i\lambda)/B(i\lambda) \in L^2(\mu_Y)$ and that B has no zeros on the imaginary axis. Write the partial fraction expansion of $A(z)/B(z)$ and use the formulas

$$\frac{1}{(i\lambda - r)^n} = \int_0^{\infty} \frac{t^{n-1}}{(n-1)!}\, e^{rt} e^{-i\lambda t}\, dt, \qquad \mathrm{Re}\ r < 0$$

$$\frac{1}{(i\lambda - r)^n} = (-1)^n \int_0^{\infty} \frac{t^{n-1}}{(n-1)!}\, e^{-rt} e^{i\lambda t}\, dt, \qquad \mathrm{Re}\ r > 0$$

to obtain a time domain formula for $\{X(t)\}$ in terms of $\{Y(t)\}$. Observe that if B has any roots with positive real parts, then $X(t)$ depends on $Y(s)$, $s > t$ as well as on $Y(s)$, $s < t$.

2.7   Continuous Parameter Prediction in Special Cases; Yaglom's Method

In this section we describe a method of solving the prediction problem due to Yaglom (1962). First recall from the discussion at the beginning of Section 2.4 (and also Problem 3 of that section) that if $\varphi_\tau(\lambda)$ is the projection of $e^{it\lambda}$ on the subspace of $L^2(\mu_X)$ spanned by the functions $e^{is\lambda}$, $s \leq 0$, the best predictor of $X(t + \tau)$ based on $X(s)$, $s \leq t$, is given by

$$P_t X(t + \tau) = \int_{-\infty}^{\infty} e^{it\lambda} \varphi_\tau(\lambda)\, dZ_X(\lambda).$$

If $\{X(t)\}$ has a spectral density $f(\lambda)$, $L^2(\mu_X)$ is the collection of all $g: R \to C$ such that $\int_{-\infty}^{\infty} |g(\lambda)|^2 f(\lambda) \, d\lambda < \infty$; this space will be denoted by $L^2(f)$. By the projection theorem, the function φ_τ is specified by the following requirements.

2.7.1 Conditions Determining φ_τ

(1) $\varphi_\tau \in L^2(f)$;

(2) φ_τ is the limit in $L^2(f)$ of a sequence of finite linear combinations of $e^{is\lambda}$, $s \leq 0$;

(3) $\int_{-\infty}^{\infty} e^{is\lambda}[e^{it\lambda} - \varphi_\tau(\lambda)]f(\lambda) \, d\lambda = 0$ for all $s \geq 0$.

The following theorem gives sufficient conditions for φ_τ to satisfy these requirements.

2.7.2 Theorem

Let f be a bounded spectral density, and assume that we have found a function $\varphi_\tau(\lambda)$ of the *complex* variable λ with the following properties:

(a) $\int_{-\infty}^{\infty} |\varphi_\tau(\lambda)|^2 f(\lambda) \, d\lambda < \infty$;

(b) φ_τ is analytic for Im $\lambda \leq 0$, and for some $r > 0$, $\varphi_\tau(\lambda) = O(|\lambda|^r)$ as $|\lambda| \to \infty$ with Im $\lambda \leq 0$;

(c) $\psi_\tau(\lambda) = [e^{it\lambda} - \varphi_\tau(\lambda)]f(\lambda)$ is analytic for Im $\lambda \geq 0$, and for some $\varepsilon > 0$, $\psi_\tau(\lambda) = O(|\lambda|^{-1-\varepsilon})$ as $|\lambda| \to \infty$ with Im $\lambda \geq 0$.

Then φ_τ (restricted to real values) satisfies the conditions of 2.7.1, and hence is the transfer function of the optimal predictor. Furthermore, the prediction error is given by

$$\sigma_\tau^2 = \int_{-\infty}^{\infty} |e^{it\lambda} - \varphi_\tau(\lambda)|^2 f(\lambda) \, d\lambda = \int_{-\infty}^{\infty} [1 - |\varphi_\tau(\lambda)|^2]f(\lambda) \, d\lambda.$$

PROOF Let Γ_R be the semicircle $|z| = R$ in the upper half-plane, oriented counterclockwise. If γ_R is the closed curve formed by Γ_R and $[-R, R]$, then by (c) and Cauchy's theorem, $\int_{\gamma_R} \psi_\tau(\lambda)e^{is\lambda} \, d\lambda = 0$. But if $s \geq 0$, then on Γ_R the integrand is bounded in absolute value by $|\psi_\tau(\lambda)| = O(R^{-1-\varepsilon})$, and it follows that $\int_{\Gamma_R} \psi_\tau(\lambda)e^{is\lambda} \, d\lambda \to 0$ as $R \to \infty$. Therefore $\int_{-\infty}^{\infty} \psi_\tau(\lambda)e^{is\lambda} \, d\lambda = 0$ for all $s \geq 0$, proving (3).

Since condition (1) holds by (a), it remains to prove (2). If $\int_{-\infty}^{\infty} |\varphi_\tau(\lambda)|^2 \, d\lambda < \infty$, φ_τ may be represented as a Fourier transform:

$$\varphi_\tau(\lambda) = \int_{-\infty}^{\infty} e^{-is\lambda}a(s) \, ds$$

where

$$a(s) = \frac{1}{2\pi} \int_{-\infty}^{\infty} e^{is\lambda} \varphi_\tau(\lambda) \, d\lambda$$

and the integrals converge in L^2 [Lebesgue measure] (see Appendix 2, in particular Theorem A2.6). Now by (b), $\varphi_\tau(\lambda)$ is analytic for Im $\lambda \le 0$; if in addition $\varphi_\tau(\lambda) = O(|\lambda|^{-1-\varepsilon})$ as $|\lambda| \to \infty$, Im $\lambda \le 0$ (this implies that $\int_{-\infty}^{\infty} |\varphi_\tau(\lambda)|^2 \, d\lambda < \infty$), an argument similar to the above shows that $a(s) = 0$ for $s \le 0$, so that

$$\varphi_\tau(\lambda) = \int_0^\infty e^{-is\lambda} a(s) \, ds$$

in the sense that as $T \to \infty$, $\int_0^T e^{-is\lambda} a(s) \, ds$ converges to $\varphi_\tau(\lambda)$ in L^2 [Lebesgue measure], and hence in $L^2(f)$ because f is bounded. Consideration of approximating sums to $\int_0^T e^{-is\lambda} a(s) \, ds$ shows that (2) holds, under our assumption that $\varphi_\tau(\lambda) = O(|\lambda|^{-1-\varepsilon})$ as $|\lambda| \to \infty$, Im $\lambda \le 0$. To deal with this difficulty we approximate φ_τ by

$$\varphi_{\tau,n}(\lambda) = \frac{\varphi_\tau(\lambda)}{(1 + i\lambda/n)^q}$$

where q is an integer greater than $r + 1 + \varepsilon$. Then by (b), $\varphi_{\tau,n}$ is analytic for Im $\lambda \le 0$ and $\varphi_{\tau,n}(\lambda) = O(|\lambda|^{-1-\varepsilon})$, so (2) holds for $\varphi_{\tau,n}$. But $|\varphi_{\tau,n}| \le |\varphi_\tau| \in L^2(f)$, and $\varphi_{\tau,n}(\lambda) \to \varphi_\tau(\lambda)$ as $n \to \infty$; thus by the dominated convergence theorem, $\varphi_{\tau,n} \to \varphi_\tau$ in $L^2(f)$. It follows that (2) holds for φ_τ.

Finally, the first expression for σ_τ^2 follows from the definition of prediction error; to establish the second formula, note that $e^{it\lambda}$ is the sum of the orthogonal functions $e^{it\lambda} - \varphi_\tau(\lambda)$ and $\varphi_\tau(\lambda)$, and apply the Pythagorean relation. ‖

Theorem 2.7.2 is especially useful in the case of *rational spectral densities*, namely, spectral densities of the form

$$f(\lambda) = A \left| \frac{(\lambda - \alpha_1) \cdots (\lambda - \alpha_m)}{(\lambda - \beta_1) \cdots (\lambda - \beta_n)} \right|^2, \qquad m < n.$$

We assume that Im $\alpha_j > 0$ for all j, and Im $\beta_k > 0$ for all k. In order for $\psi_\tau(\lambda) = [e^{it\lambda} - \varphi_\tau(\lambda)]f(\lambda)$ to be analytic for Im $\lambda \ge 0$, φ_τ can have singularities only at $\alpha_1, \ldots, \alpha_m$; singularities elsewhere would not be canceled by zeros of f. Thus we may write $\varphi_\tau(\lambda) = C_\tau(\lambda)/(\lambda - \alpha_1) \cdots (\lambda - \alpha_m)$ where C_τ is entire. For (a) to hold we require that C_τ be a polynomial of degree at most $n - 1$; since Im $\alpha_j > 0$, (b) is satisfied as well, as is the $O(|\lambda|^{-1-\varepsilon})$ condition of (c). Now we write

$$\psi_\tau(\lambda) = \frac{A[e^{it\lambda}(\lambda - \alpha_1) \cdots (\lambda - \alpha_m) - C_\tau(\lambda)](\lambda - \bar\alpha_1) \cdots (\lambda - \bar\alpha_m)}{(\lambda - \beta_1) \cdots (\lambda - \beta_n)(\lambda - \bar\beta_1) \cdots (\lambda - \bar\beta_n)}.$$

For this to be analytic for Im $\lambda \geq 0$, the factors $(\lambda - \beta_1) \cdots (\lambda - \beta_n)$ in the denominator must be canceled by zeros of $e^{it\lambda}(\lambda - \alpha_1) \cdots (\lambda - \alpha_m) - C_\tau(\lambda)$ in the numerator. If β_k is of multiplicity m_k, we must choose the n coefficients of C_τ to satisfy the following n equations.

2.7.3 Solution of the Prediction Problem for Rational Spectral Densities

$$\frac{d^j}{d\lambda^j}[e^{it\lambda}(\lambda - \alpha_1) \cdots (\lambda - \alpha_m) - C_\tau(\lambda)] = 0 \qquad \text{at} \quad \lambda = \beta_k,$$

for $j = 0, 1, \ldots, m_k - 1$ and all k. (The zeroth derivative of a function is interpreted as the function itself; also, if $m = 0$, the product $(\lambda - \alpha_1) \cdots (\lambda - \alpha_m)$ is replaced by 1.)

It is always possible to choose $C_\tau(\lambda) = c_0 + c_1 \lambda + \cdots + c_{n-1} \lambda^{n-1}$ to satisfy these conditions (see Problem 5). The transfer function of the optimal predictor is given by

$$\varphi_\tau(\lambda) = \frac{C_\tau(\lambda)}{(\lambda - \alpha_1) \cdots (\lambda - \alpha_m)}.$$

Let us look at some examples.

2.7.4 Example

Suppose the covariance function is given by $K(t) = e^{-\alpha|t|}$ where $\alpha > 0$; the corresponding spectral density is

$$f(\lambda) = \frac{1}{2\pi} \int_{-\infty}^{\infty} K(t)e^{-i\lambda t} \, dt = \frac{\alpha}{\pi} \frac{1}{\lambda^2 + \alpha^2} = \frac{\alpha}{\pi} \frac{1}{(\lambda - i\alpha)(\lambda + i\alpha)}.$$

In this case $C_\tau(\lambda)$ is a polynomial of degree 0, in other words a constant c_0. The equations of 2.7.3 become simply $e^{it\lambda} - c_0 = 0$ at $\lambda = i\alpha$; thus $c_0 = e^{-\alpha\tau}$. Therefore the best linear predictor of $X(t + \tau)$ based on $X(s)$, $s \leq t$ is

$$\int_{-\infty}^{\infty} e^{it\lambda} e^{-\alpha\tau} \, dZ_X(\lambda) = e^{-\alpha\tau} X(t)$$

and the prediction error is

$$\sigma_\tau^2 = \int_{-\infty}^{\infty} [1 - e^{-2\alpha\tau}] f(\lambda) \, d\lambda = 1 - e^{-2\alpha\tau}.$$

2.7.5   Example

Suppose the spectral density is given by

$$f(\lambda) = \frac{1}{(\lambda^2 + \alpha^2)^2} \qquad \text{where} \quad \alpha > 0.$$

In this case $C_\tau(\lambda)$ is of the form $c_0 + c_1 \lambda$, and the equations 2.7.3 become

$$e^{it\lambda} - (c_0 + c_1\lambda) = 0 \qquad \text{at} \quad \lambda = i\alpha$$

$$i\tau e^{it\lambda} - c_1 = 0 \qquad \text{at} \quad \lambda = i\alpha.$$

Thus $c_1 = i\tau e^{-\alpha\tau}$, $c_0 = e^{-\alpha\tau}(1 + \alpha\tau)$.

The best linear predictor of $X(t + \tau)$ based on $X(s)$, $s \leq t$, is

$$\int_{-\infty}^{\infty} e^{it\lambda}[e^{-\alpha\tau}(1 + \alpha\tau) + e^{-\alpha\tau}i\tau\lambda]\, dZ_X(\lambda)$$

$$= e^{-\alpha\tau}[(1 + \alpha\tau)X(t) + \tau X'(t)]$$

and the prediction error is

$$\sigma_\tau^2 = \int_{-\infty}^{\infty} [1 - |e^{-\alpha\tau}(1 + \alpha\tau + i\tau\lambda)|^2]f(\lambda)\, d\lambda$$

$$= \int_{-\infty}^{\infty} (1 - e^{-2\alpha\tau}[(1 + \alpha\tau)^2 + \tau^2\lambda^2])f(\lambda)\, d\lambda.$$

Integration yields

$$\sigma_\tau^2 = \sigma^2[1 - e^{-2\alpha\tau}(1 + 2\alpha\tau + 2\alpha^2\tau^2)]$$

where $\sigma^2 = \int_{-\infty}^{\infty} f(\lambda)\, d\lambda = \pi/2\alpha^3$; note that σ^2 is $K(0)$, the variance of the process.

Problems

1.   If the covariance function of $\{X(t)\}$ is a sum of exponentials, then as in 2.7.4, the spectral density can be expressed as

$$f(\lambda) = \frac{R(\lambda)}{S(\lambda)} = A\frac{\prod_j (\lambda - \alpha_j)}{\prod_k (\lambda - \beta_k)}$$

where R and S are relatively prime polynomials and A is constant; since f is integrable, none of the β_k is real.

If f is a nonnegative Lebesgue integrable function expressible in the above form, we are going to show (as in Problem 4, Section 2.4) that f is a rational spectral density.

(a) Show that z is a zero of $R(z)/S(z)$ iff $\bar{z}$ is a zero of $R(z)/S(z)$, and similarly for poles.

(b) Show that $f(\lambda)$ can be written in the form

$$A \frac{|\prod_{\mathrm{Im}\,\alpha_j > 0}(\lambda - \alpha_j)|^2}{|\prod_{\mathrm{Im}\,\beta_k > 0}(\lambda - \beta_k)|^2} \prod_{\mathrm{Im}\,\alpha_j = 0}(\lambda - \alpha_j).$$

(c) Show that if α_j is real, then α_j is a root of even multiplicity. Conclude that f can be expressed as

$$f(\lambda) = |P(\lambda)/Q(\lambda)|^2$$

where P and Q are relatively prime polynomials.

2. Find the best linear predictor for $X(t + \tau)$ based on $X(s)$, $s \le t$, for processes with the following spectral densities ($\alpha > 0$):

(a) $\dfrac{1}{\lambda^4 + \alpha^4}$,   (b) $\dfrac{1}{(\lambda^2 + 2\alpha^2)(\lambda^2 + \frac{1}{2}\alpha^2)}$,   (c) $\dfrac{\lambda^2 + \alpha^2}{\lambda^4 + \alpha^4}$.

Also find the prediction error in each case.

3. Yaglom's method can also be used to solve the filtering problem in certain cases. Suppose we observe $X(t) = S(t) + N(t)$ where $\{S(t)\}$ is a stationary signal process with spectral density f_S and $\{N(t)\}$ is a stationary noise process with spectral density f_N. Assume that $\{S(t)\}$, $\{N(t)\}$ are orthogonal processes; that is, $S(t)$ and $N(t')$ are orthogonal for all t, t'.

(a) Show that the best linear estimate of $S(t)$ based on $X(s)$, $-\infty < s < \infty$, is given by

$$\int_{-\infty}^{\infty} e^{it\lambda} \frac{f_S(\lambda)}{f_S(\lambda) + f_N(\lambda)} dZ_X(\lambda)$$

(assuming $f_S/(f_S + f_N) \in L^2(\mu_X)$) and the estimation error by

$$\int_{-\infty}^{\infty} \frac{f_S(\lambda) f_N(\lambda)}{f_S(\lambda) + f_N(\lambda)} d\lambda.$$

Thus perfect estimation is possible iff the signal and noise have disjoint spectra.

(b) Show that the best linear estimate of $S(t + \tau)$ based on $X(s)$, $s \le t$, is given by $\int_{-\infty}^{\infty} e^{it\lambda} \varphi_\tau(\lambda) dZ_X(\lambda)$, where φ_τ satisfies the following conditions:

(1) $\varphi_\tau \in L^2(f_X)$;

(2) φ_τ is the limit in $L^2(f_X)$ of a sequence of finite linear combinations of $e^{is\lambda}$, $s \le 0$;

(3) $\int_{-\infty}^{\infty} e^{is\lambda}[e^{it\lambda}f_S(\lambda) - \varphi_\tau(\lambda)f_X(\lambda)]\, d\lambda = 0$ for all $s \ge 0$.

(c)   Show that for the conditions of (b) to hold, it is sufficient that:

(i) $\varphi_\tau \in L^2(f_X)$;

(ii) φ_τ is analytic for Im $\lambda \le 0$, and for some $r > 0$, $\varphi_\tau(\lambda) = O(|\lambda|^r)$ as $|\lambda| \to \infty$ with Im $\lambda \le 0$;

(iii) $\psi_\tau(\lambda) = e^{it\lambda}f_S(\lambda) - \varphi_\tau(\lambda)f_X(\lambda)$ is analytic for Im $\lambda \ge 0$, and for some $\varepsilon > 0$, $\psi_\tau(\lambda) = O(|\lambda|^{-1-\varepsilon})$ as $|\lambda| \to \infty$ with Im $\lambda \ge 0$.

(d)   Show that the estimation error is

$$\sigma_\tau^2 = \int_{-\infty}^{\infty} f_S(\lambda)\, d\lambda - \int_{-\infty}^{\infty} |\varphi_\tau(\lambda)|^2 f_X(\lambda)\, d\lambda$$

$$= \int_{-\infty}^{\infty} [1 - |\varphi_\tau(\lambda)|^2] f_S(\lambda)\, d\lambda + \int_{-\infty}^{\infty} |\varphi_\tau(\lambda)|^2 f_N(\lambda)\, d\lambda.$$

4.   In the filtering problem, let

$$f_S(\lambda) = \frac{A}{\lambda^2 + \alpha^2}, \qquad f_N(\lambda) = \frac{B}{\lambda^2 + \beta^2},$$

where α, β, A, $B > 0$. Show that for $\tau \ge 0$,

$$\varphi_\tau(\lambda) = \frac{A}{A + B} \frac{\alpha + \beta}{\alpha + \gamma} \frac{\beta + i\lambda}{\gamma + i\lambda} e^{-\alpha\tau}$$

where

$$\gamma^2 = \frac{A\beta^2 + B\alpha^2}{A + B}.$$

If $\tau \le 0$, show that

$$\varphi_\tau(\lambda) = \frac{A}{A + B} \frac{(\alpha + \gamma)(\lambda^2 + \beta^2)e^{it\lambda} + (\beta - \gamma)(\lambda - i\alpha)(\lambda - i\beta)e^{\gamma\tau}}{(\alpha + \gamma)(\lambda^2 + \gamma^2)}.$$

5.   In connection with 2.7.3, we wish to show the existence of n constants $a_0, \ldots, a_{n-1}$ such that the polynomial $a_0 + a_1 x + \cdots + a_{n-1} x^{n-1}$ satisfies the n conditions

$$p^{(i)}(x_1) = w_{i1}, \qquad i = 0, 1, \ldots, m_1 - 1$$
$$\vdots$$
$$p^{(i)}(x_r) = w_{ir}, \qquad i = 0, 1, \ldots, m_r - 1$$

where the w_{ij} are given constants, $m_1 + \cdots + m_r = n$ and $x_1, \ldots, x_r$ are distinct. Define a linear transformation T from the space of polynomials of degree at most $n - 1$ to the space R^n by

$$Tp(x) = (p(x_1), p'(x_1), \ldots, p^{(m_1 - 1)}(x_1), \ldots, p(x_r), p'(x_r), \ldots, p^{(m_r - 1)}(x_r)).$$

Our problem is solvable for arbitrary w_{ij} if and only if T maps onto all of R^n. Show that T is one-to-one; since T is a linear transformation on a finite-dimensional space, it must therefore be onto.

2.8   Some Stochastic Differential Equations

In 2.6.3 we studied a class of stochastic differential equations of the form

$$\sum_{j=0}^{m} b_j X^{(j)}(t) = \sum_{k=0}^{n} a_k Y^{(k)}(t)$$

where the L^2 derivatives are assumed to exist. We now drop the requirement that the forcing function be L^2-differentiable, and consider the equation

$$b_n X^{(n)}(t) + \cdots + b_1 X'(t) + b_0 X(t) = Y'(t) \tag{1}$$

where the b_j are given constants and $\{Y(t)\}$ is an L^2-continuous process with 0 mean and orthogonal increments. Since $Y'(t)$ need not exist, a *solution* to (1) on an interval I will mean a process $\{X(t), t \in I\}$, with $n - 1$ L^2-derivatives on I, satisfying an integrated version of (1):

$$b_n[X^{(n-1)}(t) - X^{(n-1)}(t_0)] + \cdots + b_1[X(t) - X(t_0)]$$

$$+ b_0 \int_{t_0}^{t} X(s)\, ds = Y(t) - Y(t_0) \tag{2}$$

where t_0 is a point of I.

The following theorem is analogous to a standard result about ordinary differential equations.

2.8.1   Theorem

The unique solution to (1) on $[t_0, \infty)$, subject to initial conditions $X(t_0) = X'(t_0) = \cdots = X^{(n-1)}(t_0) = 0$, is given by

$$X(t) = \int_{t_0}^{t} h(t - s)\, dY(s)$$

where h is the *impulse response* (or *Green's function*) of (1); that is, $h(t) = 0$ for $t < 0$, and for $t \geq 0$, $h(t)$ is the unique solution to the homogeneous initial value problem

$$b_n x^{(n)}(t) + \cdots + b_1 x'(t) + b_0 x(t) = 0 \qquad (3)$$

$$x(0) = \cdots = x^{(n-2)}(0) = 0, \qquad x^{(n-1)}(t) = 1/b_n .$$

(The integral is of the type discussed in 2.1.6.) If $\varphi_0, \ldots, \varphi_{n-1}$ are the unique solutions to (3) such that $\varphi_i^{(j)}(t_0) = \delta_{ij}$, $i, j = 0, 1, \ldots, n - 1$, the general solution to (1) on $[t_0, \infty)$ is given by

$$X(t) = C_0 \varphi_0(t) + \cdots + C_{n-1} \varphi_{n-1}(t) + \int_{t_0}^{t} h(t - s) \, dY(s)$$

where $C_0, \ldots, C_{n-1}$ are arbitrary L^2 random variables. Furthermore, $C_j = X^{(j)}(t_0)$, $0 \leq j \leq n - 1$, so that the solution is unique subject to specification of $X(t_0), \ldots, X^{(n-1)}(t_0)$.

PROOF   Set $X(t) = \int_{t_0}^{t} h(t - s) \, dY(s)$. An integration by parts (Problem 1) gives

$$X(t) = -h(t - t_0)Y(t_0) + \int_{t_0}^{t} h'(t - s)Y(s) \, ds.$$

Now apply Leibniz's rule (Problem 2) and the fact that $h^{(j)}(0) = 0$, $1 \leq j \leq n - 2$, to obtain

$$X^{(j)}(t) = -h^{(j)}(t - t_0)Y(t_0) + \int_{t_0}^{t} h^{(j+1)}(t - s)Y(s) \, ds,$$

$$1 \leq j \leq n - 2$$

$$X^{(n-1)}(t) = -h^{(n-1)}(t - t_0)Y(t_0)$$

$$+ \int_{t_0}^{t} h^{(n)}(t - s)Y(s) \, ds + h^{(n-1)}(0)Y(t).$$

It   follows   immediately   that $X(t_0) = X'(t_0) = \cdots = X^{(n-1)}(t_0) = 0$. Furthermore,

$$\int_{t_0}^{t} X(s) \, ds = \int_{t_0}^{t} \left[-h(s - t_0)Y(t_0) + \int_{t_0}^{s} h'(s - u)Y(u) \, du \right] ds$$

$$= -\int_{t_0}^{t} h(s - t_0)Y(t_0) \, ds + \int_{t_0}^{t} \left[\int_{u}^{t} h'(s - u) \, ds \right] Y(u) \, du$$

$$= -\int_{t_0}^{t} h(s - t_0)Y(t_0) \, ds + \int_{t_0}^{t} h(t - u)Y(u) \, du.$$

Therefore, the left side of (2) is

$$-Y(t_0)\left[b_n h^{(n-1)}(t-t_0) + \cdots + b_1 h(t-t_0) + b_0 \int_{t_0}^t h(s-t_0)\,ds\right]$$

$$+ \int_{t_0}^t [b_n h^{(n)}(t-s) + \cdots + b_1 h'(t-s) + b_0 h(t-s)] Y(s)\,ds$$

$$+ b_n h^{(n-1)}(0) Y(t).$$

Since h is a solution for the homogeneous equation, the derivative with respect to t of the first term in brackets is 0; the second term in brackets is 0 for the same reason. Since $b_n h^{(n-1)}(0) = 1$, the first term in brackets, which is constant for all t, must be 1. Thus the left side of (2) reduces to $Y(t) - Y(t_0)$, so that $\{X(t)\}$ is a solution, as desired.

To prove uniqueness, consider the vector differential equation

$$A_0 X'(t) + A_1 X(t) = Y'(t) \tag{4}$$

where A_0 and A_1 are n by n matrices with A_0 nonsingular, and $X(t)$ and $Y(t)$ are n by 1 random column vectors. If $X(t) = \text{col}\,(X_1(t),\ldots,X_n(t))$, then $X'(t)$ is defined as col $(X'_1(t), \ldots, X'_n(t))$, where the derivatives are in the L^2 sense. The integrated form of (4) (recall that $Y'(t)$ may not be defined) is

$$A_0(X(t) - X(t_0)) + A_1 \int_{t_0}^t X(s)\,ds = Y(t) - Y(t_0)$$

which we may rewrite as

$$e^{At}\left[X(t) + A\int_{t_0}^t X(s)\,ds\right] = e^{At}[X(t_0) + A_0^{-1}(Y(t) - Y(t_0))]$$

where $A = A_0^{-1}A_1$. This in turn may be expressed as

$$\frac{d}{dt}\left[e^{At}\int_{t_0}^t X(s)\,ds\right] = e^{At}[X(t_0) - A_0^{-1}Y(t_0)] + e^{At}A_0^{-1}Y(t).$$

Integrating from t_0 to t and multiplying by e^{-At} yields

$$\int_{t_0}^t X(s)\,ds = e^{-At}[X(t_0) - A_0^{-1}Y(t_0)] \int_{t_0}^t e^{As}\,ds$$

$$+ e^{-At}\int_{t_0}^t e^{As}A_0^{-1}Y(s)\,ds.$$

Finally, differentiation with respect to t gives an explicit expression for $X(t)$ in terms of A_0, A_1, $X(t_0)$, and $\{Y(t)\}$; consequently, there is at most one vector-valued L^2 process $\{X(t), t \geq t_0\}$ satisfying (4) with $X(t_0)$ specified. We apply this result with

$$A_0 = \begin{bmatrix} b_n & b_{n-1} & b_{n-2} & \cdots & b_2 & b_1 \\ 0 & 1 & 0 & \cdots & 0 & 0 \\ 0 & 0 & 1 & \cdots & 0 & 0 \\ \vdots & & & & & \\ 0 & 0 & 0 & \cdots & 1 & 0 \\ 0 & 0 & 0 & \cdots & 0 & 1 \end{bmatrix},$$

$$A_1 = \begin{bmatrix} 0 & 0 & 0 & \cdots & 0 & b_0 \\ -1 & 0 & 0 & \cdots & 0 & 0 \\ 0 & -1 & 0 & \cdots & 0 & 0 \\ \vdots & & & & & \\ 0 & 0 & 0 & \cdots & -1 & 0 \end{bmatrix},$$

$$\mathbf{Y}(t) = \begin{bmatrix} Y(t) \\ 0 \\ 0 \\ \vdots \\ 0 \end{bmatrix}, \qquad \mathbf{X}(t) = \begin{bmatrix} X_1(t) \\ \vdots \\ X_n(t) \end{bmatrix}.$$

If we set $X_n(t) = X(t)$, then (4) requires that

$$X_{n-1}(t) = X'(t), \qquad X_{n-2}(t) = X''(t), \qquad \ldots, \qquad X_1(t) = X^{(n-1)}(t),$$

and

$$b_n X^{(n)}(t) + \cdots + b_1 X'(t) + b_0 X(t) = Y'(t).$$

Therefore, Eq. (1) has a unique solution with $X(t_0), \ldots, X^{(n-1)}(t_0)$ specified.

To prove the assertion about the general solution to (1), we observe that if $\{X(t), t \geq t_0\}$ satisfies (1), and if

$$Z(t) = X(t) - X(t_0)\varphi_0(t) - X'(t_0)\varphi_1(t) - \cdots - X^{(n-1)}(t_0)\varphi_{n-1}(t)$$

then $\{Z(t), t \geq t_0\}$ satisfies (1) and the initial conditions $Z(t_0) = Z'(t_0) = \cdots = Z^{(n-1)}(t_0) = 0$. By what we have already proved, $Z(t)$ has the form $\int_{t_0}^{t} h(t - s)\, dY(s)$. ‖

If the forcing function is derived from Brownian motion, we may obtain additional results.

2.8.2   Theorem

In 2.8.1, assume all roots of the characteristic equation $b_n \lambda^n + \cdots + b_1 \lambda + b_0 = 0$ have negative real parts. Let $\{Y(t)\}$ be extended Brownian motion $\{B(t), -\infty < t < \infty\}$ (see Section 2.6, Problem 4), and let

$$X_0(t) = \int_{-\infty}^{t} h(t - s)\, dB(s), \qquad -\infty < t < \infty.$$

Then

(a) $\{X_0(t),\ -\infty < t < \infty\}$ is a stationary Gaussian process with 0 mean and covariance function

$$K_0(t) = \sigma^2 \int_0^\infty h(u)h(u + t)\,du.$$

(b) The general solution to (1) on any interval is given by

$$X(t) = C_0\varphi_0(t) + \cdots + C_{n-1}\varphi_{n-1}(t) + X_0(t)$$

where $C_0, \ldots, C_{n-1}$ are arbitrary L^2 random variables.

(c) $\{X_0(t),\ -\infty < t < \infty\}$ is the only stationary process that is a solution to (1) on all bounded intervals.

(d) As $s, t \to \infty$, we have

$$X(t) \to X_0(t) \quad \text{a.s.,} \qquad EX(t) \to EX_0(t),$$
$$\text{Cov}\,[X(s), X(t)] \to \text{Cov}\,[X_0(s), X_0(t)].$$

In this sense, any solution of (1) approaches the stationary solution as $t \to \infty$.

PROOF (a) By hypothesis, $h(t)$ is 0 for $t < 0$, and is a linear combination of "negative exponentials" for $t \geq 0$. Therefore the function $s \to h(t - s)$ belongs to $L^2(m)$, m is Lebesgue measure, and hence X_0 is well defined (see 2.1.6). Since $E[B(t)] \equiv 0$, we have $E[X_0(t)] \equiv 0$. Now since $h(t) = 0$ for $t < 0$, we may write

$$X_0(t) = \int_{-\infty}^\infty h(t - u)\,dB(u) = L^2 \lim_{T \to \infty} \int_{-T}^T h(t - u)\,dB(u).$$

Thus

$$\text{Cov}\,[X_0(s), X_0(t)] = E[X_0(s)X_0(t)]$$
$$= \lim_{T \to \infty} E\left[\int_{-T}^T h(s - u)\,dB(u) \int_{-T}^T h(t - u)\,dB(u)\right]$$
$$= \lim_{T \to \infty} \int_{-T}^T h(s - u)h(t - u)\sigma^2\,du$$
$$= \sigma^2 \int_{-\infty}^\infty h(-v)h(t - s - v)\,dv.$$

It follows that $\{X_0(t)\}$ is stationary, with covariance function

$$K_0(t) = \sigma^2 \int_{-\infty}^{\infty} h(-v)h(t-v)\, dv = \sigma^2 \int_0^{\infty} h(u)h(t+u)\, du.$$

To prove that $\{X_0(t)\}$ is Gaussian, observe that a linear combination $\sum_{i=1}^m a_i X_0(t_i)$ can be written as $\int_{-\infty}^{\infty} \sum_{i=1}^m a_i h(t_i - s)\, dB(s)$, which is a Gaussian random variable by Problem 4d, Section 2.6. By Problem 8, Section 1.2, $\{X_0(t)\}$ is Gaussian.

(b)   Integration by parts gives

$$X_0(t) = \int_{-\infty}^{t} h'(t-s)B(s)\, ds.$$

(Note that as $s \to -\infty$, $h(t-s)B(s) \to 0$ in L^2 since $E[\,|h(t-s)B(s)|^2] = h^2(t-s)\sigma^2 s \to 0$; remember that h is a combination of negative exponentials.) By Leibniz's rule,

$$X_0^{(j)}(t) = \int_{-\infty}^{t} h^{(j+1)}(t-s)B(s)\, ds,$$

$$j = 0, 1, \ldots, n-2,$$

$$X_0^{(n-1)}(t) = h^{(n-1)}(0)B(t) + \int_{-\infty}^{t} h^{(n)}(t-s)B(s)\, ds.$$

Also,

$$\int_{-\infty}^{t} X_0(s)\, ds = \int_{-\infty}^{t} \int_{-\infty}^{s} h'(s-u)B(u)\, du\, ds$$

$$= \int_{-\infty}^{t} \int_{u}^{t} h'(s-u)\, ds B(u)\, du$$

$$= \int_{-\infty}^{t} h(t-u)B(u)\, du.$$

Thus

$$b_n X_0^{(n-1)}(t) + \cdots + b_1 X_0(t) + b_0 \int_{-\infty}^{t} X_0(s)\, ds$$

$$= \int_{-\infty}^{t} [b_n h^{(n)}(t-s) + \cdots + b_1 h'(t-s) + b_0 h(t-s)]B(s)\, ds + B(t)$$

$$= B(t).$$

If we replace t by t_0 and then subtract, we see that $\{X_0(t)\}$ is a particular solution of (1) on any interval. If $\{X(t)\}$ is any solution, then $X(t) - X_0(t)$ is a solution of the homogeneous equation ($Y' = 0$ in (1)). By 2.8.1, $X(t) - X_0(t)$ is expressible as $C_0 \varphi_0(t) + \cdots + C_{n-1} \varphi_{n-1}(t)$.

(c)   This is immediate from (a) and (b) if we observe that if any of the C_i are nonzero, $\{X(t)\}$ cannot be stationary.

(d)   By (b), $X(t) - X_0(t) = C_0 \varphi_0(t) + \cdots + C_{n-1} \varphi_{n-1}(t)$, which converges to 0, in L^1, L^2 and a.s., again because the $\varphi_i(t)$ are linear combinations of negative exponentials. It follows that $X(t) \to X_0(t)$ a.s. and $E[X(t)] \to E(X_0(t))$. The covariance assertion follows from 1.3.1.   ‖

Problems

1.   (*Integration by Parts*)   Let $\{Y(t), \ a \leq t \leq b\}$ be an L^2-continuous process with orthogonal increments, and let f be a complex-valued function with a continuous derivative on $[a, b]$. Show that

$$\int_a^b f(t) \, dY(t) = [f(t)Y(t)]_a^b - \int_a^b f'(t)Y(t) \, dt.$$

2.   (*Leibniz's Rule*)   Let $f = f(s, t)$ be a complex-valued function on $I \times I$, I an open interval of reals. Assume f and $\partial f / \partial t$ are continuous on $I \times I$. Let $\{Y(t), s \in I\}$ be an L^2-continuous process with 0 mean. If $t_0, t \in I$, show that

$$\frac{d}{dt} \int_{t_0}^t f(s, t)Y(s) \, ds = \int_{t_0}^t \frac{\partial f}{\partial t}(s, t)Y(s) \, ds + f(t, t)Y(t).$$

3.   (*Uhlenbeck–Ornstein Process*)   Consider the stochastic differential equation

$$a_0 X''(t) + a_1 X'(t) = B'(t), \qquad t \geq 0,$$

with initial conditions

$$X(0) = x_0, \qquad X'(0) = v_0.$$

If $V(t) = X'(t)$, then $\{V(t), t \geq 0\}$ is called the *Langevin velocity process* and $\{X(t), t \geq 0\}$ is called the *Uhlenbeck–Ornstein process*; note that $X(t) = x_0 + \int_0^t V(s) \, ds$. Express each of the processes $\{V(t)\}$ and $\{X(t)\}$ in terms of $\{B(t)\}$, and find the mean and covariance functions.

4.   (*Prediction Problem for Solutions of Stochastic Differential Equations*) Consider the stochastic differential equation

$$b_n X^{(n)}(t) + \cdots + b_0 X(t) = B'(t), \qquad t \geq 0,$$

with $X(0) = c_0, \ldots, X^{(n-1)}(0) = c_{n-1}$ (the c_j constant). Let $\varphi_0, \ldots, \varphi_{n-1}$ be the unique solutions of the ordinary differential equation $b_n x^{(n)}(t) + \cdots + b_0 x(t) = 0$, subject to $\varphi_i^{(j)}(0) = \delta_{ij}$, $i, j = 0, \ldots, n-1$.

(a)   If $0 < t_1 < t_2$, show that the best linear predictor of $X(t_2)$ based on $X(t)$, $0 \le t \le t_1$, is

$$\hat{X}(t_2) = X(t_1)\varphi_0(t_2 - t_1) + \cdots + X^{(n-1)}(t_1)\varphi_n(t_2 - t_1).$$

(Thus only values of $X(t)$ for $t_1 - \varepsilon \le t \le t_1$, $\varepsilon > 0$ arbitrary, are used in prediction.)

(b)   Show that $\{\hat{X}(t), t \ge t_1\}$ is the unique solution to the homogeneous equation

$$b_n \hat{X}^{(n)}(t) + \cdots + b_0 \hat{X}(t) = 0, \qquad t \ge t_1,$$

subject to the initial conditions

$$\hat{X}(t_1) = X(t_1), \ldots, \hat{X}^{(n-1)}(t_1) = X^{(n-1)}(t_1).$$

(c)   Show that the prediction error is

$$E[\,|\hat{X}(t_2) - X(t_2)|^2] = \sigma^2 \int_0^{t_2-t_1} h^2(s) \, ds.$$

2.9   Continuous Parameter Prediction: Remarks on the General Solution

The development of the general case of continuous parameter prediction closely parallels that of discrete parameter prediction. The two main differences are (1) the complex analysis results are for the half-plane instead of the unit disk, and (2) the concept of white noise in the continuous case is not as straightforward as in the discrete case. We summarize the main results for the continuous case.

Let $\{X(t), -\infty < t < \infty\}$ have spectral measure μ, and let f be the density of the absolutely continuous component of μ in the decomposition with respect to Lebesgue measure. Then $\{X(t)\}$ is deterministic if and only if

$$\int_{-\infty}^{\infty} \frac{\ln f(\lambda)}{1 + \lambda^2} \, d\lambda = -\infty.$$

If $\{X(t)\}$ is not deterministic, let

$$C(\lambda) = \exp\left\{ \frac{1}{2\pi} \int_{-\infty}^{\infty} \frac{\ln f(\eta)}{1 + \eta^2} \frac{i + \eta\lambda}{\eta + i\lambda} \, d\eta \right\}.$$

Then C can be expressed as the inverse Fourier transform of a function c vanishing on $(-\infty, 0)$:

$$C(\lambda) = \frac{1}{2\pi} \int_0^\infty e^{-it\lambda} c(t) \, dt.$$

Let

$$C_\tau(\lambda) = \frac{1}{2\pi} \int_\tau^\infty e^{-it\lambda} c(t) \, dt.$$

Then the best linear predictor of $X(t + \tau)$, $\tau > 0$, based on $X(s)$, $s \le t$, is

$$P_t X(t + \tau) = \int_{-\infty}^\infty e^{it\lambda} \left[e^{it\lambda} \frac{C_\tau(i\lambda)}{C(i\lambda)} \right] dZ_X(\lambda)$$

and the prediction error is

$$\delta(\tau) = \int_0^\tau |c(t)|^2 \, dt.$$

2.10   Notes

There is a large body of literature on L^2 processes and prediction; we mention only a few. Introductory treatments are given by Yaglom (1962) and Cramér and Leadbetter (1967); Papoulis (1967) and Wong (1971) give presentations oriented toward engineering applications. Gikhman and Skorokhod (1969) discuss the general prediction problem. Rozanov (1967) and Hannan (1970) treat the prediction problem for vector-valued stationary processes, and Hannan includes broad coverage of the statistical theory of these processes.

Chapter 3

Ergodic Theory

3.1 Introduction

In Chapters 1 and 2, we considered the concept of wide sense stationarity of an L^2 process; in this chapter our main interest is in strictly stationary sequences of random variables. Our purpose is to sharpen the intuition by considering some reasonable examples of stochastic processes before developing the general theory of Chapter 4. Ergodic theory provides a convenient vehicle for the study of stationary sequences, and we shall see that the basic "pointwise ergodic theorem" is a generalization of the strong law of large numbers (see RAP, Chapter 7).

The starting point for ergodic theory is the notion of a measure-preserving transformation, defined as follows.

3.1.1 Definition

Let $(\Omega, \mathscr{F}, \mu)$ be a measure space, and T a *measurable transformation* on $(\Omega, \mathscr{F}, \mu)$, that is, $T: (\Omega, \mathscr{F}) \to (\Omega, \mathscr{F})$.

The transformation T is said to be *measure-preserving* (we also say that T is *μ-preserving* or that T *preserves* μ iff $\mu(T^{-1}A) = \mu(A)$ for all $A \in \mathscr{F}$.

(This implies that $\mu(T^{-k}A) = \mu(A)$ for all $k = 1$, 2, ..., where $T^{-k}A = \{\omega : T^k\omega \in A\}$ and T^k is the composition of T with itself k times.)

The physical concept of a *flow* may be used to motivate the study of measure-preserving transformations. A flow may be regarded as a process in which a system of particles of a fluid (each point of the container corresponding to a particle) moves about under the action of an externally applied force. The force is assumed to be independent of time, so that at least at discrete times $t = 0, 1, 2, \ldots$, the flow can be described by a single (measurable) function T. If x is a point of the container, Tx is the position of the particle at x after one second has elapsed; thus $T^2 x = T(Tx)$ is the position after two seconds, and so on. If A is a (Borel) subset of the container, then $T^{-1}(A)$ corresponds to the set of particles that will be in A after one application of T. If μ is Lebesgue measure (volume in this case) and the fluid is incompressible, it is reasonable to expect that $\mu(T^{-1}A) = \mu(A)$.

We consider some mathematical examples.

3.1.2   Examples

(a)   Let Ω be a finite set $\{x_1, \ldots, x_n\}$, $n \geq 2$, with $\mathcal{F}$ consisting of all subsets of Ω. Let T be a cyclic permutation of Ω, say $T(x_i) = x_{i+1}$, with indices reduced modulo n. Since $T^{-1}\{x_i\} = \{x_{i-1}\}$, T preserves μ iff $\mu\{x_i\}$ is constant for all i. Thus if μ is a probability measure P, then $P\{x_i\}$ must be $1/n$ for all i.

More generally, if T is any permutation of Ω, T can be expressed as a product of disjoint cycles $C_1, \ldots, C_k$. Then T preserves μ iff within each cycle, μ assigns equal weight to each point.

(b)   Let $\Omega = R$, $\mathcal{F} = \mathcal{B}(R)$, and let μ be Lebesgue measure. If $T(x) = x + c$, c constant, then T preserves μ because μ is translation-invariant.

(c)   Let Ω be the unit circle in the plane R^2 (Ω can be identified with the interval $[0, 2\pi)$ under the correspondence $e^{i\theta} \to \theta$). Take $\mathcal{F}$ as the Borel sets, and $\mu = P = \lambda/2\pi$, where λ is arc length on the circle (or Lebesgue measure on $[0, 2\pi)$). Thus if A is a Borel subset of $[0, 2\pi)$, $\mu(A) = \int_A (2\pi)^{-1} \, d\theta$.

Let α be fixed in $[0, 2\pi)$, and let T be rotation by α. Thus T is defined on Ω by $T(e^{i\theta}) = e^{i(\theta + \alpha)}$, or equivalently on $[0, 2\pi)$ by $T(\theta) = \theta + \alpha$ (modulo 2π). As in (b), T preserves μ by translation-invariance of Lebesgue measure.

(d)   Let $\Omega = R^\infty$, the collection of all sequences $(\omega_0, \omega_1, \ldots)$ of real numbers; take $\mathcal{F} = [\mathcal{B}(R)]^\infty$, and let μ be any probability measure P on $\mathcal{F}$. Define $T(\omega_0, \omega_1, \omega_2, \ldots) = (\omega_1, \omega_2, \ldots)$; T is called the *one-sided shift transformation*.

We show that T *preserves* P *iff* P *is stationary*, in other words,

$$\text{iff } P\{\omega : (\omega_0, \omega_1, \ldots, \omega_{n-1}) \in B_n\} = P\{(\omega_k, \omega_{k+1}, \ldots, \omega_{k+n-1}) \in B_n\}$$

for all $n, k = 1, 2, \ldots$ and all n-dimensional Borel sets B_n.

(If $X_k(\omega) = \omega_k$, $k = 0, 1, \ldots$, then stationarity of P is equivalent to (strict) stationarity of the random sequence $X_0, X_1, \ldots$, that is, $(X_0, \ldots,$

X_{n-1}) and $(X_k, \ldots, X_{k+n-1})$ have the same distribution for all n, $k = 1, 2, \ldots$. In this chapter, stationarity will always have the strict rather than the wide sense connotation.)

First we note that

$$T^{-k}\{\omega : (\omega_0, \ldots, \omega_{n-1}) \in B_n\} = \{\omega : (\omega_k, \ldots, \omega_{k+n-1}) \in B_n\}.$$

If T preserves P, then

$$P(A) = P(T^{-1}A) = \cdots = P(T^{-k}A), \qquad A \in \mathscr{F},$$

and it follows that P is stationary.

Conversely, if P is stationary and $A = \{\omega : (\omega_0, \ldots, \omega_{n-1}) \in B_n\}$ is a measurable cylinder, then $T^{-1}(A) = \{\omega : (\omega_1, \ldots, \omega_n) \in B_n\}$ and hence $P(A) = P(T^{-1}A)$. The class of sets $A \in \mathscr{F}$ such that $P(A) = P(T^{-1}A)$ is a monotone class containing the measurable cylinders, hence coincides with $\mathscr{F}$. The result follows.

(e)   Let Ω be the set of all doubly infinite sequences $(\ldots \omega_{-1}; \omega_0, \omega_1, \ldots)$ of real numbers, $\mathscr{F}$ the σ-field generated by the measurable cylinders

$$\{\omega : (\omega_k, \omega_{k+1}, \ldots, \omega_{k+n-1}) \in B_n\},$$

$n = 1, 2, \ldots$, $k = 0, \pm 1, \pm 2, \ldots$, $B_n \in \mathscr{B}(R^n)$. Let μ be any probability measure P on $\mathscr{F}$, and let T be the *two-sided shift*, defined by

$$T(\ldots \omega_{-1}; \omega_0, \omega_1, \ldots) = (\ldots \omega_0; \omega_1, \omega_2, \ldots).$$

In other words if $X_k(\omega) = \omega_k$, $k = 0, \pm 1, \ldots$, then $X_k(T\omega) = \omega_{k+1}$.

Exactly as in (d), T preserves P iff P is stationary, that is,

$$P\{\omega : (\omega_0, \ldots, \omega_{n-1}) \in B_n\} = P\{\omega : (\omega_k, \ldots, \omega_{k+n-1}) \in B_n\}$$

for all $n = 1, 2, \ldots$, $k = 0, \pm 1, \ldots$, $B_n \in \mathscr{B}(R^n)$.

Note that the transformations of (a), (b), (c), and (e) are *invertible* (measurable, one-to-one, onto, with T^{-1} measurable), while that of (d) is not invertible (it is not one-to-one).

We now consider a physical example to motivate the concept of an ergodic transformation. Suppose that rainfall data are collected at a very large number of observation points $a_0, a_1, \ldots$ at times $t = 0, 1, \ldots$. Assume that the statistical character of the observations at a_i is the same for all i, and is represented by a stationary sequence of random variables $X_0, X_1, \ldots$, with X_n the amount of rainfall at time n. Assume also that the a_i are "independent," in other words, the a_i correspond to a sequence of independent performances of a random experiment, where a performance means an observation of the entire random sequence $\{X_0, X_1, \ldots\}$.

Suppose that the problem is to measure the average rainfall. Scientist A might take the following approach. He might take measurements at each observation point at a given time, say $t = 0$, and average the results. Scientist B might reason as follows. Since all observation points have the same statistical character, we can simply go to one observation point, take a large number of observations, say at $t = 0, 1, \ldots, n$, and average the results. Scientist A is using what might be called a vertical measuring scheme, and scientist B a horizontal scheme, as illustrated below.

Observation point	Measurements			
	$t = 0$	1	2	$\cdots$
a_0	X_{00}	X_{01}	X_{02}	$\cdots$
a_1	X_{10}	X_{11}	X_{12}	$\cdots$
a_2	X_{20}	X_{21}	X_{22}	$\cdots$
$\vdots$				

A's observations correspond to the first column, B's to the first row.

Now A and B will not necessarily obtain the same result (not even "essentially" the same). For example, suppose that "nature" flips an unbiased coin at each observation point. If the coin comes up heads, the rain is one inch at each observation time; if the coin comes up tails, there is no rain at any time. Roughly half of A's observers will measure one inch of rainfall, and half will measure none. Thus A will arrive at an average rainfall of one-half inch. But B will either measure an average of one inch or no rain at all, and thus will not get the same answer.

Mathematically, we are in the situation described in Example 3.1.2(d), and what B is computing is a *time average*, namely,

$$\frac{1}{n} \sum_{k=0}^{n-1} X_k(\omega) \qquad \text{for a } particular \ \omega.$$

(Note also that if T is the one-sided shift, $X_k(\omega) = X_0(T^k\omega)$.) But A is observing an *ensemble average* at a *particular time*, namely, $n^{-1}(Y_0 + Y_1 + \cdots + Y_{n-1})$, where the Y_j are independent random variables, all having the same distribution as X_0. Thus A's result would approximate $E(X_0) = \int_\Omega X_0 \, dP$. For A and B to get the same answer, we must have

$$\frac{1}{n} \sum_{k=0}^{n-1} X_0(T^k\omega) \to \int_\Omega X_0 \, dP.$$

More generally, we might ask when it will be true that for each integrable function f on $(\Omega, \mathscr{F}, P)$, we have

$$\frac{1}{n} \sum_{k=0}^{n-1} f(T^k\omega) \to \int_\Omega f \, dP \tag{1}$$

at least for almost every ω. In particular, if f is an indicator I_A, the property to be verified is simply the convergence of the relative frequency of visits to A in the first n steps to the probability of A.

Now suppose that A is an "almost invariant" set, in other words, A and $T^{-1}A$ differ only by a set of measure 0. (In the case where nature flips an unbiased coin to determine rainfall, we may take $A = \{\omega : X_0(\omega) = 1\}$, so that $T^{-1}A = \{\omega : X_1(\omega) = 1\}$.) Then we have, almost everywhere,

$$\frac{1}{n} \sum_{k=0}^{n-1} I_A(T^k\omega) = I_A(\omega) \qquad \text{for all} \quad n.$$

Thus the relative frequency of visits to A cannot converge to $P(A)$, except when $P(A) = 0$ or 1. Conversely, the pointwise ergodic theorem, to be proved in Section 3.3, implies that if every almost invariant set has probability 0 or 1, the convergence result (1) holds. In the next section we shall prepare for the proof of this basic result.

One comment on terminology. In this chapter, we deal exclusively with real as opposed to complex-valued functions; thus the L^p spaces are to be regarded as real vector spaces. There would be no difficulty in working with complex-valued functions, but for our purposes, nothing is to be gained in doing so.

3.2 Ergodicity and Mixing

The following definitions are motivated by the analysis at the end of Section 3.1.

3.2.1 Definitions

Let T be a measure-preserving transformation on $(\Omega, \mathcal{F}, \mu)$. A set $A \in \mathcal{F}$ is said to be *invariant* (under T) iff $A = T^{-1}A$, that is, $\omega \in A$ iff $T\omega \in A$; *almost invariant* iff A and $T^{-1}A$ differ by a set of measure 0, in other words, $\mu(A \triangle T^{-1}A) = 0$.

It is easily checked that the invariant sets form a σ-field, as do the almost invariant sets.

If $g \colon (\Omega, \mathcal{F}) \to (R, \mathcal{B}(R))$, g is said to be *invariant* iff $g(T\omega) = g(\omega)$ for all ω, *almost invariant* iff $g(T\omega) = g(\omega)$ for almost all ω. Note that a set is invariant (respectively almost invariant) iff its indicator is invariant (respectively almost invariant).

The measure-preserving transformation T is said to be *ergodic* iff for every invariant set A, either $\mu(A) = 0$ or $\mu(\Omega - A) = 0$. In the case of a probability space, ergodicity means that each invariant set has probability 0 or 1.

Invariance may be replaced by almost invariance in the definition of ergodicity, as the following result shows.

3.2.2 Lemma

(a) If A is an almost invariant set, there is a (strictly) invariant set B such that $\mu(A \triangle B) = 0$.

(b) A measure-preserving transformation T is ergodic iff for each almost invariant set A, either $\mu(A) = 0$ or $\mu(\Omega - A) = 0$.

PROOF Take $B = \limsup_n T^{-n}A$. Then $T^{-1}B = \limsup_n T^{-(n+1)}A = B$, hence B is invariant.

Now $A \triangle B \subset \bigcup_{k=0}^{\infty} (T^{-k}A \triangle T^{-(k+1)}A)$; for if $\omega \in A - B$ then $\omega \in T^{-n}A$ for only finitely many n, including $n = 0$. Thus $\omega \in T^{-k}A - T^{-(k+1)}A$ for some k. If $\omega \in B - A$, then $T^n\omega \in A$ for infinitely many n, but $\omega \notin A$. If $k + 1$ is the smallest integer such that $T^{k+1}\omega \in A$, then $\omega \in T^{-(k+1)}A - T^{-k}A$.

Since $\mu(T^{-k}A \triangle T^{-(k+1)}A) = \mu(A \triangle T^{-1}A) = 0$, it follows that $\mu(A \triangle B) = 0$, proving (a). To prove (b), let T be ergodic, and let A be almost invariant. If B is invariant and $\mu(A \triangle B) = 0$, then $\mu(A) = \mu(B)$ and $\mu(\Omega - A) = \mu(\Omega - B)$; hence $\mu(A) = 0$ or $\mu(\Omega - A) = 0$. The converse is clear since every invariant set is almost invariant. ‖

We give another way of expressing ergodicity.

3.2.3 Lemma

Let T be a measure-preserving transformation on $(\Omega, \mathscr{F}, \mu)$. The following conditions are equivalent:

(a) T is ergodic.
(b) Every almost invariant function is a.e. constant.
(c) Every invariant function is a.e. constant.

PROOF (a) *implies* (b): Let g be an almost invariant function. Then for each real λ, $A_\lambda = \{\omega : g(\omega) \le \lambda\}$ is an almost invariant set since $g(\omega) = g(T\omega)$ a.e. By (a) and 3.2.2(b), $\mu(A_\lambda) = 0$ or $\mu(A_\lambda^c) = 0$. Let $c = \sup \{\lambda : \mu(A_\lambda) = 0\}$; c is finite (ignoring the trivial case $\mu \equiv 0$) since $A_\lambda \uparrow \Omega$ as $\lambda \uparrow \infty$, $A_\lambda^c \uparrow \Omega$ as $\lambda \downarrow -\infty$. Then

$$\mu\{\omega : g(\omega) < c\} = \mu\left[\bigcup_{n=1}^{\infty} \left\{\omega : g(\omega) \le c - \frac{1}{n}\right\}\right] = 0,$$

and similarly $\mu\{\omega : g(\omega) > c\} = 0$. Thus $g = c$ a.e.

(b) *implies* (c): Every invariant function is almost invariant.

(c) *implies* (a): If A is an invariant set, I_A is an invariant function, hence I_A is a.e. constant. If $I_A = 0$ a.e., then $\mu(A) = \int_\Omega I_A \, d\mu = 0$, and if $I_A = 1$ a.e. then $\mu(A^c) = \int_\Omega (1 - I_A) \, d\mu = 0$. ‖

Invariance and almost invariance can be defined just as above for *extended* real-valued Borel measurable functions. Lemma 3.2.3 holds in this case also, with essentially the same proof.

The following characterization of almost invariance is often useful.

3.2.4 Lemma

Assume μ is finite. A set $A \in \mathscr{F}$ is almost invariant iff either $\mu(T^{-1}A - A) = 0$ or $\mu(A - T^{-1}A) = 0$, that is, $T\omega \in A$ essentially implies $\omega \in A$, or $\omega \in A$ essentially implies $T\omega \in A$.

PROOF We may write

$$\mu(A - T^{-1}A) = \mu(A) - \mu(A \cap T^{-1}A) = \mu(T^{-1}A) - \mu(A \cap T^{-1}A)$$

$$= \mu(T^{-1}A - A).$$

Thus $\mu(A \,\triangle\, T^{-1}A) = 2\mu(A - T^{-1}A) = 2\mu(T^{-1}A - A)$. ‖

Nonergodicity, that is, the existence of a nontrivial invariant set, indicates that T does not completely stir up the space. This concept of "stirring" may be further developed as follows.

3.2.5 Definition

If T is a measure-preserving transformation on the *probability* space $(\Omega, \mathscr{F}, P)$, T is said to be *mixing* iff for all $A, B \in \mathscr{F}$,

$$\lim_{n \to \infty} P(A \cap T^{-n}B) = P(A)P(B).$$

The restriction to a probability measure is essential here. If, for example, T has the mixing property with respect to the measure μ, let $A = B = \Omega$ to obtain $\mu(\Omega) = [\mu(\Omega)]^2$. Thus if $\mu(\Omega)$ is finite it must be 1; if $\mu(\Omega) = \infty$, and A is a set in $\mathscr{F}$ with finite, strictly positive measure, take $B = \Omega$. Then $\mu(A \cap T^{-n}B) = \mu(A) < \infty$, but $\mu(A)\mu(B) = \infty$, a contradiction.

The mixing property has the following intuitive interpretation (this example is due to Halmos). Regard the transformation T as defining a flow, as in the discussion after 3.1.1. Suppose, for example, that initially the container is filled with a liquid that is 90 percent gin, 10 percent vermouth, the "vermouth particles" occupying the set A, the "gin particles" the set $\Omega - A$. The externally applied force is due to a swizzle stick. The condition of the container is observed at times $t = 0, 1, 2, \ldots$. If B is any Borel subset of the container, let $P(B)$ be the volume of B divided by the volume of the container (so that $P(A) = .1$). It is reasonable to expect that if the mixing process is continued long enough, the percentage of vermouth in B should be approximately the same as the percentage in the entire container, namely 10 percent. To translate this into mathematical terms note that if ω is a point of the container, and a particle is initially at ω, then $T^n\omega$ is the position of the particle n seconds later. Thus the set of vermouth particles that are in B at time $t = n$ is $\{\omega \in A : T^n\omega \in B\} = A \cap T^{-n}B$. The fraction of vermouth in B at time $t = n$ is $P(A \cap T^{-n}B)/P(B)$, and the mixing property is expressed by saying that $P(A \cap T^{-n}B)/P(B) \to P(A) = .1$.

Mixing is a stronger property than ergodicity, as we now prove.

3.2.6 Theorem

Let T be a mixing transformation on $(\Omega, \mathscr{F}, P)$. Then T is ergodic.

PROOF   Let B be an invariant set. If $A \in \mathscr{F}$, then since $B = T^{-n}B$, we have $P(A \cap T^{-n}B) = P(A \cap B)$ for all n. If we let $n \to \infty$ and invoke the mixing property, we obtain $P(A \cap B) = P(A)P(B)$. But since A is an arbitrary set in $\mathscr{F}$, we may take $A = B$, and thus $P(B) = [P(B)]^2$, hence $P(B) = 0$ or 1.   ‖

It is useful to observe that it is not necessary to verify the mixing condition for all sets $A, B \in \mathscr{F}$, but only for A, B in a field $\mathscr{F}_0$ whose minimal σ-field is $\mathscr{F}$.

3.2.7 Theorem

Let T be a measure-preserving transformation on $(\Omega, \mathscr{F}, P)$. Let $\mathscr{F}_0$ be a field of subsets of Ω such that the σ-field generated by $\mathscr{F}_0$ is $\mathscr{F}$. If the mixing condition holds for all $A, B \in \mathscr{F}_0$, it holds for all $A, B \in \mathscr{F}$ and hence T is mixing.

PROOF   Let $A, B \in \mathscr{F}$, and find sets $A_k, B_k \in \mathscr{F}_0$ $(k = 1, 2, \ldots)$ such that $P(A \triangle A_k)$ and $P(B \triangle B_k) \to 0$ as $k \to \infty$ (see RAP, 1.3.11, p. 20). Now

$$(A \cap T^{-n}B) \triangle (A_k \cap T^{-n}B_k) \subset (A \triangle A_k) \cup (T^{-n}(B \triangle B_k)),$$

so the probability of the set on the left is at most $P(A \triangle A_k) + P(B \triangle B_k)$, which approaches 0 as $k \to \infty$, uniformly in n. Thus $P(A_k \cap T^{-n}B_k) \to$

$P(A \cap T^{-n}B)$ as $k \to \infty$, uniformly in n. By hypothesis, $P(A_k \triangle T^{-n}B_k) \to P(A_k)P(B_k)$ as $n \to \infty$. Therefore by the standard double limit theorem,

$$\lim_{n \to \infty} P(A \cap T^{-n}B) = \lim_{n \to \infty} \lim_{k \to \infty} P(A_k \cap T^{-n}B_k)$$

$$= \lim_{k \to \infty} \lim_{n \to \infty} P(A_k \cap T^{-n}B_k)$$

$$= \lim_{k \to \infty} P(A_k)P(B_k) = P(A)P(B),$$

the desired result. ‖

The following basic measure-theoretic result will often be used; for the proof, see RAP, 1.6.12, p. 50.

3.2.8   Theorem

Let $T: (\Omega, \mathscr{F}) \to (\Omega_0, \mathscr{F}_0)$ be a measurable mapping, and let μ be a measure on $\mathscr{F}$. Define a measure $\mu_0 = \mu T^{-1}$ on $\mathscr{F}_0$ by

$$\mu_0(A) = \mu(T^{-1}A), \qquad A \in \mathscr{F}_0.$$

(If $\Omega_0 = \Omega$, $\mathscr{F}_0 = \mathscr{F}$, and T is μ-preserving, then $\mu_0 = \mu$.) If $f: (\Omega_0, \mathscr{F}_0) \to (\bar{R}, \mathscr{B}(\bar{R}))$, and $A \in \mathscr{F}_0$, then

$$\int_{T^{-1}A} f(T\omega)\, d\mu(\omega) = \int_A f(\omega)\, d\mu_0(\omega)$$

in the sense that if one of the integrals exists so does the other, and the integrals are equal.

3.2.9   Examples

We consider again the examples of 3.1.2.

(a)   Let T be a permutation of $\Omega = \{x_1, \ldots, x_n\}$, $n \geq 2$, μ any measure on all subsets of Ω that assigns equal weight to each point within a given cycle of T. (When talking about the mixing property we assume μ is a probability measure; assume also that $\mu\{x_i\} > 0$ for all i.)

We claim that T is ergodic iff T has only one cycle; this follows because the only invariant sets are unions of cycles of T (and the empty set).

But T is never mixing. For suppose that $\{x_1, \ldots, x_k\}$ is a cycle of T and $Tx_i = x_{i+1}$, with indices reduced mod k. Assume $k \geq 2$; if $k = 1$, $\{x_1\}$ is a nontrivial invariant set.

Let $A = B = \{x_i\}$. Then $A \cap T^{-n}B$ coincides with A if n is a multiple of k, and is the empty set otherwise. Thus $\lim_{n \to \infty} \mu(A \cap T^{-n}B)$ does not exist.

(b)   Let $T(x) = x + c$ on R, with Borel sets and Lebesgue measure. Then T *is not ergodic*; $A = \bigcup_{n=-\infty}^{\infty} (nc, nc + c/2)$ is a nontrivial invariant set.

(c)   Let T be rotation by α on the unit circle (or $T(\theta) = \theta + \alpha \pmod{2\pi}$ on $[0, 2\pi)$).

We claim that T *is ergodic iff* $\alpha/2\pi$ *is irrational*, that is, iff $e^{i\alpha}$ is not a root of unity.

Assume $\alpha/2\pi$ irrational, and let A be an invariant set in $\mathscr{F}$. Let a_n be the nth Fourier coefficient of the indicator function I_A, that is,

$$a_n = \frac{1}{2\pi} \int_0^{2\pi} I_A(\omega) e^{-in\omega} \, d\omega = \int_A e^{-in\omega} \, dP(\omega).$$

By 3.2.8,

$$a_n = \int_{T^{-1}A} e^{-in(\omega+\alpha)} \, dP(\omega) = e^{-in\alpha} \int_A e^{-in\omega} \, dP(\omega) = e^{-in\alpha} a_n$$

by invariance of A. Since $\alpha/2\pi$ is irrational, $e^{-in\alpha} \neq 1$ for $n \neq 0$, and it follows that $a_n = 0$ for $n \neq 0$. But $I_A \in L^2$, so that the Fourier series $\sum_{n=-\infty}^{\infty} a_n e^{in\omega}$ converges in L^2 to $I_A(\omega)$ (see RAP, p. 127, Problem 9). It follows that $I_A = a_0$ a.e., and therefore $P(A) = 0$ or 1. Thus T is ergodic.

Conversely, if $e^{i\alpha}$ is a root of unity, say $e^{in\alpha} = 1$, then α is an integral multiple of $2\pi/n$. Let A be the union of the sectors $0 \leq \theta \leq \pi/n, 2\pi/n \leq \theta \leq 3\pi/n, 4\pi/n \leq \theta \leq 5\pi/n, \ldots, (2n-2)\pi/n \leq \theta \leq (2n-1)\pi/n$. Then A is invariant, but $P(A) = \frac{1}{2}$, so that T is not ergodic.

Now T *is never mixing*; to establish this we may assume that $\alpha/2\pi$ is irrational. Let $A = B = \{\theta : 0 \leq \theta \leq \pi\}$, corresponding to the upper semicircle. Given $\varepsilon > 0$, $e^{in\alpha}$ is within distance ε of $e^{i0} = 1$ for infinitely many n. (Extract a convergent subsequence $\{z_k\}$ from $\{e^{in\alpha}\}$, and select z_i and z_{i+j} such that dist $(z_i, z_{i+j}) < \varepsilon$; i can be chosen larger than any preassigned positive integer. Then it is possible to form a chain that eventually goes entirely around the circle, with the distance between successive points less than ε.) It follows that A and $T^n A$ overlap except for a set of measure less than ε. Thus

$$P(A \cap T^{-n}B) = P(A \cap T^{-n}A)$$

$$= P(T^n A \cap A) \qquad \text{since } T \text{ is}$$

$$\text{measure-preserving and invertible}$$

$$\geq P(A) - \varepsilon$$

$$= \tfrac{1}{2} - \varepsilon > \tfrac{3}{8} \qquad \text{if } \varepsilon < \tfrac{1}{8}.$$

But $P(A)P(B) = [P(A)]^2 = \frac{1}{4}$, so the mixing property fails.

(d)   (and (e)).   Let T be the one-sided (or two-sided) shift transforma-tion on the space of all infinite (or doubly infinite) sequences of real numbers.

The problem of finding conditions on P such that T will be ergodic (or mixing) is largely unsolved. At the present time, we consider only the case in which the coordinate random variables X_k are independent, that is, the measure P has the property that

$$P\{\omega : X_i(\omega) \in A_i, \ i = 1, 2, \ldots, n\} = \prod_{i=1}^{n} P\{\omega : X_i(\omega) \in A_i\}$$

for all real Borel sets $A_1, \ldots, A_n$ and all $n = 1, 2, \ldots$. In this case, T *is mixing* (hence ergodic).

Let

$$A = \{\omega : X_0(\omega), \ldots, X_{k-1}(\omega)) \in B_k\}$$

and

$$B = \{\omega : (X_0(\omega), \ldots, X_{r-1}(\omega)) \in B_r'\}$$

be measurable cylinders. For sufficiently large n we have $n > k - 1$, and therefore the indices defining the sets A and $T^{-n}B$ are distinct. Thus by independence,

$$P(A \cap T^{-n}B) = P\{\omega : (\omega_0, \ldots, \omega_{k-1}) \in B_k, (\omega_n, \ldots, \omega_{n+r-1}) \in B_r'\}$$

$$= P(A)P(T^{-n}B) = P(A)P(B).$$

Thus the mixing condition holds for all measurable cylinders, and hence by 3.2.7, the mixing condition holds for all $A, B \in \mathscr{F}$, so that T is mixing.

Problems

1.   If A is an almost invariant set, show that for almost every ω, $\omega \in A$ iff $T^n\omega \in A$ for all $n = 1, 2, \ldots$.

2.   If μ is counting measure on the integers, show that $\omega \to \omega + 1$ is er-godic, but $\omega \to \omega + 2$ is not ergodic.

3.   Let T be a measure-preserving transformation on $(\Omega, \mathscr{F}, \mu)$. Give examples to show that the following results are possible:

(a) $A \in \mathscr{F}$ does not imply $T(A) \in \mathscr{F}$.
(b)   If $A \in \mathscr{F}$ and $T(A) \in \mathscr{F}$, $P(A)$ need not equal $P(TA)$.
(c)   If T is one-to-one onto, it need not be invertible (that is, T^{-1} need not be measurable).

4. (Jacobs, 1962)   Let T be a measurable (but not necessarily measure-preserving) transformation on $(\Omega, \mathscr{F}, \mu)$. We say that T is *recurrent* iff for every $A \in \mathscr{F}$ and almost every $\omega \in A$, $T^n\omega \in A$ for some $n \geq 1$; T is *infinitely recurrent* iff for every $A \in \mathscr{F}$ and almost every $\omega \in A$, $T^n\omega \in A$ for infinitely many n. (In these definitions, the exceptional sets of measure 0 are allowed to depend on A.) A set $B \in \mathscr{F}$ is *wandering* iff $B, T^{-1}B, T^{-2}B, \ldots$ are disjoint; T is said to be *conservative* iff all wandering sets have measure 0. Finally, T is *incompressible* iff $A \subset T^{-1}A$ implies $\mu(T^{-1}A - A) = 0$. (Show that equivalently, $T^{-1}A \subset A$ implies $\mu(A - T^{-1}A) = 0$.)

(a)   Show that the following are equivalent:

1. T is incompressible;
2. T is conservative;
3. T is recurrent;
4. T is infinitely recurrent.

One possible scheme is $1 \Rightarrow 2 \Rightarrow 3 \Rightarrow 1, 4 \Rightarrow 3, 1 \Rightarrow 4$.)

(b)   Show that $T(x) = x + 1$ on R (with Borel sets and Lebesgue measure) violates all four conditions of (a).

(c)   (Poincaré)   If T is measure-preserving and μ is finite, show that T is (infinitely) recurrent.

3.3   The Pointwise Ergodic Theorem

We are going to prove the pointwise ergodic theorem, which states that if T is a measure-preserving transformation on $(\Omega, \mathscr{F}, \mu)$ and $f \in L^1(\Omega, \mathscr{F}, \mu)$, then $n^{-1}(f(\omega) + f(T\omega) + \cdots + f(T^{n-1}\omega))$ converges to an integrable function $\hat{f}(\omega)$, for almost all ω.

It is possible to prove this result in a somewhat more general form (see the comments at the end of the chapter). The generalization is based on the fact that associated with a measure-preserving transformation is a positive contraction operator, as follows.

3.3.1   Theorem

If f is an extended real-valued Borel measurable function on $(\Omega, \mathscr{F}, \mu)$, and T is μ-preserving, let $\hat{T}f$ denote the function $f \circ T$. Then for every $p \in (0, \infty]$ we have $\|\hat{T}f\|_p = \|f\|_p$. Thus if we consider $\hat{T}$ as a linear operator on the Banach space $L^p(\Omega, \mathscr{F}, \mu)$, $1 \leq p \leq \infty$, then $\hat{T}$ is an *isometry* (a one-to-one, linear, norm-preserving map), in particular, $\hat{T}$ is a *contraction*, that is, $\|\hat{T}\| \leq 1$ (in this case, $\|\hat{T}\| = 1$). Furthermore, $\hat{T}$ is *positive*, that is, if $f \geq 0$ a.e., then $\hat{T}f \geq 0$ a.e.

PROOF By 3.2.8, $\int_\Omega |f(T\omega)|^p \, d\mu(\omega) = \int_\Omega |f(\omega)|^p \, d\mu(\omega)$, hence $\|\hat{T}f\|_p = \|f\|_p$, $0 < p < \infty$. If $p = \infty$, we have

$$\|\hat{T}f\|_\infty = \inf \{c : \mu\{|\hat{T}f| > c\} = 0\}$$

$$= \inf \{c : \mu\{\omega : |f(T\omega)| > c\} = 0\}$$

$$= \inf \{c : \mu T^{-1}\{\omega : |f(\omega)| > c\} = 0\}$$

$$= \inf \{c : \mu\{|f| > c\} = 0\} \text{ since } T \text{ preserves } \mu$$

$$= \|f\|_\infty .$$

If $\mu(N) = 0$, and $f \geq 0$ on N^c, then $\mu(T^{-1}N) = 0$ and $\hat{T}f \geq 0$ on $(T^{-1}N)^c$, proving positivity. ∥

The sequence of averages $f^{(n)}(\omega) = n^{-1}[f(\omega) + f(T\omega) + \cdots + f(T^{n-1}\omega)]$ can now be expressed in terms of $\hat{T}$ as

$$f^{(n)} = n^{-1}(f + \hat{T}f + \cdots + \hat{T}^{n-1}f)$$

where $\hat{T}^0 f = f$ and $\hat{T}^k$ is the composition of $\hat{T}$ with itself k times, $k \geq 1$.

In the three results to follow, T is a measure-preserving transformation on $(\Omega, \mathcal{F}, \mu)$, and $f: (\Omega, \mathcal{F}) \to (\bar{R}, \mathcal{B}(\bar{R}))$. Inspection of the proofs will show, however, that the results hold if $\hat{T}$ is replaced by an arbitrary positive contraction on L^1, not necessarily arising from a measure-preserving transformation.

3.3.2 Lemma

If $f \in L^1(\Omega, \mathcal{F}, \mu)$, let $f_0 = f$, and

$$f_n = \max (f, f + \hat{T}f, \ldots, f + \hat{T}f + \cdots + \hat{T}^n f), \qquad n \geq 1.$$

Then

$$f_{n+1} \leq f + \hat{T}f_n^+, \qquad n = 0, 1, \ldots .$$

PROOF If $0 \leq m \leq n$, then $\sum_{k=0}^{m+1} \hat{T}^k f = f + \hat{T}(\sum_{k=0}^m \hat{T}^k f)$. Since $\sum_{k=0}^m \hat{T}^k f \leq f_n$ and $\hat{T}$ is positive, we have

$$\hat{T}\left(\sum_{k=0}^m \hat{T}^k f\right) \leq \hat{T}f_n \leq \hat{T}f_n^+ .$$

Thus

$$\sum_{k=0}^{m+1} \hat{T}^k f \leq f + \hat{T}f_n^+, \qquad 0 \leq m \leq n.$$

Since $f \leq f + \hat{T}f_n^+$, we have $f_{n+1} \leq f + \hat{T}f_n^+$, as desired. ∥

3.3.3   Lemma

Let f_n be defined as in 3.3.2, and assume $f \in L^1(\Omega, \mathcal{F}, \mu)$. If $A_n = \{f_n > 0\}$, then $\int_{A_n} f \, d\mu \geq 0$.

PROOF   By 3.3.2,

$$\int_{A_n} f \, d\mu \geq \int_{A_n} f_{n+1} \, d\mu - \int_{A_n} (\hat{T} f_n^+) \, d\mu$$

$$\geq \int_{A_n} f_n \, d\mu - \int_{A_n} (\hat{T} f_n^+) \, d\mu$$

$$= \int_{\Omega} f_n^+ \, d\mu - \int_{A_n} (\hat{T} f_n^+) \, d\mu$$

$$\geq \int_{\Omega} f_n^+ \, d\mu - \int_{\Omega} (\hat{T} f_n^+) \, d\mu$$

$$= \|f_n^+\|_1 - \|\hat{T} f_n^+\|_1 \geq 0$$

since $\hat{T}$ is a contraction.   ‖

3.3.4   Maximal Ergodic Theorem

If $f \in L^1(\Omega, \mathcal{F}, \mu)$ and

$$A = \left\{ \omega : \sup_{n \geq 1} \sum_{k=0}^{n-1} (\hat{T}^k f)(\omega) > 0 \right\}$$

$$\left(= \left\{ \omega : \sup_{n \geq 1} f^{(n)}(\omega) > 0 \right\}, \qquad \text{where} \quad f^{(n)} = n^{-1} \sum_{k=0}^{n-1} \hat{T}^k f \right)$$

then $\int_A f \, d\mu \geq 0$.

PROOF   The sets A_n of 3.3.3 increase to A.   ‖

It will be convenient to isolate some of the technical difficulties in the proof of the pointwise ergodic theorem. Let $f \in L^1(\Omega, \mathcal{F}, \mu)$, and let $f^{(n)}(\omega) = n^{-1} \sum_{k=0}^{n-1} f(T^k \omega)$, $n = 1, 2, \ldots$, where T is μ-preserving. If $a < b$, define

$$C_{ab} = C_{ab}(f) = \left\{ \omega : \liminf_{n \to \infty} f^{(n)}(\omega) < a < b < \limsup_{n \to \infty} f^{(n)}(\omega) \right\},$$

$$N_b = \left\{ \omega : \sup_{n \geq 1} f^{(n)}(\omega) > b \right\};$$

note that C_{ab} is a subset of N_b.

We can establish a.e. convergence of the sequence $f^{(n)}$ if we show that $\mu(C_{ab})$ is always 0. To do this, we may assume without loss of generality that $b > 0$. For note that

$$C_{ab}(f) = \{\omega : \lim \inf -f^{(n)}(\omega) < -b < -a < \lim \sup -f^{(n)}(\omega)\}$$

$$= C_{-b, -a}(-f).$$

If $b \leq 0$, then $-a > 0$, and the argument below will show that $C_{-b, -a}(-f)$ has measure 0, and thus $\mu(C_{ab}) = 0$.

3.3.5 Lemma

(a)   The set C_{ab} is almost invariant, and

(b) $\mu(C_{ab}) < \infty$; in fact,

(c) $\mu(C_{ab}) = 0$.

PROOF   (a)   We may write

$$f^{(n)}(T\omega) = \frac{1}{n} \sum_{k=0}^{n-1} f(T^{k+1}\omega)$$

$$= \frac{n+1}{n} \left[\frac{1}{n+1} \sum_{k=0}^{n} f(T^k\omega) \right] - \frac{f(\omega)}{n}$$

$$= \frac{n+1}{n} f^{(n+1)}(\omega) - \frac{f(\omega)}{n}.$$

Since $f \in L^1$, $f(\omega)$ is finite for almost all ω, hence

$$\lim_{n \to \infty} \inf f^{(n)}(\omega) = \lim_{n \to \infty} \inf f^{(n)}(T\omega),$$

and similarly for lim sup, except possibly on a set of measure 0. Thus, outside a set of measure 0, $\omega \in C_{ab}$ iff $T\omega \in C_{ab}$, and the result follows.

(b)   Let C be any set in $\mathscr{F}$ such that $C \subset C_{ab}$ and $\mu(C) < \infty$, and define $F_b = \{\omega : \sup_{n \geq 1} (f - bI_C)^{(n)}(\omega) > 0\}$. Note that if $\omega \in C_{ab}$, then $\omega \in N_b$, hence

$$\frac{1}{n} \sum_{k=0}^{n-1} f(T^k\omega) > b \qquad \text{for some} \quad n$$

$$\geq \frac{1}{n} \sum_{k=0}^{n-1} bI_C(T^k\omega).$$

Thus $\omega \in F_b$; in particular, C is a subset of F_b. By the maximal ergodic theorem 3.3.4,

$$\int_{F_b} (f - bI_C) \, d\mu \geq 0.$$

Thus

$$\int_\Omega |f| \, d\mu \geq \int_{F_b} f \, d\mu \geq b \int_{F_b} I_C \, d\mu = b\mu(C \cap F_b) = b\mu(C)$$

so that

$$\mu(C) \leq b^{-1} \int_\Omega |f| \, d\mu < \infty.$$

Now note that C_{ab} is a subset of

$$\bigcup_{n=0}^{\infty} \{\omega : |f(T^n\omega)| > 0\}.$$

By 3.2.8,

$$\int_\Omega |f(T^n\omega)| \, d\mu(\omega) = \int_\Omega |f(\omega)| \, d\mu(\omega) < \infty,$$

and it follows that $\{\omega : |f(T^n\omega)| > 0\}$ is a countable union of sets of finite measure, and therefore so is C_{ab} (see RAP, p. 69, Problem 2). Thus

$$\mu(C_{ab}) = \sup \{\mu(C) : C \in \mathscr{F}, C \subset C_{ab}, \mu(C) < \infty\}$$

$$\leq b^{-1} \int_\Omega |f| \, d\mu < \infty.$$

(c) Since C_{ab} is almost invariant, T is a well-defined measure-preserving transformation on $(C_{ab}, \mathscr{F}_{ab}, \mu_{ab})$, where

$$\mathscr{F}_{ab} = \{A \in \mathscr{F} : A \subset C_{ab}\} = \{B \cap C_{ab} : B \in \mathscr{F}\},$$

and $\mu_{ab} = \mu$ restricted to C_{ab}. (Strictly speaking, if $D = C_{ab} \triangle T^{-1}C_{ab}$, then T is well defined on $C_{ab} - D$. Since $\mu(D) = 0$, this causes no difficulty. For example, we may redefine T as the identity on D, and then it will be well defined and measure-preserving on C_{ab}.)

The argument of part (b) may now be applied to T on C_{ab}. In particular, the equation $\int_{F_b} (f - bI_C) \, d\mu \geq 0$ now becomes

$$\int_{C_{ab} \cap F_b} (f - bI_C) \, d\mu \geq 0.$$

Since C_{ab} has finite measure, we may set $C = C_{ab}$, and since $C_{ab} \subset F_b$, we obtain

$$\int_{C_{ab}} (f - b) \, d\mu \geq 0.$$

Now let

$$F'_{ab} = \left\{\omega \in C_{ab} : \sup_{n \geq 1} (a - f)^{(n)}(\omega) > 0\right\}.$$

If $\omega \in C_{ab}$, then $f^{(n)}(\omega) < a$ for at least one n, hence $\omega \in F'_{ab}$; thus $C_{ab} = F'_{ab}$. Since C_{ab} has finite measure, constant functions are integrable on C_{ab}, and we may therefore apply the maximal ergodic theorem to obtain

$$\int_{C_{ab}} (a - f) \, d\mu \geq 0.$$

Thus we have, for $a < b$,

$$b\mu(C_{ab}) \leq \int_{C_{ab}} f \, d\mu \leq a\mu(C_{ab}),$$

a contradiction unless $\mu(C_{ab}) = 0$.   ‖

We may now prove the main result.

3.3.6   Pointwise Ergodic Theorem

Let T be a measure-preserving transformation on $(\Omega, \mathscr{F}, \mu)$, and let $f \in L^1(\Omega, \mathscr{F}, \mu)$. Then there is a function $\hat{f} \in L^1$ such that we have $n^{-1} \sum_{k=0}^{n-1} f(T^k\omega) \to \hat{f}(\omega)$ almost everywhere.

PROOF   Let $D = \{\omega : f^{(n)}(\omega)$ does not converge to a finite or infinite limit$\}$. Then $D = \bigcup \{C_{ab}(f) : a < b, a, b \text{ rational}\}$. By 3.3.5, $\mu(D) = 0$, and hence $f^{(n)}(\omega)$ converges for almost all ω; we call the limit $\hat{f}(\omega)$. (Define $\hat{f} = 0$ on the exceptional set.) By Fatou's lemma,

$$\int_\Omega |\hat{f}| \, d\mu = \int_\Omega \lim_{n \to \infty} |f^{(n)}(\omega)| \, d\mu(\omega) \leq \liminf_{n \to \infty} \int_\Omega |f^{(n)}(\omega)| \, d\mu(\omega).$$

But

$$\int_\Omega |f^{(n)}| \, d\mu \leq \frac{1}{n} \sum_{k=0}^{n-1} \int_\Omega |f(T^k\omega)| \, d\mu(\omega)$$

$$= \frac{1}{n} \sum_{k=0}^{n-1} \int_\Omega |f| \, d\mu \qquad \text{by 3.2.8}$$

$$= \int_\Omega |f| \, d\mu < \infty,$$

and the theorem is proved.   ‖

We now look more closely at the convergence of the sequence $\{f^{(n)}\}$.

3.3.7   Theorem

If $\mu(\Omega) < \infty$ and $f \in L^p$ $(1 \le p < \infty)$, then $\hat{f} \in L^p$ and $f^{(n)} \xrightarrow{L^p} \hat{f}$.

PROOF   Since the finite-valued simple functions are dense in L^p, for each $\varepsilon > 0$ there is a bounded measurable function g such that $\|f - g\|_p < \varepsilon$. If $f_k(\omega) = f(T^k\omega)$, $g_k(\omega) = g(T^k\omega)$, and $|g| \le M$, then

$$\|f^{(n)} - \hat{f}\|_p \le \left\|\frac{1}{n}\sum_{k=0}^{n-1}(f_k - g_k)\right\|_p$$

$$+ \left\|\left(\frac{1}{n}\sum_{k=0}^{n-1}g_k\right) - \hat{g}\right\|_p + \|\hat{g} - \hat{f}\|_p. \tag{1}$$

Since $|n^{-1}\sum_{k=0}^{n-1}g_k| \le M$ (hence $|\hat{g}| \le M$ a.e.), the second term on the right in (1) approaches zero as n approaches ∞, by the dominated convergence theorem. By 3.2.8, $\|f_k - g_k\|_p = \|f - g\|_p < \varepsilon$, hence the first term is less than ε. Now

$$\int_\Omega |\hat{f} - \hat{g}|^p \, d\mu = \int_\Omega \lim_n \left|\frac{1}{n}\sum_{k=0}^{n-1}(f_k - g_k)\right|^p \, d\mu$$

$$\le \liminf_n \int_\Omega \left|\frac{1}{n}\sum_{k=0}^{n-1}(f_k - g_k)\right|^p \, d\mu$$

$$= \liminf_n \left\|\frac{1}{n}\sum_{k=0}^{n-1}(f_k - g_k)\right\|_p^p < \varepsilon^p.$$

Thus $\|f^{(n)} - \hat{f}\|_p < 2\varepsilon$ for large enough n, and the result follows. ‖

If $p = 1$ in 3.3.7, the hypothesis that $\mu(\Omega) < \infty$ cannot be dropped (Problem 1). Also, the result fails for $p = \infty$, even if $\mu(\Omega) < \infty$ (Problem 2).

We now identify the limit function $\hat{f}$. Theorem 3.3.9 indicates that although the pointwise ergodic theorem can be presented without reference to probability, some insight is lost in doing so.

3.3.8   Lemma

If $f \in L^1$, then $\hat{f}$ is almost invariant. Thus if $\mathscr{G}$ is the σ-field of almost invariant sets, then $\hat{f}: (\Omega, \mathscr{G}) \to (\bar{R}, \mathscr{B}(\bar{R}))$.

PROOF   If $f^{(n)}(\omega) \to \hat{f}(\omega)$ for $\omega \notin N$, where $\mu(N) = 0$, then $f^{(n)}(T\omega) \to \hat{f}(T\omega)$ for $\omega \notin T^{-1}N$, where $\mu(T^{-1}N) = 0$. But

$$f^{(n)}(T\omega) = \left(\frac{n+1}{n}\right)f^{(n+1)}(\omega) - \frac{f(\omega)}{n},$$

and $f(\omega)/n \to 0$ a.e. since $f \in L^1$. Thus $f^{(n)}(T\omega) \to \hat{f}(\omega)$ a.e. hence $\hat{f}(\omega) = \hat{f}(T\omega)$ a.e., proving $\hat{f}$ almost invariant.

If $B \in \mathcal{B}(\bar{R})$ and $C = \{\omega : \hat{f}(\omega) \in B\}$, then if $\hat{f}(\omega) = \hat{f}(T\omega)$ we have $\omega \in C$ iff $T\omega \in C$. Thus C is almost invariant, and the proof is complete.   ‖

3.3.9   Theorem

If $f \in L^1$, and if A is an almost invariant set of finite measure, then $\int_A f \, d\mu = \int_A \hat{f} \, d\mu$. Thus in a probability space, $\hat{f} = E(f \mid \mathcal{G})$, where $\mathcal{G}$ is the σ-field of almost invariant sets.

PROOF   Restrict T to A as in 3.3.5(c). Since $\mu(A) < \infty$, we have L^1 convergence (on A) by 3.3.7, hence $\int_A f^{(n)} \, d\mu \to \int_A \hat{f} \, d\mu$. But

$$\int_A f^{(n)} \, d\mu = \int_A f \, d\mu \qquad \text{by 3.2.8.} \quad ‖$$

In the ergodic case, $\hat{f}$ assumes a very special form.

3.3.10   Theorem

If T is ergodic and $f \in L^1$, then $\hat{f}$ is constant a.e. If $\mu(\Omega) = \infty$, the constant is $c = 0$; if $\mu(\Omega) < \infty$, we have

$$c = \frac{1}{\mu(\Omega)} \int_\Omega f \, d\mu.$$

Thus on a probability space, $\hat{f} = E(f)$ a.e.

PROOF   By 3.3.8, $\hat{f}$ is almost invariant, so by 3.2.3(b), $\hat{f} = c$ a.e. If $\mu(\Omega) = \infty$, then c must be 0 because $\hat{f} \in L^1$ by 3.3.6. If $\mu(\Omega) < \infty$, then $c = [\mu(\Omega)]^{-1} \int_\Omega f \, d\mu$ by 3.3.9 (with $A = \Omega$).   ‖

If T is ergodic and μ is finite, consider the case $f = I_A$. Then by 3.3.10, $\hat{f} = \mu(A)/\mu(\Omega)$ a.e., so that $n^{-1} \sum_{k=0}^{n-1} I_A(T^k\omega)$, the relative frequency of visits to A, converges a.e. to the relative mass of A (the probability of A if $\mu(\Omega) = 1$). Apparently we have a version of the strong law of large numbers, and in fact the pointwise ergodic theorem can be regarded as a generalization of this result. Let T be the one-sided shift transformation (see 3.1.2(d) and 3.2.9(d)), with coordinate random variables X_k. If Z is an integrable random variable on $(\Omega, \mathcal{F}, P)$, then

$$Z^{(n)}(\omega) = n^{-1}[Z(\omega_0, \omega_1, \ldots) + Z(\omega_1, \omega_2, \ldots) + \cdots + Z(\omega_{n-1}, \omega_n, \ldots)],$$

in other words,

$$Z^{(n)} = n^{-1}[Z(X_0, X_1, \ldots) + Z(X_1, X_2, \ldots) + \cdots + Z(X_{n-1}, X_n, \ldots)].$$

By 3.3.9,

$$Z^{(n)} \stackrel{\text{a.e.}}{\to} E(Z \mid \mathscr{G})$$

where $\mathscr{G}$ is the σ-field of almost invariant sets; if T is ergodic, the limit is $E(Z)$ by 3.3.10. In particular, let $X_0, X_1, \ldots$ be iid random variables with finite expectation. If $Z(\omega) = \omega_0$, that is, $Z = X_0$, we obtain

$$n^{-1}(X_0 + \cdots + X_{n-1}) \to E(X_0) \quad \text{a.e.,}$$

the iid case of the strong law of large numbers (RAP, 7.2.5, p. 275).

In the next section, it will be necessary to consider probability measures P_1 and P_2 that are each preserved by a fixed measurable transformation T on $(\Omega, \mathscr{F})$; P_i is said to be *ergodic* (relative to T) iff T is ergodic on $(\Omega, \mathscr{F}, P_i)$. If P_1 and P_2 are both ergodic, they must be identical or mutually singular.

3.3.11 Theorem

If P_1 and P_2 are ergodic probability measures relative to T, then either $P_1 \equiv P_2$ or $P_1 \perp P_2$.

PROOF   Suppose that $P_1(A) \neq P_2(A)$ for some $A \in \mathscr{F}$, and let

$$A_i = \{\omega : I_A^{(n)}(\omega) \to P_i(A)\}, \qquad i = 1, 2.$$

By 3.3.6 and 3.3.10, $P_1(A_1) = P_2(A_2) = 1$. But A_1 and A_2 are disjoint since $P_1(A) \neq P_2(A)$, hence $A_1 \perp A_2$. ‖

Theorem 3.3.11 gives us a criterion for ergodicity (unfortunately impractical).

3.3.12 Theorem

The probability measure P is ergodic relative to T iff there is no probability measure P_1 preserved by T such that P_1 is absolutely continuous with respect to P but not identical to P.

PROOF   If $P_1 \ll P$, $P_1 \not\equiv P$, and P is ergodic, then so is P_1. For if A is an invariant set, then $P(A) = 0$ or $P(A^c) = 0$ by ergodicity, hence $P_1(A) = 0$ or $P_1(A^c) = 0$ by absolute continuity. By 3.3.11, $P_1 \perp P$. But then P_1 is both absolutely continuous and singular with respect to P, hence P_1 is the zero measure, a contradiction.

Conversely, if P is not ergodic, let A be a T-invariant set with $0 < P(A) < 1$. Define $P_1(B) = P(B \mid A) = P(A \cap B)/P(A)$, $B \in \mathcal{F}$; then $P_1 \ll P$, and since $P_1(A) = 1 \neq P(A)$, $P_1 \not\equiv P$. Now

$$P_1(T^{-1}E) = \frac{P(T^{-1}E \cap A)}{P(A)}$$

$$= \frac{P(T^{-1}E \cap T^{-1}A)}{P(A)} \qquad \text{since } A \text{ is invariant}$$

$$= \frac{P(E \cap A)}{P(A)} \qquad \text{since } T \text{ preserves } P$$

$$= P_1(E).$$

Thus P_1 is preserved by T.   ‖

Problems

1. Let $T(\omega) = \omega + 1$ on R (with Borel sets and Lebesgue measure). If f is the indicator of $(0, 1]$, show that $f^{(n)}$ does not converge in L^1 to $\hat{f}$.

2. Let $\Omega = R^\infty$, $\mathcal{F} = [\mathcal{B}(R)]^\infty$, $X_n(\omega_0, \omega_1, \dots) = \omega_n$, P the unique probability measure making the X_n independent with $P\{X_n = 0\} = P\{X_n = 1\} = \frac{1}{2}$. (In other words, consider an infinite sequence of Bernoulli trials with probability $\frac{1}{2}$ of success on a given trial.) If T is the one-sided shift and $f(\omega) = \omega_0$, show that $n^{-1} \sum_{k=0}^{n-1} f(T^k\omega)$ does not converge in L^∞ to $\hat{f}(\omega) = \frac{1}{2}$.

3. Let T be a measure-preserving transformation on $(\Omega, \mathcal{F}, \mu)$. If T is ergodic, we know that for every $f \in L^1$, $f^{(n)}$ converges a.e. to a constant. Conversely, if $\mu(\Omega) < \infty$ and if for every $f \in L^1$ there is a constant $c = c(f)$ such that $f^{(n)} \to c$ a.e., show that T is ergodic. Give a counterexample to this result if $\mu(\Omega) = \infty$.

4. Let T be an ergodic measure-preserving transformation on $(\Omega, \mathcal{F}, \mu)$ with $\mu(\Omega) < \infty$. Let f be a real-valued Borel measurable function such that $\int_\Omega f \, d\mu$ exists. If $f \in L^1$, we know that $f^{(n)}$ converges a.e. to $[\mu(\Omega)]^{-1} \int_\Omega f \, d\mu$. Conversely, if $f^{(n)}$ converges a.e. to a finite limit, show that $f \in L^1$. (A special case of this result was considered in RAP, p. 280, Problem 1. Note also that the result fails when $\mu(\Omega) = \infty$; take $f \equiv c$.)

5. (*Mean Ergodic Theorem in a Hilbert Space*)   Let U be a bounded linear operator on the Hilbert space H, and let U^* be the adjoint of U, defined by the requirement that $\langle Uf, g \rangle = \langle f, U^*g \rangle$ for all $f, g \in H$. (If we follow the general procedure for defining the adjoint of a bounded linear operator on a normed linear space (see RAP, p. 149, Problem 4), we obtain an operator U^* defined on H^*, the space of continuous linear functionals on H. We then identify H^* with H by means of a conjugate isometry (RAP, 3.3.4(a), p. 130), to obtain U^* defined on H. This makes U^* conjugate linear, that is, $U^*(af + bg) = \bar{a}U^*(f) + \bar{b}U^*(g)$.)

Establish the following results.

(a)   The following conditions are equivalent (and define a unitary operator, that is, an invertible isometry).

   (1) $UU^* = U^*U = I$, the identity operator on H.
   (2) U is one-to-one onto, and $\langle f, g \rangle = \langle Uf, Ug \rangle$ for all $f, g \in H$.
   (3) U is one-to-one onto, and $\|Uf\| = \|f\|$ for all $f \in H$.

(b)   The following conditions are equivalent (and define an isometry).

   (1) $U^*U = I$.
   (2) $\langle f, g \rangle = \langle Uf, Ug \rangle$ for all $f, g \in H$.
   (3) $\|Uf\| = \|f\|$ for all $f \in H$.

For the remainder of the problem, U is an isometry of H.

(c)   If $f \in H$, then $Uf = f$ iff $U^*f = f$.

(d)   Define $A_n = n^{-1}(I + U + U^2 + \cdots + U^{n-1})$; note that

$$\|A_n\| \le n^{-1}\sum_{k=0}^{n-1} \|U\|^k = 1.$$

If $E = \{f \in H : \lim_{n\to\infty} A_n f \text{ exists (in } H)\}$, E is a closed subspace of H.

(e)   Let M be the set of elements of H that are *invariant* under U, that is, $M = \{f \in H : Uf = f\}$; note that M is a closed subspace of H, by continuity of U. Let $N_0 = \{g - Ug : g \in H\}$. If we define $\hat{f} = \lim_{n\to\infty} A_n f$ (where the limit exists) then:

   $f \in M$ implies $f \in E$ and $\hat{f} = f$;
   $f \in N_0$ implies $f \in E$ and $\hat{f} = 0$.

(f)   If $N = \bar{N}_0$, the closure of N_0, then H is the orthogonal direct sum of M and N (see RAP, p. 121). Thus by (d) and (e), $E = H$.

(g)   (*Mean Ergodic Theorem*)   Let U be an isometry of the Hilbert space H, and let P be the projection of H on the space M of all elements invariant under U. For every $f \in H$,

$$\frac{1}{n}\sum_{k=0}^{n-1} U^k f \to Pf.$$

If T is a measure-preserving transformation and $U = \hat{T}$, we obtain L^2 convergence of $f^{(n)}$ to $\hat{f}$.

(h)   If in addition, T is invertible and $S = T^{-1}$, then U is a unitary operator and $U^* = U^{-1} = \hat{S}$.

6.   Let T be a measurable transformation on $(\Omega, \mathcal{F})$. Within the linear space of finite signed measures on $\mathcal{F}$, let K be the convex set of probability measures preserved by T. Show that P is ergodic relative to T iff P is an extreme point of K, that is, P cannot be expressed as $\lambda_1 P_1 + \lambda_2 P_2$, with

$$\lambda_1, \lambda_2 > 0, \lambda_1 + \lambda_2 = 1, P_1, P_2 \in K, P_1 \not\equiv P_2 .$$

7.   This problem gives many conditions equivalent to ergodicity; in particular, some of the conditions involve convergence in probability rather than almost everywhere convergence.

Let T be a measure-preserving transformation on the probability space $(\Omega, \mathcal{F}, P)$, and let $\mathcal{F}_0$ be a field of sets whose minimal σ-field is $\mathcal{F}$. Show that the following conditions are equivalent:

(a) T is ergodic.
(b) $I_A^{(n)} \to P(A)$ a.e. for each $A \in \mathcal{F}$, where

$$I_A^{(n)}(\omega) = n^{-1} \sum_{k=0}^{n-1} I_A(T^k \omega).$$

(c) $I_A^{(n)} \to P(A)$ in probability for each $A \in \mathcal{F}$.
(d) $I_A^{(n)} \to P(A)$ in probability for each $A \in \mathcal{F}_0$.
(e) $I_A^{(n)} \to P(A)$ a.e. for each $A \in \mathcal{F}_0$.
(f) $n^{-1} \sum_{k=0}^{n-1} P(A \cap T^{-k}B) \to P(A)P(B)$ for all $A, B \in \mathcal{F}_0$.
(g) $n^{-1} \sum_{k=0}^{n-1} P(A \cap T^{-k}B) \to P(A)P(B)$ for all $A, B \in \mathcal{F}$.

If T is a one-sided shift transformation, we may take $\mathcal{F}_0$ to be the field of measurable cylinders. Furthermore, if the coordinate random variables take on only finitely many possible values, a measurable cylinder is a finite disjoint union of sets of the form $\{X_0 = i_0, \ldots, X_m = i_m\}$. Thus condition (d) is equivalent to the following statement:

For each $\alpha = (i_0, \ldots, i_m)$, $m = 0, 1, \ldots$, with the i_k belonging to the coordinate space, let N_α^n be the number of times that $i_0, \ldots, i_m$ occur in sequence in the first $n + m$ coordinates, that is,

$$N_\alpha^n(\omega) = \sum_{k=0}^{n-1} I_A(T^k \omega), \qquad \text{where} \quad A = \{\omega : \omega_0 = i_0, \ldots, \omega_m = i_m\};$$

then $n^{-1} N_\alpha^n$ converges in probability to $p(\alpha) = P(A)$.

Note also that by the Kolmogorov extension theorem, given any one-sided shift with coordinate random variables X_n, there is a two-sided shift with coordinate random variables X_n', such that $(X_n', n \geq 0)$ and $(X_n, n \geq 0)$ have the same distribution. [For example, specify that (X_{-8}', X_{-3}', X_6') have the same distribution as (X_0, X_5, X_{14}).] Condition (d) shows that the one-sided shift is ergodic iff the corresponding two-sided shift is ergodic; by 3.2.7, the same is true if "ergodic" is replaced by "mixing."

3.4   Applications to Real Analysis

In this section we establish a significant connection between probability and real analysis. Suppose that a number in $[0, 1]$ is generated by selecting the digits $i_1, i_2, \ldots$ of its r-adic expansion at random. The resulting number $Y = .i_1 i_2 \cdots = \sum_{n=1}^{\infty} i_n r^{-n}$ is a random variable, and the distribution of Y will depend on the way the digits are selected. For example, let the digits X_1, $X_2, \ldots$ be chosen independently, with $P\{X_k = i\} = p_i, i = 0, 1, \ldots, r - 1$. If all $p_i = 1/r$, it turns out that $Y = \sum_{n=1}^{\infty} X_n r^{-n}$ is uniformly distributed, so that the induced probability measure P_Y is Lebesgue measure. If $p_i \neq 1/r$ for some i, then P_Y is singular with respect to Lebesgue measure.

We formalize as follows. Let $S = \{0, 1, \ldots, r - 1\}$, and let $\mathscr{S}$ consist of all subsets of S; take $\Omega' = S^{\infty}$, the set of all sequences $(\omega_1, \omega_2, \ldots)$ with values in S. Let $\mathscr{F}'$ be the product σ-field $\mathscr{S}^{\infty}$, and define $X_n(\omega_1, \omega_2, \ldots) = \omega_n, n = 1$, $2, \ldots$. Finally, let $p_0, p_1, \ldots, p_{r-1}$ be nonnegative numbers that add to 1, and let P' be the unique probability measure on $\mathscr{F}'$ making $X_1, X_2, \ldots$ independent, with $P'\{X_n = i\} = p_i, i = 0, 1, \ldots, r - 1$ (P' exists by the Kolmogorov extension theorem). To avoid degeneracy assume that all $p_i < 1$.

First we note that $P'\{X_n = i$ for sufficiently large $n\} = 0, i = 0, 1, \ldots$, $r - 1$. For

$$P'\{X_n = i \text{ for sufficiently large } n\} = P'\left(\liminf_n \{X_n = i\}\right)$$

$$= P'\left(\bigcup_{n=1}^{\infty} \bigcap_{k=n}^{\infty} \{X_k = i\}\right)$$

$$\leq \sum_{n=1}^{\infty} P'\left(\bigcap_{k=n}^{\infty} \{X_k = i\}\right)$$

$$= 0 \qquad \text{since} \quad p_i < 1.$$

Now remove $D = \{X_n = r - 1$ for sufficiently large $n\}$ from the space, that is, let $\Omega = \Omega' - D$, $\mathscr{F} = \{A \in \mathscr{F}' : A \subset \Omega\} = \{B \cap \Omega : B \in \mathscr{F}'\}$, $P = P'$ restricted to $\mathscr{F}$. For the remainder of the discussion, our probability space will be $(\Omega, \mathscr{F}, P)$.

Let $Y(\omega) = .\omega_1 \omega_2 \cdots = \sum_{n=1}^{\infty} \omega_n r^{-n}$; then Y is a one-to-one map of Ω onto $[0, 1)$. Furthermore, Y and Y^{-1} are both measurable (relative to $\mathscr{F}$ and $\mathscr{B}[0, 1)$). To see this, let $kr^{-n} = .i_1 i_2 \cdots i_n = \sum_{j=1}^{n} i_j r^{-j}$ be an r-adic rational number ($k = 0, 1, \ldots, r^n - 1$). To avoid ambiguity, we always choose the r-adic expansion that does not end with all $(r - 1)$'s. Then

$$Y^{-1}[kr^{-n}, (k + 1)r^{-n}) = \{\omega \in \Omega : \omega_1 = i_1, \ldots, \omega_n = i_n\}.$$

The sets $[kr^{-n}, (k + 1)r^{-n})$ generate $\mathscr{B}[0, 1)$, and the sets $\{\omega \in \Omega : \omega_1 = i_1,$ $\ldots, \omega_n = i_n\}$ generate $\mathscr{F}$, proving measurability of Y and Y^{-1}.

Now if $A \in \mathscr{F}$, then

$$P(A) = P\{\omega : \omega \in A\} = P\{\omega : Y(\omega) \in Y(A)\} = P_Y(Y(A)).$$

To summarize: Y is a one-to-one correspondence between Ω and $[0, 1)$, and if $A \subset \Omega$, we have $A \in \mathscr{F}$ iff $Y(A) \in \mathscr{B}[0, 1)$, and in this case $P(A) = \mu(Y(A))$, where $\mu = P_Y$. Such a mapping Y is called a *measure-theoretic isomorphism* between $(\Omega, \mathscr{F}, P)$ and $([0, 1), \mathscr{B}[0, 1), \mu)$. The isomorphism allows us to establish a connection between real analysis on the interval $[0, 1)$ and the properties of the random sequence $\{X_n\}$.

Let F be the distribution function of Y, so that P_Y is the Lebesgue–Stieltjes measure corresponding to F. The assumption that all $p_i < 1$ is still in force.

3.4.1 Theorem

(a)   The distribution function F is continuous.

(b)   If all $p_i > 0$, then F is strictly increasing on $[0, 1]$.

(c)   If P_1 is the probability measure on $\mathscr{F}$ corresponding to $(p_0, \ldots, p_{r-1})$, and P_2 corresponds to $(p_0', \ldots, p_{r-1}') \neq (p_0, \ldots, p_{r-1})$, then $P_1 \perp P_2$.

(d)   If μ_1 is the induced probability measure P_Y corresponding to $(p_0, \ldots, p_{r-1})$, and μ_2 corresponds to $(p_0', \ldots, p_{r-1}') \neq (p_0, \ldots, p_{r-1})$, then $\mu_1 \perp \mu_2$. Furthermore, if all $p_i = 1/r$, then μ_1 is Lebesgue measure, that is, Y is uniformly distributed.

PROOF   (a)   If $y = \sum_{n=1}^{\infty} i_n r^{-n} \in [0, 1)$, then

$$P_Y\{y\} = P\left(\bigcap_{n=1}^{\infty} \{X_n = i_n\}\right) = \prod_{n=1}^{\infty} P\{X_n = i_n\} = 0$$

since all $p_i < 1$.

(b)   If $y_1 < y_2$ let $I = [kr^{-n}, (k + 1)r^{-n})$, with $kr^{-n} = .i_1, \ldots, i_n$, be an r-adic interval included in $(y_1, y_2]$. Then

$$F(y_2) - F(y_1) \geq P\{Y \in I\} = P\{X_1 = i_1, \ldots, X_n = i_n\}$$

$$= p_{i_1} \cdots p_{i_n} > 0.$$

(c)   The probability measures P_1 and P_2 are ergodic, that is, the one-sided shift T is ergodic on $(\Omega, \mathscr{F}, P_i)$, $i = 1, 2$, because of the independence of the X_i's (see 3.2.9(d)). Now $P_1 \neq P_2$; to see this, observe that if $A = \{X_1 = j\}$ and $p_j \neq p_j'$, then $P_1(A) \neq P_2(A)$. By 3.3.11, $P_1 \perp P_2$.

(d)   By (c), there are disjoint sets B_1, $B_2 \in \mathcal{F}$ such that $P_1(B_1) = P_2(B_2) = 1$. Then $Y(B_1)$ and $Y(B_2)$ are disjoint, and

$$\mu_j(Y(B_j)) = P_j\{Y \in Y(B_j)\} = P_j(B_j) = 1, j = 1, 2;$$

thus $\mu_1 \perp \mu_2$. If all $p_i = 1/r$, then if $kr^{-n} = .i_1 i_2 \cdots i_n$, we have

$$\mu_1[kr^{-n}, (k+1)r^{-n}) = P_1\{kr^{-n} \leq Y < (k+1)r^{-n}\}$$
$$= P\{X_1 = i_1, \ldots, X_n = i_n\} = p_{i_1} \cdots p_{i_n} = r^{-n}.$$

Thus $\mu_1[a, b) = b - a$ for all r-adic intervals, hence for all intervals in $[0, 1)$. Thus μ_1 is Lebesgue measure.   ‖

3.4.2 Calculation of F

We now show how to calculate the distribution function F explicitly (or at least to as close an approximation as desired). We shall construct a sequence of distribution functions F_n converging uniformly to F.

Define $F_n(y) = F(y)$ if $y = kr^{-n}$, $k = 0, 1, \ldots, r^n$, and interpolate F_n linearly between these points. (Take $F_n(y) = F(y) = 0$, $y < 0$, and $F_n(y) = F(y) = 1$, $y > 1$.) If $kr^{-n} \leq y \leq (k+1)r^{-n}$, with $kr^{-n} = .i_1 \cdots i_n$, then

$$F(kr^{-n}) \leq F(y) \leq F((k+1)r^{-n})$$

and

$$F(kr^{-n}) = F_n(kr^{-n}) \leq F_n(y) \leq F_n((k+1)r^{-n}) = F((k+1)r^{-n}).$$

It follows that

$$|F(y) - F_n(y)| \leq F((k+1)r^{-n}) - F(kr^{-n})$$
$$= P_Y(kr^{-n}, (k+1)r^{-n}]$$
$$= P_Y[kr^{-n}, (k+1)r^{-n})$$
$$= p_{i_1} \cdots p_{i_n}$$
$$\leq \varepsilon^n \to 0,$$

where $\varepsilon = \max\{p_i : i = 0, 1, \ldots, r - 1\} < 1$; thus $F_n \to F$ uniformly. Since $F((k+1)r^{-n}) - F(kr^{-n}) = p_{i_1} \cdots p_{i_n}$, the F_n are easily calculated. As an illustration, assume $r = 3$, $p_0 = \frac{1}{2}$, $p_1 = p_2 = \frac{1}{4}$, and abbreviate $F(b) - F(a)$ by $F(a, b]$. The computation of F_1 and F_2 is as follows:

Let $n = 1$; $F(0, \frac{1}{3}] = P\{X_1 = 0\} = \frac{1}{2}$

$\frac{1}{3} = .1$ $F(\frac{1}{3}, \frac{2}{3}] = P\{X_1 = 1\} = \frac{1}{4}$

$\frac{2}{3} = .2$ $F(\frac{2}{3}, 1] = P\{X_1 = 2\} = \frac{1}{4}$

Let $n = 2$;
$$F(0, \tfrac{1}{9}] = P\{X_1 = X_2 = 0\} = \tfrac{1}{4}$$

$\tfrac{1}{9} = .01$	$F(\tfrac{1}{9}, \tfrac{2}{9}] = P\{X_1 = 0, X_2 = 1\} = \tfrac{1}{8}$
$\tfrac{2}{9} = .02$	$F(\tfrac{2}{9}, \tfrac{3}{9}] = P\{X_1 = 0, X_2 = 2\} = \tfrac{1}{8}$
$\tfrac{3}{9} = .10$	$F(\tfrac{3}{9}, \tfrac{4}{9}] = P\{X_1 = 1, X_2 = 0\} = \tfrac{1}{8}$
$\tfrac{4}{9} = .11$	$F(\tfrac{4}{9}, \tfrac{5}{9}] = P\{X_1 = 1, X_2 = 1\} = \tfrac{1}{16}$
$\tfrac{5}{9} = .12$	$F(\tfrac{5}{9}, \tfrac{6}{9}] = P\{X_1 = 1, X_2 = 2\} = \tfrac{1}{16}$
$\tfrac{6}{9} = .20$	$F(\tfrac{6}{9}, \tfrac{7}{9}] = P\{X_1 = 2, X_2 = 0\} = \tfrac{1}{8}$
$\tfrac{7}{9} = .21$	$F(\tfrac{7}{9}, \tfrac{8}{9}] = P\{X_1 = 2, X_2 = 1\} = \tfrac{1}{16}$
$\tfrac{8}{9} = .22$	$F(\tfrac{8}{9}, 1] = P\{X_1 = 2, X_2 = 2\} = \tfrac{1}{16}.$

Since all distribution functions are 0 at the origin, the specification of F_1 and F_2 is complete; see Figure 3.1.

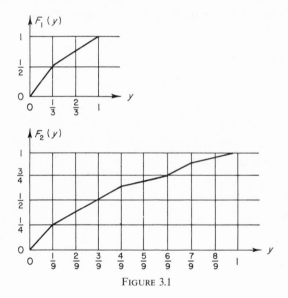

FIGURE 3.1

By 3.4.1(d), if $p_i \neq 1/r$ for some i, then F is singular with respect to Lebesgue measure, that is, the associated Lebesgue–Stieltjes measure $\mu = P_Y$ is singular with respect to Lebesgue measure. By 3.4.1(b), if all $p_i > 0$, then F is strictly increasing; the following result gives some information about the other case.

3.4.3 Theorem

If $p_j = 0$ for some j, then $F' = 0$ a.e. [Lebesgue measure].

PROOF   If $p_j = 0$, then $F(kr^{-n}, (k + 1)r^{-n}] = 0$ for any r-adic interval such that kr^{-n} has at least one j in its r-adic expansion.

We call an open r-adic interval $I = (kr^{-n}, (k + 1)r^{-n}) \subset [0, 1)$ a j-interval iff kr^{-n} (hence each $y \in I$) has at least one j in its r-adic expansion. If I is a j-interval, then $F(kr^{-n}, (k + 1)r^{-n}] = 0$; thus F is constant, hence $F' = 0$, on I.

Therefore $F' = 0$ on D, the union of all j-intervals. If λ denotes Lebesgue measure and $C = [0, 1) - D$, then $\lambda(C) = 0$. For if $y \in C$ and y is not an r-adic rational, then the r-adic expansion of y contains no j's. If P^* corresponds to all $p_i = 1/r$, then

$$\lambda(C) = P^*\{Y \in C\}$$

$$\leq P^*\{Y \text{ is an } r\text{-adic rational}\} + P^*\{X_n \text{ never equals } j\}$$

$$= P^*\{X_n \text{ never equals } j\} \text{ since the } r\text{-adic rationals are countable}$$

$$= \prod_{n=1}^{\infty} \left(1 - \frac{1}{r}\right) = 0.$$

Therefore $F' = 0$ a.e. $[\lambda]$. ‖

When $r = 3$, $p_0 = \frac{1}{2}$, $p_1 = 0$, $p_2 = \frac{1}{2}$, F is the *Cantor function*, so named because in this case the set C in the proof of 3.4.3 is the Cantor set. For the classical approach to this function, see RAP, p. 77.

Problems

1. (*Borel's Normal Number Theorem*)   A number $y \in [0, 1)$ is said to be *normal* to the base r iff for each $b = (i_1, \ldots, i_m)$, $m = 1, 2, \ldots, i_1, \ldots, i_m \in \{0, 1, \ldots, r - 1\}$, the relative frequency of b in the first $n + m - 1$ digits of the r-adic expansion of y converges as $n \to \infty$ to r^{-m}. All rational numbers are nonnormal; if $.i_1 i_2 \cdots$ is any irrational number, then $.i_1 0 i_2 0 i_3 0 \cdots$ is irrational and nonnormal. The irrational number $.12345678910111213 \cdots$ (base 10) can be shown to be normal.

Show that almost every number in $[0, 1)$ is normal to all bases, that is, if B is the set of numbers that are nonnormal to at least one base $r = 2, 3, \ldots$, then B has Lebesgue measure 0.

2. Consider the interval $[0, 1)$, with Borel sets and Lebesgue measure. Define $Tx = 2x \pmod 1$, that is, $Tx = 2x$, $0 \leq x < \frac{1}{2}$; $Tx = 2x - 1$, $\frac{1}{2} \leq x < 1$; T is called the *dyadic transformation*. Show that T may essentially be regarded (make this precise) as a one-sided shift transformation on the space of all infinite sequences of 0's and 1's.

3. As in Problem 2, show that the transformation $Tx = rx \pmod 1$ can be regarded as a one-sided shift transformation on the space of all infinite sequences with values in $\{0, 1, \ldots, r - 1\}$. The precise definition of T is

$$Tx = rx, \qquad\qquad\qquad 0 \le x < 1/r;$$

$$Tx = rx - 1, \qquad\qquad 1/r \le x < 2/r; \ldots;$$

$$Tx = rx - (r - 1), \qquad (r - 1)/r \le x < 1.$$

In Problems 2 and 3, the shift transformations have independent coordinate random variables, taking values in a finite set. Such transformations are called *Bernoulli shifts*.

3.5   Applications to Markov Chains

If T is a one-sided shift transformation and the coordinate random variables X_n are independent, we have seen that T is ergodic (we also say that the sequence $\{X_n\}$ is ergodic). In this section we exhibit a large class of examples in which T is ergodic, but the X_n are not independent.

First we must consider some general properties of shift transformations. If $X_0, X_1, \ldots$ are the coordinate random variables of a one-sided shift, recall (RAP, 7.2.6, p. 278) that the *tail σ-field* of the X_n is defined by $\mathscr{F}_\infty = \bigcap_{n=0}^\infty \mathscr{F}_n$, where $\mathscr{F}_n = \mathscr{F}(X_n, X_{n+1}, \ldots)$, the smallest σ-field that makes X_i measurable for all $i \ge n$. We prove that the σ-field of almost invariant sets is essentially included in $\mathscr{F}_\infty$.

3.5.1   Theorem

If A is an almost invariant set, there is a set $B \in \mathscr{F}_\infty$ such that $P(A \bigtriangleup B) = 0$.

PROOF   By 3.2.2(a), there is a strictly invariant set B with $P(A \bigtriangleup B) = 0$. Now if $C \in \mathscr{F}$ then $T^{-n}C \in \mathscr{F}_n$; for if $C = \{(X_0, X_1, \ldots) \in C'\}$, then $T^{-n}C = \{(X_n, X_{n+1}, \ldots) \in C'\}$. (Actually, $C = C'$ here since the X_n are coordinate random variables.) But $B = T^{-n}B$, so that $B \in \mathscr{F}_n$ for all n, hence $B \in \mathscr{F}_\infty$.   ‖

The proof of 3.5.1 shows that all strictly invariant sets belong to $\mathscr{F}_\infty$; however, an almost invariant set might not be in $\mathscr{F}_\infty$. For example, let $A = \{\omega : X_n(\omega) = X_0(\omega) \text{ for all } n\}$; then $A \subset T^{-1}A$, so that A is almost invariant by 3.2.4; but $A \notin \mathscr{F}_\infty$.

The following fact about conditional probabilities will be used.

3.5.2 Lemma

If T is a measure-preserving transformation on $(\Omega, \mathscr{F}, P)$, and $\mathscr{G}$ is a sub σ-field of $\mathscr{F}$, then for any $A \in \mathscr{F}$,

$$P(A \mid \mathscr{G})(T\omega) = P(T^{-1}A \mid T^{-1}\mathscr{G})(\omega) \quad \text{a.e.}$$

In particular, if T is a shift transformation, then

$$P(A \mid X_n)(T\omega) = P(T^{-1}A \mid X_{n+1})(\omega) \quad \text{a.e.}$$

PROOF   If $T^{-1}B \in T^{-1}\mathscr{G}$, then

$$P(T^{-1}A \cap T^{-1}B) = \int_{T^{-1}B} P(T^{-1}A \mid T^{-1}\mathscr{G}) \, dP.$$

But since T is measure-preserving,

$$P(T^{-1}A \cap T^{-1}B) = P(A \cap B) = \int_B P(A \mid \mathscr{G}) \, dP$$

$$= \int_{T^{-1}B} P(A \mid \mathscr{G})(T\omega) \, dP(\omega) \qquad \text{by 3.2.8.}$$

Since $P(A \mid \mathscr{G})(T\omega)$ is $T^{-1}\mathscr{G}$-measurable, the result follows.   ‖

We may give an intuitive interpretation of 3.5.2. If T is a shift transformation and $\omega^* = (\omega_0^*, \omega_1^*, \ldots) \in R^\infty$, then $P(T^{-1}A \mid X_{n+1})(\omega^*)$ is the probability that $\omega \in T^{-1}A$, given that the $(n+1)$th coordinate of ω is $X_{n+1}(\omega^*) = \omega_{n+1}^*$; $P(A \mid X_n)(T\omega^*)$ is the probability that $T\omega \in A$, given that the nth coordinate of $T\omega$ is $X_n(T\omega^*) = \omega_{n+1}^*$; these two expressions agree.

Our interest will be in sequences having the Markov property, defined as follows.

3.5.3 Definition

A sequence of random variables $\{X_n, n \geq 0\}$ is said to have the *Markov property* iff for each $B \in \mathscr{F}(X_{n+1}, X_{n+2}, \ldots)$, $n = 0, 1, \ldots$,

$$P(B \mid X_0, \ldots, X_n) = P(B \mid X_n) \quad \text{a.e.}$$

If $\{X_n\}$ is a Markov chain (see RAP, Section 5.11, and Ash, 1970, Chapter 7), then

$$P\{X_{n+1} = i_{n+1}, \ldots, X_{n+k} = i_{n+k} \mid X_0 = i_0, \ldots, X_n = i_n\}$$

$$= P\{X_{n+1} = i_{n+1}, \ldots, X_{n+k} = i_{n+k} \mid X_n = i_n\}$$

for all $i_0, \ldots, i_{n+k}$ in the state space. Thus the Markov property holds when B is a measurable cylinder, hence for all $B \in \mathscr{F}(X_{n+1}, X_{n+2}, \ldots)$ by the monotone class theorem.

In the Markov case, almost invariant sets assume a special form.

3.5.4 Theorem

Let the coordinate random variables X_n of a one-sided shift have the Markov property. If A is an almost invariant set, there is a set B of the form $\{X_0 \in C\}$, $C \in \mathscr{B}(R)$, such that $A = B$ a.e., that is, $P(A \triangle B) = 0$.

PROOF By 3.5.1, there is a set B in the tail σ-field $\mathscr{F}_\infty$ such that $P(A \triangle B) = 0$. Since $B \in \mathscr{F}(X_{n+1}, X_{n+2}, \ldots)$ for all n, $P(B \mid X_0, \ldots, X_n) = P(B \mid X_n)$ a.e. Consequently, $P(A \mid X_0, \ldots, X_n) = P(A \mid X_n)$ a.e. Now

$$P(A \mid X_0, \ldots, X_n) \to P(A \mid X_0, X_1, \ldots) \quad \text{a.e.}$$

(see RAP, 7.6.4, p. 299); but

$$P(A \mid X_0, X_1, \ldots) = E(I_A \mid X_0, X_1, \ldots) = I_A \quad \text{a.e.}$$

since $A \in \mathscr{F}(X_0, X_1, \ldots) = \mathscr{F}$. Therefore $P(A \mid X_n) \to I_A$ a.e.

Given $\varepsilon > 0$, let $H_n = \{\omega : |P(A \mid X_n)(\omega) - I_A(\omega)| \geq \varepsilon\}$. Then $P(H_n) \to 0$ since $P(A \mid X_n) \to I_A$ a.e., hence in probability. Now

$$T^{-1}(H_n) = \{\omega : |P(A \mid X_n)(T\omega) - I_A(T\omega)| \geq \varepsilon\}$$
$$= \{\omega : |P(A \mid X_{n+1})(\omega) - I_A(\omega)| \geq \varepsilon\} \quad \text{a.e.}$$

by 3.5.2 and the almost invariance of A. Thus $T^{-1}(H_n) = H_{n+1}$ a.e. Since T preserves P, $P(H_n) = PT^{-1}(H_n) = P(H_{n+1})$ for all n. Since $P(H_n) \to 0$, we must have $P(H_n) \equiv 0$, so $P(A \mid X_n) = I_A$ a.e. for all n, in particular, $I_A = P(A \mid X_0)$ a.e.

Now $P(A \mid X_0)$ is $\mathscr{F}(X_0)$-measurable, hence can be expressed as $f(X_0)$ for some Borel measurable $f: R \to R$ (see RAP, 6.4.2(c), p. 251). Thus (a.e.)

$$\omega \in A \quad \text{iff} \quad I_A(\omega) = 1$$
$$\text{iff} \quad f(X_0(\omega)) = 1$$
$$\text{iff} \quad X_0(\omega) \in C, \quad \text{where} \quad C = f^{-1}\{1\}. \quad \|$$

3.5.5 Corollary

Under the hypothesis of 3.5.4, a set $A \in \mathscr{F}$ is almost invariant iff A is of the form $\{X_n \in C \text{ for all } n\}$ for some $C \in \mathscr{B}(R)$.

PROOF Let A be almost invariant. By 3.5.4, A is of the form $\{X_0 \in C\}$. But $A = T^{-1}A$ a.e., so $\{X_0 \in C\} = \{X_1 \in C\}$ a.e. Inductively, $A = \{X_n \in C \text{ for}$

all n}. Conversely, every set A of this form has $A \subset T^{-1}A$, hence is almost invariant by 3.2.4. $\parallel$

We therefore have the following criterion for ergodicity of a Markov sequence.

3.5.6   Theorem

If $\{X_n\}$ has the Markov property, then $\{X_n\}$ is not ergodic iff there is a set $C \in \mathscr{B}(R)$ such that $0 < P\{X_n \in C$ for all $n\} < 1$.

PROOF   Apply 3.5.5 and 3.2.2(b). $\parallel$

We now apply these results to Markov chains. Let $\{X_n\}$ be a Markov chain; assume the initial distribution $\{v_i\}$ is a stationary distribution, so that $\{X_n\}$ is a stationary sequence and the machinery of ergodic theory is applicable. (See Ash, 1970, pp. 236–240, for the appropriate background material on Markov chains.)

3.5.7   Theorem

(a)   If there is exactly one positive recurrent class C, then $\{X_n\}$ is ergodic.

(b)   If there are at least two positive recurrent classes C_1 and C_2, and $\sum_{i \in C_1} v_i > 0$, $\sum_{i \in C_2} v_i > 0$, then $\{X_n\}$ is not ergodic.

PROOF   (a)   Since the stationary distribution assigns probability 1 to C, we may as well assume that C is the entire state space. Let D be a nonempty proper subset of C, and let $i \in D, j \in C - D$. By recurrence, if the initial state is i, then j must be visited; hence

$$P\{X_n \in D \text{ for all } n\} = \sum_i v_i P\{X_n \in D \text{ for all } n \mid X_0 = i\} = 0.$$

By 3.5.6, $\{X_n\}$ is ergodic.

(b)   It is impossible to exit from a recurrent class, hence

$$P\{X_n \in C_1 \text{ for all } n\} = P\{X_0 \in C_1\} = \sum_{i \in C_1} v_i \in (0, 1).$$

By 3.5.6, $\{X_n\}$ is not ergodic. $\parallel$

The case in which there are no positive recurrent classes is not discussed in 3.5.7 because in this case, there is no stationary distribution for the chain. Note also that if there is exactly one positive recurrent class, the stationary distribution is unique.

We now have many examples of ergodic sequences $\{X_n\}$ where the X_n need not be independent. For example, consider a finite Markov chain such that every state is reachable from every other state. The state space then forms a single equivalence class, necessarily recurrent positive. Thus if the initial distribution is the unique stationary distribution, the sequence $\{X_n\}$ is ergodic.

Now suppose $\{X_n\}$ is an ergodic Markov chain, and assume that the entire state space forms a positive recurrent class. Then by the pointwise ergodic theorem, if $f \in L^1(\Omega, \mathscr{F}, P)$, then

$$\frac{1}{n} \sum_{k=0}^{n-1} f(T^k\omega) \to E(f) \quad \text{a.e.}$$

We have been assuming that the initial distribution $\{v_i\}$ is stationary, but this result holds regardless of the initial distribution. For

$$0 = P\left\{\frac{1}{n} \sum_{k=0}^{n-1} f(T^k\omega) \nrightarrow E(f)\right\} = \sum_i v_i P\left\{\frac{1}{n} \sum_{k=0}^{n-1} f(T^k\omega) \nrightarrow E(f) \,|\, X_0 = i\right\}.$$

Since $v_i > 0$ for all i, we have $P\{n^{-1} \sum_{k=0}^{n-1} f(T^k\omega) \nrightarrow E(f) \,|\, X_0 = i\} = 0$ for all i, and the result follows.

If a Markov chain has exactly one positive recurrent class, and therefore a unique stationary distribution $\{v_j\}$, then the mean recurrence time of state j, that is, the average number of steps required to return to j when the initial state is j, is the reciprocal of the probability v_j that the chain will be in state j at any particular time. This is intuitively reasonable; if, say, $v_j = \frac{1}{4}$, then in the long run, we are in state j on one out of four trials, so on the average it should take four steps to return to j.

We are going to prove a more general result of this type.

3.5.8 Theorem

Let T be a measure-preserving transformation on the probability space $(\Omega, \mathscr{F}, P)$. If $A \in \mathscr{F}$, let

$$A_k = \{\omega \in A : T^n\omega \notin A, n = 1, \ldots, k-1, T^k\omega \in A\};$$

thus A_k is the set of points $\omega \in A$ such that $T^n\omega$ returns to A for the first time at $n = k$.

Define the *recurrence time* of A by $r_A(\omega) = k$ if $\omega \in A_k$, $k = 1, 2, \ldots$; $r_A(\omega) = \infty$ if $\omega \in A - \bigcup_{k=1}^{\infty} A_k$ (define r_A arbitrarily on A^c). Then

$$\int_A r_A(\omega) \, dP(\omega) = P\left(\bigcup_{n=0}^{\infty} T^{-n}A\right).$$

Before giving the proof, let us go into more detail on the meaning of the theorem. If T is ergodic and $P(A) > 0$, let $E = \bigcup_{n=0}^{\infty} T^{-n}A$. Since $T^{-1}E \subset E$, E is almost invariant by 3.2.4, and since $P(E) \geq P(A) > 0$, $P(E)$ must be 1. Thus if $Q(B) = P(B \mid A) = P(B \cap A)/P(A)$, $B \in \mathscr{F}$, we have

$$\int_A r_A(\omega) \, dQ(\omega) = \frac{1}{P(A)}.$$

If T is a one-sided shift and $A = \{X_0 \in C\}$, then $\int_A r_A \, dQ$ is the average length of time required for X_n to return to the set C, given that the initial value X_0 belongs to C. Thus the mean recurrence time of C is the reciprocal of the probability that the process will be in C at any particular time.

PROOF   Let $C_k = T^{-k}A$, $B_k = C_k^c$, $k = 0, 1, \ldots$ (take $C_0 = A$). Then, with intersections written as products,

$$
\begin{aligned}
P(A_{k+1}) &= P(C_0 B_1 \cdots B_k C_{k+1}), \qquad k \geq 1 \\
&= P(C_0 B_1 \cdots B_k) - P(C_0 B_1 \cdots B_{k+1}) \\
&= P(B_1 \cdots B_k) - P(B_0 B_1 \cdots B_k) - P(B_1 \cdots B_{k+1}) \\
&\quad + P(B_0 B_1 \cdots B_{k+1}) \\
&= P(B_0 \cdots B_{k-1}) - 2P(B_0 \cdots B_k) + P(B_0 \cdots B_{k+1}) \\
&\qquad \text{since } T \text{ preserves } P \\
&= b_k - 2b_{k+1} + b_{k+2}
\end{aligned}
$$

where $b_k = P(B_0 \cdots B_{k-1})$, $k \geq 1$. When $k = 0$ we have

$$
\begin{aligned}
P(A_1) &= P(C_0 C_1) \\
&= P(C_0) - P(C_0 B_1) \\
&= 1 - P(B_0) - P(B_1) + P(B_0 B_1) \\
&= 1 - 2P(B_0) + P(B_0 B_1),
\end{aligned}
$$

hence we have $P(A_{k+1}) = b_k - 2b_{k+1} + b_{k+2}$ for all $k \geq 0$, if we take $b_0 = 1$. Now

$$
\begin{aligned}
\int_A r_A \, dP &= \sum_{k=0}^{\infty} (k+1)P(A_{k+1}) \\
&= \lim_{n \to \infty} \left[\sum_{k=0}^{n-1} (k+1)b_k - 2 \sum_{k=1}^{n} k b_k + \sum_{k=2}^{n+1} (k-1)b_k \right] \\
&= \lim_{n \to \infty} \left[1 - n(b_n - b_{n+1}) - b_n \right].
\end{aligned}
$$

Now $b_n - b_{n+1} = P(B_0 \cdots B_{n-1} C_n) \geq 0$, and

$$b_n \to P\left(\bigcap_{k=0}^{\infty} B_k\right) = 1 - P\left(\bigcup_{k=0}^{\infty} C_k\right) = 1 - P\left(\bigcup_{n=0}^{\infty} T^{-n}A\right).$$

Since $1 - n(b_n - b_{n+1}) - b_n$ has a limit and b_n approaches a finite limit, $n(b_n - b_{n+1})$ must approach a finite limit. But the sets $B_0 \cdots B_{n-1} C_n$ are disjoint, so that $\sum_n (b_n - b_{n+1}) < \infty$; this implies that $n(b_n - b_{n+1}) \to 0$. ‖

Problems

1. The one-sided shift transformation T is said to be a *Kolmogorov shift* (or to be *tail trivial*) iff the tail σ-field $\mathscr{F}_\infty$ consists only of sets with probability 0 or 1. It follows from 3.5.1 that a Kolmogorov shift is ergodic; show that, in fact, every Kolmogorov shift is mixing.

2. Let T be the shift transformation associated with a finite Markov chain where the initial distribution is stationary. Assume that $v_i > 0$ for all i. (If $v_i = 0$, then $P\{X_n = i \text{ for some } n\} \leq \sum_n P\{X_n = i\} = 0$, so that i may as well be removed from the state space.)

Show that T is mixing iff a steady state distribution exists, that is, iff $p_{ij}^{(n)} \to q_j$ (independent of i) as $n \to \infty$, where $\sum_j q_j = 1$. (Necessarily, $q_j = v_j$, and the stationary distribution is unique; see Ash, 1970, pp. 236–237.)

3. Let $X_1, X_2, \ldots$ be iid random variables, and let $S_n = \sum_{k=1}^n X_k, n = 1, 2,$ ....

(a) If $R_n(\omega)$ is the number of distinct points in the set $\{S_1(\omega), \ldots, S_n(\omega)\}$, and A is the event that S_n never returns to 0, that is, $A = \{S_1 \neq 0, S_2 \neq 0, \ldots\}$, show that $n^{-1} E(R_n) \to P(A)$. (*Hint*: Express R_n as $1 + \sum_{k=2}^n I_{B_k}$, where

$$B_k = \{S_k \neq S_{k-1}, S_k \neq S_{k-2}, \ldots, S_k \neq S_1\}.)$$

(b) For a fixed positive integer N, let $Z_k(\omega)$ be the number of distinct points in $\{S_{(k-1)N+1}(\omega), \ldots, S_{kN}(\omega)\}, k = 1, 2, \ldots$. Use the strong law of large numbers to show that

$$\limsup_{n \to \infty} \frac{R_{nN}}{nN} \leq \frac{E(Z_1)}{N} \quad \text{a.e.}$$

(c) Show that $\limsup_{n \to \infty} n^{-1} R_n \leq P(A)$ a.e.

(d) Let V_k be the indicator of the set $\{S_{k+1} \neq S_k, S_{k+2} \neq S_k, \ldots\}$; thus $V_k = 1$ iff S_k is never revisited. Use the pointwise ergodic theorem to show that $n^{-1}(V_1 + \cdots + V_n) \to E(V_1)$ a.e.

(e) Show that $\liminf_{n \to \infty} n^{-1} R_n \geq E(V_1)$ a.e., and conclude that $n^{-1} R_n \to P(A)$ a.e.

3.6   The Shannon–McMillan Theorem

We are going to apply the pointwise ergodic theorem to prove a basic result of information theory. Consider a shift transformation with coordinate random variables X_n taking values in a countable set. We define a rather unusual sequence of random variables $p(X_0, X_1, \ldots, X_{n-1})$ as follows. If $X_0(\omega) = i_0, \ldots, X_{n-1}(\omega) = i_{n-1}$, let

$$p(X_0(\omega), \ldots, X_{n-1}(\omega)) = P\{\omega' : X_0(\omega') = i_0, \ldots, X_{n-1}(\omega') = i_{n-1}\}.$$

For example, if $P\{X_0 = 1\} = P\{X_0 = 2\} = \frac{1}{6}$, $P\{X_0 = 3\} = \frac{2}{3}$, then $p(X_0)$ has the value $\frac{1}{6}$ with probability $\frac{1}{3}$, namely, when $X_0 = 1$ or 2, and $p(X_0) = \frac{2}{3}$ with probability $\frac{2}{3}$, that is, when $X_0 = 3$.

Similarly, define the random variable $p(X_n \mid X_{n-1}, X_{n-2}, \ldots, X_{n-r})$ by specifying that if $X_n = i_n, X_{n-1} = i_{n-1}, \ldots, X_{n-r} = i_{n-r}$, then

$$p(X_n \mid X_{n-1}, \ldots, X_{n-r}) = P\{X_n = i_n \mid X_{n-1} = i_{n-1}, \ldots, X_{n-r} = i_{n-r}\}$$

assuming $P\{X_{n-1} = i_{n-1}, \ldots, X_{n-r} = i_{n-r}\} > 0$.

The Shannon–McMillan theorem is an assertion about the convergence of $-n^{-1} \log p(X_0, \ldots, X_{n-1})$; in particular, if T is ergodic, we have convergence to a constant H, almost everywhere and in L^1. Before turning to the proof, let us look at the intuitive interpretation of the convergence statement. For this, we require only that $-n^{-1} \log p(X_0, \ldots, X_{n-1})$ converge to H in probability. (It is traditional in information theory to use logs to the base 2, and we shall follow this practice here. Switching to natural logs involves only a multiplicative constant.)

Given $\varepsilon > 0$, $\delta > 0$, let

$$A_n = \left\{ \left| -\frac{1}{n} \log p(X_0, \ldots, X_{n-1}) - H \right| \le \delta \right\};$$

thus if $\omega \in A_n$, then $p(X_0(\omega), \ldots, X_{n-1}(\omega))$ is between $2^{-n(H+\delta)}$ and $2^{-n(H-\delta)}$. If $-n^{-1} \log p(X_0, \ldots, X_{n-1}) \xrightarrow{P} H$, then $P(A_n) \ge 1 - \varepsilon$ for sufficiently large n.

Now let S_n be the set of all sequences of length n, with values in the coordinate space, corresponding to points in A_n, that is, S_n is the set of all sequences $(X_0(\omega), \ldots, X_{n-1}(\omega))$, $\omega \in A_n$. If $(i_0, \ldots, i_{n-1}) \in S_n$, then

$$2^{-n(H+\delta)} \le P\{X_0 = i_0, \ldots, X_{n-1} = i_{n-1}\} \le 2^{-n(H-\delta)};$$

furthermore, $P\{(X_0, \ldots, X_{n-1}) \in S_n\} = P(A_n)$, so that

$$1 - \varepsilon \le P\{(X_0, \ldots, X_{n-1}) \in S_n\} \le 1.$$

Thus each sequence in S_n has a probability between $2^{-n(H+\delta)}$ and $2^{-n(H-\delta)}$, and the total probability assigned to S_n is between $1 - \varepsilon$ and 1. Consequently, the maximum number of sequences in S_n is $1/2^{-n(H+\delta)} = 2^{n(H+\delta)}$, and the minimum number is $(1 - \varepsilon)/2^{-n(H-\delta)} = (1 - \varepsilon)2^{n(H-\delta)}$.

Thus, roughly, for large n there are approximately 2^{nH} sequences of length n, each with probability approximately 2^{-nH}; the remaining sequences are negligible, i.e., have total probability at most ε. In information theory, this is referred to as the *asymptotic equipartition property*. If $\{X_n\}$ represents the output of an "information source" (such as a language) with r symbols, then of all the $r^n = 2^{n\log r}$ possible sequences of length n, only 2^{nH} can reasonably be expected to appear.

The number H will turn out to be the entropy of the sequence $\{X_n\}$, and we must discuss this concept before going any further.

3.6.1 Definitions

If X is a discrete random variable (or random vector), define the *entropy* (also called the *uncertainty*) of X as

$$H(X) = -\sum_x p(x) \log p(x)$$

where $p(x) = P\{X = x\}$. If X and Y are discrete, define the *conditional entropy* of Y given $X = x$ as

$$H(Y \mid X = x) = -\sum_y p(y \mid x) \log p(y \mid x)$$

where $p(y \mid x) = P\{Y = y \mid X = x\}$; also define the *conditional entropy of Y given X* as a weighted average of the $H(Y \mid X = x)$, namely,

$$H(Y \mid X) = \sum_x p(x)H(Y \mid X = x)$$

$$= -\sum_{x,y} p(x, y) \log p(y \mid x).$$

The *joint entropy* of X and Y is defined by

$$H(X, Y) = -\sum_{x,y} p(x, y) \log p(x, y);$$

since $H(X, Y)$ is the entropy of the random vector (X, Y), nothing new is involved.

Note that entropy is always nonnegative; if the random variables are allowed to have a countably infinite set of values, an entropy of $+\infty$ may be obtained. Difficulties with events of probability zero are avoided by defining $0 \log 0 = 0$, $-\log 0 = +\infty$.

It is often convenient to express entropy in terms of the random variables $p(X)$ introduced at the beginning of the section; we have

$$H(X) = E[-\log p(X)], \qquad H(Y \mid X) = E[-\log p(Y \mid X)].$$

We now establish a few properties of entropy.

3.6.2   Theorem

Let X, Y, and Z be discrete random vectors with finite entropy.

(a)   If p_1, p_2, $\ldots$, q_1, q_2, $\ldots$ are nonnegative numbers with $\sum_i p_i = \sum_i q_i = 1$, and either $-\sum_i p_i \log p_i < \infty$ or $-\sum_i p_i \log q_i < \infty$, then

$$-\sum_i p_i \log p_i \le -\sum_i p_i \log q_i,$$

with equality iff $p_i = q_i$ for all i.

(b) $H(X, Y) \le H(X) + H(Y)$, with equality iff X and Y are independent.

(c) $H(X, Y) = H(X) + H(Y \mid X) = H(Y) + H(X \mid Y)$.

(d) $H(Y \mid X) \le H(Y)$, with equality iff X and Y are independent.

(e)   If X takes on r possible values $x_1, \ldots, x_r$, then $H(X) \le \log r$, with equality iff $p(x_i) = 1/r$, $i = 1, \ldots, r$.

(f) $H(Y, Z \mid X) \le H(Y \mid X) + H(Z \mid X)$, with equality iff Y and Z are conditionally independent given X, that is,

$$p(y, z \mid x) = p(y \mid x) p(z \mid x) \qquad \text{for all} \quad x, y, z.$$

(g) $H(Y, Z \mid X) = H(Y \mid X) + H(Z \mid X, Y)$.

(h) $H(Z \mid X, Y) \le H(Z \mid X)$, with equality iff Y and Z are conditionally independent given X.

PROOF   (a)   For convenience we switch to natural logs. Since $x - 1$ is the tangent to $\log x$ at $x = 1$, we have $\log x \le x - 1$, with equality iff $x = 1$. Thus $\log (q_i/p_i) \le (q_i/p_i) - 1$, with equality iff $p_i = q_i$; hence, even if p_i or $q_i = 0$,

$$p_i \log q_i = p_i \log (q_i/p_i) + p_i \log p_i \le q_i - p_i + p_i \log p_i$$

with equality iff $p_i = q_i$. Sum over i to obtain the desired result. (Note that if $\sum_i p_i \log q_i = \sum_i p_i \log p_i$, then the sums are finite by hypothesis, so that $p_i = q_i$ for all i.)

(b) By (a), $-\sum_{x,y} p(x, y) \log p(x, y) \leq -\sum_{x,y} p(x, y) \log p(x)p(y)$, with equality iff $p(x, y) = p(x)p(y)$ for all x, y.

(c) This follows from the fact that $p(x, y) = p(x)p(y \mid x) = p(y)p(x \mid y)$.

(d) This is immediate from (b) and (c).

(e) Apply (a) with $p_i = p(x_i)$, $q_i = 1/r$.

(f) Since $H(Y \mid X) < \infty$ by (d), $H(Y \mid X = x) < \infty$ for each x (such that $p(x) > 0$), and similarly for $H(Z \mid X = x)$. The proof of (b) shows that

$$H(Y, Z \mid X = x) \leq H(Y \mid X = x) + H(Z \mid X = x),$$

with equality iff $p(y, z \mid x) = p(y \mid x)p(z \mid x)$ for all y, z. Multiply by $p(x)$ and sum over x to complete the proof.

(g) This follows from $p(y, z \mid x) = p(y \mid x)p(z \mid x, y)$.

(h) Apply (f) and (g). ‖

We also need an entropy concept for stochastic processes.

3.6.3   Definitions and Comments

Let $X_0, X_1, \ldots$ be a stationary sequence of discrete random variables, and assume that $H(X_0) < \infty$. Define the *entropy* of the sequence (also called the entropy of the associated shift transformation T) as

$$H\{X_n\} = \lim_{n \to \infty} H(X_n \mid X_0, X_1, \ldots, X_{n-1}).$$

Now

$$H(X_{n+1} \mid X_0, \ldots, X_n) \leq H(X_{n+1} \mid X_1, \ldots, X_n) \qquad \text{by 3.6.2(h)}$$

$$= H(X_n \mid X_0, \ldots, X_{n-1})$$

$$\text{by stationarity;}$$

also by 3.6.2(h), $H(X_n \mid X_0, \ldots, X_{n-1}) \leq H(X_n) = H(X_0) < \infty$. Thus the limit defining H exists and is finite.

(To apply 3.6.2(h), it must be verified that $H(X_1, \ldots, X_n) < \infty$, but the proof of 3.6.2(b) shows that if $H(X_i) < \infty$, $i = 1, \ldots, n$, then

$$H(X_1, \ldots, X_n) \leq H(X_1) + \cdots + H(X_n),$$

with equality iff $X_1, \ldots, X_n$ are independent. In the present case, $H(X_i) = H(X_0) < \infty$ for all i.)

The entropy of $\{X_n\}$ may also be expressed as

$$H\{X_n\} = \lim_{n \to \infty} \frac{1}{n} H(X_0, \ldots, X_{n-1}).$$

To see this, observe that by induction using 3.6.2(c),

$$H(X_0, \ldots, X_{n-1}) = H(X_0) + H(X_1 \mid X_0) + H(X_2 \mid X_0, X_1)$$
$$+ \cdots + H(X_{n-1} \mid X_0, \ldots, X_{n-2}).$$

Thus $n^{-1}H(X_0, \ldots, X_{n-1})$ is the arithmetic average of a sequence converging to $H\{X_n\}$, hence converges to $H\{X_n\}$.

We now begin the development of the Shannon–McMillan theorem. It will be convenient to consider a two-sided shift transformation T with coordinate random variables X_n, where the X_n take values in a countable set. It is assumed throughout that $H(X_0) < \infty$.

Martingale theory will be significant, as the following result suggests.

3.6.4   Theorem

Let $Y_0 = -\log p(X_0)$, $Y_k = -\log p(X_0 \mid X_{-1}, \ldots, X_{-k})$, $k \geq 1$. Then $\{Y_0, Y_1, \ldots\}$ is a nonnegative supermartingale, hence converges a.e. to an integrable limit function Y.

PROOF   Since $E(Y_k) = H(X_0 \mid X_{-1}, \ldots, X_{-k}) \leq H(X_0) < \infty$, the Y_k are integrable. Now since all random variables are discrete, we may write

$$E(Y_{n+1} \mid X_0 = x_0, \ldots, X_{-n} = x_{-n})$$

$$= - \sum_{x_{-(n+1)}} p(x_{-(n+1)} \mid x_0, \ldots, x_{-n}) \log p(x_0 \mid x_{-1}, \ldots, x_{-(n+1)})$$

$$= + \sum_{x_{-(n+1)}} \frac{p(x_0, \ldots, x_{-(n+1)})}{p(x_0, \ldots, x_{-n})} \log \frac{p(x_{-1}, \ldots, x_{-(n+1)})}{p(x_0, \ldots, x_{-(n+1)})}.$$

This is of the form $\sum_i \alpha_i \log x_i$, where $\alpha_i \geq 0$, $\sum_i \alpha_i = 1$, hence is less than or equal to $\log (\sum_i \alpha_i x_i)$ by convexity. Thus

$$E(Y_{n+1} \mid X_0 = x_0, \ldots, X_{-n} = x_{-n})$$

$$\leq \log \left[\sum_{x_{-(n+1)}} \frac{p(x_{-1}, \ldots, x_{-(n+1)})}{p(x_0, \ldots, x_{-n})} \right]$$

$$= \log \frac{p(x_{-1}, \ldots, x_{-n})}{p(x_0, \ldots, x_{-n})}$$

$$= -\log p(x_0 \mid x_{-1}, \ldots, x_{-n}).$$

Therefore $E(Y_{n+1} \mid X_0, \dots, X_{-n}) \le Y_n$, and since Y_n is measurable relative to the σ-field $\mathscr{F}(X_0, \dots, X_{-n})$, the result follows. $\parallel$

We are going to show that the random variables Y_n are uniformly integrable. It will be convenient to assume that the coordinate space is a subset of the positive integers; this amounts only to a relabeling, and can be done without changing the distribution of $(Y_0, Y_1, \dots)$.

3.6.5   Lemma

If r is any fixed positive integer, the random variables $W_n = Y_n I_{\{X_0 \le r\}}$, $n = 0, 1, \dots$, are uniformly integrable.

PROOF   For any positive integer k,

$$\int_{\{W_n \ge k\}} W_n \, dP = \sum_{i=k}^{\infty} \int_{\{i \le W_n < i+1\}} W_n \, dP$$

$$\le \sum_{i=k}^{\infty} (i+1) P\{W_n \ge i\}.$$

But if $W_n \ge i$, then $Y_n \ge i$, or $p(X_0, \dots, X_{-n}) \le 2^{-i} p(X_{-1}, \dots, X_{-n})$. Thus

$$P\{W_n \ge i\} = \sum \{p(x_0, \dots, x_{-n}) : y_n \ge i, \, x_0 \le r\}$$

$$\le 2^{-i} \sum_{x_0 \le r} \sum_{x_{-1}, \dots, x_{-n}} p(x_{-1}, \dots, x_{-n})$$

$$\le r 2^{-i}.$$

Consequently,

$$\int_{\{W_n \ge k\}} W_n \, dP \le \sum_{i=k}^{\infty} r(i+1) 2^{-i}$$

$$\to 0 \qquad \text{as} \quad k \to \infty, \text{ uniformly in } n.   \parallel$$

3.6.6   Theorem

The random variables Y_n, $n = 0, 1, \dots$, are uniformly integrable; thus in 3.6.4 we also have $Y_n \to Y$ in L^1.

PROOF   For any positive integer k,

$$\int_{\{Y_n \ge k\}} Y_n \, dP = \int_{\{Y_n \ge k, \, X_0 > r\}} Y_n \, dP + \int_{\{Y_n \ge k, \, X_0 \le r\}} Y_n \, dP$$

$$\le \int_{\{X_0 > r\}} Y_n \, dP + \int_{\{W_n \ge k\}} W_n \, dP.$$

By 3.6.5, the second integral on the right approaches 0 as $k \to \infty$, uniformly in n for a fixed r. Now

$$
\begin{aligned}
\int_{\{X_0 > r\}} Y_n \, dP &= - \sum_{x_0 > r} \sum_{x_{-1}, \ldots, x_{-n}} p(x_0, \ldots, x_{-n}) \\
&\quad \times \log p(x_0 \mid x_{-1}, \ldots, x_{-n}) \\
&= - \sum_{x_{-1}, \ldots, x_{-n}} p(x_{-1}, \ldots, x_{-n}) \\
&\quad \times \sum_{x_0 > r} p(x_0 \mid x_{-1}, \ldots, x_{-n}) \\
&\quad \times \log p(x_0 \mid x_{-1}, \ldots, x_{-n}).
\end{aligned}
$$

Write $\log p(x_0 \mid x_{-1}, \ldots, x_{-n}) = \log p(x_0) + \log\left[p(x_0 \mid x_{-1}, \ldots, x_{-n}) / p(x_0) \right]$; the contribution due to $\log p(x_0)$ is

$$
\begin{aligned}
&- \sum_{x_0 > r} \sum_{x_{-1}, \ldots, x_{-n}} p(x_0, \ldots, x_{-n}) \log p(x_0) \\
&= - \sum_{x_0 > r} p(x_0) \log p(x_0)
\end{aligned}
$$

and the remaining contribution is

$$
\begin{aligned}
&\sum_{x_{-1}, \ldots, x_{-n}} p(x_{-1}, \ldots, x_{-n}) \\
&\quad \times \sum_{x_0 > r} p(x_0 \mid x_{-1}, \ldots, x_{-n}) \log \frac{p(x_0)}{p(x_0 \mid x_{-1}, \ldots, x_{-n})}.
\end{aligned}
$$

If we switch to natural logs for convenience and use $\log x \le x - 1$, we may upper bound this expression by

$$
\sum_{x_0 > r} p(x_0) - \sum_{x_0 > r} p(x_0) = 0.
$$

Therefore

$$
\int_{\{X_0 > r\}} Y_n \, dP \le - \sum_{x_0 > r} p(x_0) \log p(x_0)
$$

$$
\to 0 \qquad \text{as} \quad r \to \infty, \text{ uniformly in } n,
$$

$$
\text{since } H(X_0) < \infty.
$$

If $\varepsilon > 0$ is given, we may choose a fixed r such that $\int_{\{X_0 > r\}} Y_n \, dP < \varepsilon/2$ for all n; then for sufficiently large k, $\int_{\{W_n \ge k\}} W_n \, dP < \varepsilon/2$ for all n, hence $\int_{\{Y_n \ge k\}} Y_n \, dP < \varepsilon$ for all n. $\|$

The other basic property of the Y_n that we need is that $\sup_n Y_n$ is integrable. We prove this after a preliminary lemma.

3.6.7 Lemma

If i is a positive integer, define $Y_n^{(i)} = -\log p(X_0 = i \mid X_{-1}, \ldots, X_{-n})$, that is, if $X_0 = i$, $X_{-1} = i_{-1}, \ldots, X_{-n} = i_{-n}$, then

$$Y_n^{(i)} = -\log P\{X_0 = i \mid X_{-1} = i_{-1}, \ldots, X_{-n} = i_{-n}\}$$

(define $Y_0^{(i)} = -\log P\{X_0 = i\}$). If $\lambda \geq 0$, let

$$E_n(\lambda) = \left\{\max_{j<n} Y_j \leq \lambda < Y_n\right\}, \qquad E_n^{(i)}(\lambda) = \left\{\max_{j<n} Y_j^{(i)} \leq \lambda < Y_n^{(i)}\right\}.$$

If $A_i = \{X_0 = i\}$, then

$$P(E_n(\lambda) \cap A_i) = P(E_n^{(i)}(\lambda) \cap A_i) \leq 2^{-\lambda} P(E_n^{(i)}(\lambda)).$$

PROOF On A_i we have $Y_n = Y_n^{(i)}$, hence $E_n(\lambda) \cap A_i = E_n^{(i)}(\lambda) \cap A_i$. Now since $E_n^{(i)}(\lambda)$ belongs to the σ-field $\mathscr{F}(X_{-1}, \ldots, X_{-n})$,

$$P(A_i \cap E_n^{(i)}(\lambda)) = \int_{E_n^{(i)}(\lambda)} P(A_i \mid X_{-1}, \ldots, X_{-n})\, dP$$

$$= \int_{E_n^{(i)}(\lambda)} p(X_0 = i \mid X_{-1}, \ldots, X_{-n})\, dP$$

$$= \int_{E_n^{(i)}(\lambda)} 2^{-Y_n^{(i)}}\, dP$$

$$\leq 2^{-\lambda} P(E_n^{(i)}(\lambda))$$

by definition of $E_n^{(i)}(\lambda)$. ‖

3.6.8 Theorem

The random variables Y_n satisfy $E[\sup_k Y_k] < \infty$.

PROOF We may write

$$E\left[\sup_k Y_k\right] = \int_0^\infty P\left\{\sup_k Y_k > \lambda\right\} d\lambda$$

(see RAP, p. 280, Problem 2)

$$\leq \sum_{r=0}^\infty P\left\{\sup_k Y_k > r\right\}$$

$$= \sum_{r=0}^\infty \sum_{n=0}^\infty P(E_n(r))$$

$$= \sum_{r=0}^\infty \sum_{n=0}^\infty \sum_i P(E_n^{(i)}(r) \cap A_i) \qquad \text{by 3.6.7.}$$

If the coordinate space is finite, there is no difficulty; by 3.6.7,

$$E\left[\sup_k Y_k\right] \le \sum_{r,n=0}^{\infty} 2^{-r} \sum_i P(E_n^{(i)}(r)) < \infty$$

since the $E_n^{(i)}(r)$, $n = 0, 1, \ldots$, are disjoint and the sum over i is finite.

In the general case, assume without loss of generality that the numbers $p_i = P(A_i)$ decrease as i increases (if necessary, relabel the elements of the coordinate space). We then have $p_i \le 1/i$ for all i, for if $p_i > 1/i$, then $p_1, \ldots, p_i$ are all greater than $1/i$, a contradiction. Let f be a function, to be specified later, from the nonnegative integers to the nonnegative reals, such that $\{r : f(r) < i\}$ is a finite set for each i. Then by 3.6.7,

$$\sum_{n=0}^{\infty} \sum_{i \le f(r)} P(E_n^{(i)}(r) \cap A_i) \le 2^{-r} \sum_{i \le f(r)} \sum_{n=0}^{\infty} P(E_n^{(i)}(r))$$

$$\le 2^{-r} f(r)$$

by disjointness of the

$$E_n^{(i)}(r), \quad n = 0, 1, \ldots.$$

Also by disjointness,

$$\sum_{n=0}^{\infty} \sum_{i > f(r)} P(E_n^{(i)}(r) \cap A_i) \le \sum_{i > f(r)} P(A_i)$$

$$= \sum_{i > f(r)} p_i.$$

Therefore

$$E\left[\sup_k Y_k\right] \le \sum_{r=0}^{\infty} 2^{-r} f(r) + \sum_{r=0}^{\infty} \sum_{i > f(r)} p_i.$$

The second series is $\sum_{i=1}^{\infty} \sum_{f(r) < i} p_i = \sum_{i=1}^{\infty} f_0(i) p_i$, where $f_0(i)$ is the number of nonnegative integers r such that $f(r) < i$. If we set $f(r) = 2^r (r+1)^{-2}$, then the first series converges, and $f(r) < i$ iff $r < \log i + 2 \log (r+1)$.

Choose a positive integer B such that $2x^{-1} \log (x+1) < \frac{1}{2}$ for all $x \ge B$. Then:

$$f(r) < i, \quad r \ge B \quad \text{implies } r < \log i + r/2, \quad \text{that is, } r < 2 \log i,$$

and

$$f(r) < i, \quad r < B \quad \text{implies } r < B.$$

Therefore $f_0(i) \le 2 \log i + B$, $i = 1, 2, \ldots$, hence

$$\sum_{i=1}^{\infty} f_0(i)p_i \le 2 \sum_{i=1}^{\infty} p_i \log i + B \sum_{i=1}^{\infty} p_i$$

$$\le 2 \sum_{i=1}^{\infty} p_i \log \frac{1}{p_i} + B \qquad \text{since } p_i \le \frac{1}{i}$$

$$= 2H(X_0) + B < \infty. \quad \|$$

We may now give the main result.

3.6.9   Shannon–McMillan Theorem

Let T be a two-sided shift transformation with discrete coordinate random variables X_n. Assume that $H(X_0) < \infty$, and let H be the entropy of the sequence $\{X_n\}$.

If $Z_n = -n^{-1} \log p(X_0, \ldots, X_{n-1})$, there is an invariant random variable Z such that $E(Z) = H$ and $Z_n \to Z$, almost everywhere and in L^1. In particular, if T is ergodic, $Z = H$ a.e.

PROOF   If the random variables Y_n are defined as in 3.6.4, we have

$$Z_n = -\frac{1}{n} \sum_{k=0}^{n-1} \log p(X_k \mid X_{k-1}, \ldots, X_0)$$

$$= \frac{1}{n} \sum_{k=0}^{n-1} Y_k(T^k)$$

$$= \frac{1}{n} \sum_{k=0}^{n-1} Y(T^k) + \frac{1}{n} \sum_{k=0}^{n-1} [Y_k(T^k) - Y(T^k)]. \tag{1}$$

By 3.3.6 and 3.3.7, the first series in (1) converges to an invariant function $\hat{Y}$, a.e. and in L^1. By 3.3.9, $E(\hat{Y}) = E(Y)$, and by 3.6.6, $E(Y) = \underset{L^1}{\lim_{n \to \infty}} E(Y_n) = H$. The second series converges to 0 in L^1 since $Y_k \to Y$ by 3.6.6, and $\| Y_k(T^k) - Y(T^k) \|_1 = \| Y_k - Y \|_1$ by 3.2.8.

Thus if we take $Z = \hat{Y}$, all that remains is to show that $Z_n \to Z$ a.e., and this will follow if we show that the second series in (1) converges to 0 a.e. But for any positive integer N, the series is bounded in absolute value by

$$\frac{1}{n} \sum_{k=0}^{N-1} | Y_k(T^k) - Y(T^k) | + \frac{1}{n} \sum_{k=N}^{n-1} | Y_k(T^k) - Y(T^k) |. \tag{2}$$

The first series approaches 0 a.e. as $n \to \infty$ since $Y_k(T^k) - Y(T^k)$ is integrable, hence finite a.e. If we define

$$G_N = \sup \{ | Y_k - Y | : k \ge N \},$$

then $G_N \leq 2 \sup_k Y_k$, which is integrable by 3.6.8. Also, since $Y_k \to Y$ a.e., we have $G_N \to 0$ a.e. as $N \to \infty$, hence $E(G_N) \to 0$ by the dominated convergence theorem. But the second series in (2) (call it h_n) is bounded by $n^{-1} \sum_{k=N}^{n-1} G_N(T^k)$, which converges to $\hat{G}_N$, a.e. and in L^1, by 3.3.6 and 3.3.7.

Finally, given $\varepsilon > 0$, $\delta > 0$, we have $P\{\hat{G}_N \geq \varepsilon\} \leq \varepsilon^{-1} E(\hat{G}_N)$ by Chebyshev's inequality, and by 3.3.9, $E(\hat{G}_N) = E(G_N) \to 0$. If we choose N such that $\varepsilon^{-1} E(G_N) < \delta$, then $\hat{G}_N < \varepsilon$ on a set of probability greater than $1 - \delta$, hence $\lim \sup_{n\to\infty} h_n \leq \varepsilon$ on this set. Since ε and δ are arbitrary, it follows that $h_n \to 0$ a.e. ‖

Since the Shannon–McMillan theorem involves only the random variables X_n, $n \geq 0$, it holds equally well for a one-sided shift. For an indication of how to prove this formally, see the discussion at the end of Problem 7, Section 3.3.

Problems

1. (a) In the Shannon–McMillan theorem, give an example in which the limit random variable Z is a.e. constant, but T is not ergodic.

(b) In the Shannon–McMillan theorem, give an example in which Z is not a.e. constant (of course, T cannot be ergodic).

2. If the coordinate space is finite and $1 \leq p < \infty$, show that the random variables Z_n are uniformly integrable; thus $Z_n \to Z$ in L^p.

3. (A short proof of a special case of the Shannon–McMillan theorem (Gallager, 1968).) Let X_0, X_1, ... be a stationary sequence of discrete random variables, with $H(X_0) < \infty$. (In this problem, entropy will be expressed using natural logarithms for convenience.) Define an *mth order approximation* to $p(X_0, \ldots, X_{n-1})$ by

$$q_m(X_0, \ldots, X_{n-1}) = p(X_0, \ldots, X_{m-1})p(X_m \mid X_0, \ldots, X_{m-1}) \cdots$$

$$p(X_{n-1} \mid X_{n-m-1}, \ldots, X_{n-2}),$$

$$m = 1, 2, \ldots, n > m.$$

Now note that $|\ln y| = \ln^+ y + \ln^- y = 2\ln^+ y - \ln y \leq 2e^{-1}y - \ln y$, so that

$$E\left|\left|\ln\frac{q_m}{p}\right|\right| \leq \frac{2}{e}E\left(\frac{q_m}{p}\right) - E\left(\ln\frac{q_m}{p}\right)$$

where $q_m = q_m(X_0, \ldots, X_{n-1})$, $p = p(X_0, \ldots, X_{n-1})$.

(a) Show that $E(q_m/p) \leq 1$ and

$$E\left(-\ln \frac{q_m}{p}\right) = H(X_0, \ldots, X_{m-1}) + (n-m)H(X_m \mid X_0, \ldots, X_{m-1})$$
$$- H(X_0, \ldots, X_{n-1}).$$

Thus

$$E\left[\frac{1}{n}\left|\ln \frac{q_m}{p}\right|\right] \leq \frac{2}{ne} + \frac{m}{n}\frac{H(X_0, \ldots, X_{m-1})}{m}$$
$$+ \left(1 - \frac{m}{n}\right)H(X_m \mid X_0, \ldots, X_{m-1}) - \frac{H(X_0, \ldots, X_{n-1})}{n}.$$

(b) Assume $\{X_n\}$ ergodic, that is, the associated shift transformation is ergodic. Show that as $n \to \infty$, $-n^{-1} \ln q_m$ converges a.e. and in L^1 to $H(X_m \mid X_0, \ldots, X_{m-1})$.

(c) Let H be the entropy of the ergodic sequence $\{X_n\}$. Given $\varepsilon > 0$, choose m such that

$$\left|\frac{H(X_0, \ldots, X_{m-1})}{m} - H\right| < \frac{\varepsilon}{2}$$

and

$$|H(X_m \mid X_0, \ldots, X_{m-1}) - H| < \frac{\varepsilon}{2}.$$

Show that $E[|-n^{-1} \ln p(X_0, \ldots, X_{n-1}) - H|] < \varepsilon$ for sufficiently large n, thus establishing L^1 convergence in the Shannon–McMillan theorem under the hypothesis of ergodicity.

3.7 Notes

Some standard references on ergodic theory are Billingsley (1965), Halmos (1956), and Jacobs (1962). The pointwise ergodic theorem can be generalized in several ways. If T is a positive contraction operator on L^1, not necessarily arising from a measure-preserving transformation, one can investigate convergence of the sequence of averages $n^{-1}(f + Tf + \cdots + T^{n-1}f)$. In fact the arithmetic average can be replaced by a ratio of the form

$$(f + Tf + \cdots + T^{n-1}f)/(g + Tg + \cdots + T^{n-1}g),$$

where f and g belong to L^1. For results of this type (specifically, the Dunford–Schwartz ergodic theorem and the Chacon–Ornstein theorem), see Garsia

(1970). We might also mention that a long-standing conjecture has recently been settled by Ornstein (1970), who proved that if two Bernoulli shifts T and T' have the same entropy, then T and T' are isomorphic. (Essentially, this means that there is a measure-theoretic isomorphism φ between the corresponding probability spaces $(\Omega, \mathscr{F}, P)$ and $(\Omega', \mathscr{F}', P')$ such that $\varphi(T) = T'(\varphi)$; for an example, see Problems 2 and 3, Section 3.4.)

A discussion of the Shannon–McMillan theorem for a finite coordinate space may be found in Billingsley (1965). McMillan (1953) proved L^1 convergence and Breiman (1957) obtained a.e. convergence; Shannon's original paper (1948) considered convergence in probability for functions of a finite Markov chain. All these results were for finite coordinate spaces; the extension to the countable case is due to Chung (1961). For applications to information theory, see Ash (1965) and Gallager (1968).

The Shannon–McMillan theorem has been generalized to a more abstract setting. For a survey of this area and a unified approach to the various results, see Kieffer (1970).

Chapter 4

Sample Function Analysis of Continuous Parameter Stochastic Processes

4.1 Separability

Let $\{X(t), a < t < b\}$ be a continuous parameter stochastic process. We have introduced the concepts of continuity, differentiation, and integration in the L^2 sense, but we have thus far avoided the question of sample function behavior. For example, we have not considered events of the form $A = \{\omega : \sup_{a \leq t \leq b} X(t, \omega) < c\}$. The basic difficulty here is that the finite-dimensional distributions do not yield enough information to determine the probability of such events. For example, if the probability space $(\Omega, \mathscr{F}, P)$ is that given by the Kolmogorov extension theorem, the set A does not belong to $\mathscr{F}$ (see RAP, p. 194, Problem 4). Of course, one might hope that if a different probability space were taken in which A is measurable, $P(A)$ would be determined by the finite-dimensional distributions; the following example demolishes this hope.

4.1.1 Example

Let P be Lebesgue measure on $\mathscr{B}[0, 1]$. Let $X(t, \omega) \equiv 0$ for all ω and all $t \in [0, 1]$, and take $Y(t, \omega) = 0$ for $t \neq \omega$, $Y(t, \omega) = 1$ for $t = \omega$. Then for each t, $P\{\omega : X(t, \omega) = Y(t, \omega)\} = P\{\omega : \omega \neq t\} = 1$, hence $\{X(t)\}$ and $\{Y(t)\}$ have the same finite-dimensional distributions. But

$$P\left\{\omega : \sup_{0 \leq t \leq 1} X(t, \omega) < \tfrac{1}{2}\right\} = 1, \qquad P\left\{\omega : \sup_{0 \leq t \leq 1} Y(t, \omega) < \tfrac{1}{2}\right\} = 0.$$

Note also that $\{X(t)\}$ has continuous sample functions, while all sample functions of $\{Y(t)\}$ are discontinuous. Thus $P\{\omega : X(t, \omega)$ is continuous in $t\}$ is not determined by the finite-dimensional distributions.

Our basic approach will be to show that sample function analysis is possible for processes with special properties, namely separability and measurability. Then for a given process $\{X(t)\}$ we try to find another process $\{Y(t)\}$ with the same finite-dimensional distributions such that $\{Y(t)\}$ has the desired properties.

In this section, unless otherwise specified, all stochastic processes will have state space $(S, \mathscr{S})$, where S is a compact metric space and $\mathscr{S}$ is the class of Borel sets of S. The index set T may be taken for simplicity as a subset of R, but all results and proofs will apply when T is a subset of an arbitrary separable metric space M, if we replace the Euclidean metric by the given metric of T.

4.1.2   Definition

Let $\{X(t), t \in T\}$ be a stochastic process. The process is said to be *separable* iff there is a countable dense subset T_0 of T, called the *separating set*, and a set A of probability zero called the *negligible set*, such that if $\omega \notin A$ and $t \in T$, there is a sequence $t_n \in T_0$, $t_n \to t$, with $X(t_n, \omega) \to X(t, \omega)$.

Thus the behavior of the sample functions on T is determined by their behavior on T_0.

The following criterion will be useful.

4.1.3   Theorem

The process $\{X(t), t \in T\}$ is separable iff there is a countable dense set $T_0 \subset T$ and a set A of probability zero such that the following condition holds for $\omega \notin A$:

If K is a compact (equivalently, closed) subset of S and I an open interval of R, then $X(t, \omega) \in K$ for all $t \in T_0 \cap I$ implies $X(t, \omega) \in K$ for all $t \in T \cap I$. (If T is a subset of the separable metric space M, I is replaced by an open subset of M.)

PROOF   Assume separability. Let $\omega \notin A$, and assume $X(t, \omega) \in K$ for all $t \in T_0 \cap I$. If $t \in T \cap I$, let $t_n \in T_0 \cap I$, with $t_n \to t$, $X(t_n, \omega) \to X(t, \omega)$. Since $X(t_n, \omega) \in K$ for all n, and K is closed, we have $X(t, \omega) \in K$. Conversely, assume the condition of 4.1.3 is satisfied. If the process is not separable, there is an $\omega \notin A$ and $t \in T$ such that for all sequences $t_n \in T_0$, $t_n \to t$ we have $X(t_n, \omega) \nrightarrow X(t, \omega)$. Thus there is an open interval I containing t and an $\varepsilon > 0$ such that $d(X(t, \omega), X(t', \omega)) \geq \varepsilon$ for all $t' \in T_0 \cap I$ (d is the metric for S). Let K be the compact set $\{y \in S : d(y, X(t, \omega)) \geq \varepsilon\}$; then $X(t', \omega) \in K$ for all $t' \in T_0 \cap I$, hence $X(t, \omega) \in K$, a contradiction.   ‖

4.1.4 Corollary

If $\{X(t),\ t \in T\}$ is separable and $f\colon S \to S$ is continuous, then $\{f(X(t)),\ t \in T\}$ is also separable (with the same T_0 and A).

PROOF Let $f(X(t',\ \omega)) \in K$ for all $t' \in T_0 \cap I$. Then $X(t',\ \omega) \in f^{-1}(K)$ (closed) for all $t' \in T_0 \cap I$, hence $X(t,\ \omega) \in f^{-1}(K)$, that is, $f(X(t,\ \omega)) \in K$ for all $t \in T \cap I$. ‖

The behavior of a continuous function on T is determined by its values on a countable dense set T_0, and thus we expect that a process with continuous sample functions will be separable.

4.1.5 Theorem

If there is a set A of probability 0 such that for $\omega \notin A$, $X(t,\ \omega)$ is continuous in t, then $\{X(t),\ t \in T\}$ is separable, and T_0 can be taken as any countable dense subset of T.

PROOF If $t \in T$, let $t_n \in T_0$, $t_n \to t$. Then $X(t_n,\ \omega) \to X(t,\ \omega)$. ‖

If T is an interval of R, the same result holds if continuity is replaced by right continuity, except that if T has a right endpoint y, T_0 can be taken as any countable dense subset of T containing y. A similar statement holds for left continuity.

Under a mild hypothesis, a given separating set T_0 can be replaced by an arbitrary countable dense set.

4.1.6 Theorem

Let $\{X(t),\ t \in T\}$ be a real process (in other words, the $X(t)$ are real random variables) that is separable and *continuous in probability*, that is,
$$X(t) \xrightarrow{P} X(t_0) \text{ as } t \to t_0$$
(for example, a process that is L^2-continuous). Then any countable dense set $T_0 \subset T$ can be used as the separating set.

PROOF Let T_0' be the original separating set and A the negligible set. If $t \in T$, let $t_n \in T_0$, $t_n \to t$. By hypothesis, $X(t_n) \to X(t)$ in probability, hence there is a subsequence with $X(t_{n_k},\ \omega) \to X(t,\ \omega)$ for almost every ω, say for $\omega \notin A_t$. Let $B = A \cup \{A_t : t \in T_0'\}$, and let $\omega \notin B$, $t_0 \in T$. There is a sequence of points $t_n' \in T_0'$ with $t_n' \to t_0$ and $X(t_n',\ \omega) \to X(t_0,\ \omega)$. But $\omega \notin A_{t_n'}$, hence we can find $t_n'' \in T_0$ such that $|t_n' - t_n''| < 1/n$ and $|X(t_n',\ \omega) - X(t_n'',\ \omega)| < 1/n$. Thus $t_n'' \to t_0$ and $X(t_n'',\ \omega) \to X(t_0,\ \omega)$. ‖

We now apply the separability concept to sample function analysis.

4.1.7 Theorem

Let $\{X(t),\ t \in T\}$ be separable, with separating set T_0 and negligible set A. If $\omega \notin A$, t_0 is a cluster point of T, and $\lim_{t \to t_0,\ t \in T_0} X(t, \omega)$ exists, then $\lim_{t \to t_0,\ t \in T} X(t, \omega)$ exists, and the two limits are equal. (The result holds whether deleted or nondeleted neighborhoods are used in the definition of limit.)

PROOF If $\lim_{t \to t_0,\ t \in T} X(t, \omega)$ fails to exist, we can find sequences $t_n \to t_0$, $t'_n \to t_0$ and an $\varepsilon > 0$ such that $d(X(t_n, \omega), X(t'_n, \omega)) \geq \varepsilon$ for all n. But by separability we can find $u_n, u'_n \in T_0$ with $|u_n - t_n|$, $|u'_n - t'_n| < 1/n$ and $d(X(t_n, \omega), X(u_n, \omega))$, $d(X(t'_n, \omega), X(u'_n, \omega)) < 1/n$. Thus $u_n, u'_n \to t_0$ and $d(X(u_n, \omega), X(u'_n, \omega)) \geq \varepsilon/2$ for large n, contradicting the hypothesis. Since T_0 is a dense subset of T, the two limits must be equal. ‖

We may now establish a very convenient sufficient condition for sample function continuity. The result will be used in Section 4.3 to show that a separable Brownian motion process has continuous sample functions.

4.1.8 Theorem

Let $\{X(t),\ a \leq t \leq b\}$ be real and separable. Suppose that for some r, c, $\varepsilon > 0$ and all sufficiently small $h > 0$ we have

$$E[\,|X(t + h) - X(t)|^r] \leq ch^{1+\varepsilon} \qquad \text{for all}\quad t.$$

Then almost every sample function is continuous, in other words, for almost all ω, $X(\cdot, \omega)$ is continuous on $[a, b]$.

PROOF We may assume $a = 0$, $b = 1$. (Consider $Y(t) = X(a + (b - a)t)$.) By Chebyshev's inequality,

$$P\{\,|X(t + h) - X(t)| > h^k\} \leq h^{-rk}E[\,|X(t + h) - X(t)|^r]$$

$$\leq ch^{1+\varepsilon-rk}$$

$$\to 0 \qquad \text{as}\quad h \to 0 \tag{1}$$

if k is a positive number chosen so that $\varepsilon - rk > 0$. In particular, the process is continuous in probability, so by 4.1.6, T_0 may be taken as the dyadic rationals $\{j/2^n : j = 0, 1, \ldots, 2^n,\ n = 1, 2, \ldots\}$. Now

$$P\left\{ \max_{0 \leq j \leq 2^n - 1} \left| X\left(\frac{j+1}{2^n}\right) - X\left(\frac{j}{2^n}\right) \right| \geq \frac{1}{2^{nk}} \right\}$$

$$\leq \sum_{j=0}^{2^n - 1} P\left\{ \left| X\left(\frac{j+1}{2^n}\right) - X\left(\frac{j}{2^n}\right) \right| \geq \frac{1}{2^{nk}} \right\}$$

$$\leq 2^n c 2^{-n(1+\varepsilon-rk)} \qquad \text{by (1)}$$

$$= c 2^{-n(\varepsilon-rk)}.$$

By the Borel–Cantelli lemma, there is a set B of probability 0 such that for $\omega \notin B$ we have, for all $n \geq N = N(\omega)$,

$$\left| X\left(\frac{j+1}{2^n}, \omega\right) - X\left(\frac{j}{2^n}, \omega\right) \right| < \frac{1}{2^{nk}}, \qquad j = 0, 1, \ldots, 2^n - 1. \qquad (2)$$

Fix $\omega \notin B$ and let s be a dyadic rational in $[j2^{-n}, (j+1)2^{-n})$. Then $s = j2^{-n} + a_1 2^{-(n+1)} + \cdots + a_m 2^{-(n+m)}$ where the $a_j = 0$ or 1. If

$$b_r = j2^{-n} + a_1 2^{-(n+1)} + \cdots + a_r 2^{-(n+r)} \qquad (b_0 = j2^{-n}, b_m = s),$$

then

$$\left| X(s, \omega) - X(j2^{-n}, \omega) \right| \leq \sum_{r=0}^{m-1} \left| X(b_{r+1}, \omega) - X(b_r, \omega) \right|.$$

Now $[b_r, b_{r+1})$ is empty if $a_{r+1} = 0$, and is of the form $[l2^{-(n+r+1)}, (l+1)2^{-(n+r+1)})$ if $a_{r+1} = 1$. Thus by (2),

$$\left| X(s, \omega) - X(j2^{-n}, \omega) \right| \leq \sum_{r=0}^{m-1} 2^{-(n+r+1)k}$$

$$\leq 2^{-nk} \sum_{r=0}^{\infty} 2^{-(r+1)k} \leq M2^{-nk} \qquad (3)$$

where M can be assumed to be at least 1.

If $\delta > 0$ is given, choose N_1 so that $M2^{-nk} < \delta/3$ for $n \geq N_1$; since $M \geq 1$, we also have $2^{-nk} < \delta/3$, $n \geq N_1$. If $t_1, t_2 \in T_0$, $|t_1 - t_2| <$ $\min(2^{-N_1}, 2^{-N(\omega)})$, then at most one dyadic rational of rank $n = \max(N_1, N(\omega))$ can lie between t_1 and t_2. By (2) and (3), $|X(t_1, \omega) - X(t_2, \omega)| < \delta$.

Thus almost every sample function is uniformly continuous on T_0, and hence has a continuous extension to $T = [0, 1]$. By 4.1.7, the extension must coincide with the original function; therefore almost every sample function is continuous on T. ∥

A basic property of separable processes is that many sets whose definitions involve uncountably many values of the parameter t become measurable. For example, if $\{X(t), t \in T\}$ is separable, then $\{\omega : X(\cdot, \omega)$ is continuous at $t_0\}$, $t_0 \in T$, and $\{\omega : X(\cdot, \omega)$ is uniformly continuous on $T\}$ are measurable sets. Also, if the process is extended real, then for almost every ω, $\sup\{X(t, \omega) : t \in T\}$ is a measurable function of ω (see Problems 1 and 2).

In many situations, a process $\{X(t)\}$ is constructed by specifying the finite-dimensional distributions and applying the Kolmogorov extension theorem. Although $\{X(t)\}$ need not be separable, we now show that there is always a separable process $\{Y(t)\}$, defined on the same probability space, such that $\{Y(t)\}$ is a *standard modification* of $\{X(t)\}$, that is, for each t, $X(t) = Y(t)$ a.e. In particular, the processes have the same finite-dimensional distributions.

4.1.9   Separability Theorem

Let $\{X(t), t \in T\}$ be a stochastic process with state space $(S, \mathscr{S})$, where S is a compact metric space and $\mathscr{S} = \mathscr{B}(S)$. Assume $T \subset R$ (or more generally, let T be a subset of a separable metric space). Then $\{X(t), t \in T\}$ has a separable standard modification.

PROOF   We break the proof into several parts.

(a)   There is a countable set $T_0 \subset T$ and, for each $t \in T$, a set A_t of probability 0 such that if $\omega \notin A_t$, then $X(t, \omega)$ belongs to the closure of $\{X(t', \omega) : t' \in T_0\}$.

To prove this, observe that since S is second countable, there is a sequence of compact sets $K_n \subset S$ such that every compact subset of S is an intersection of some of the K_n. Let

$$\lambda_n = \inf_{\substack{t_1, \ldots, t_r \in T \\ r = 1, 2, \ldots}} P\{X(t_i) \in K_n, 1 \leq i \leq r\}.$$

By considering a sequence of approximations to the inf and taking the union of the corresponding sets $\{t_1, \ldots, t_r\}$, we find a countable set $T_n \subset T$ such that

$$\lambda_n = P\{X(t) \in K_n \text{ for all } t \in T_n\}.$$

Fix t, and let $A_n(t) = \{X(t') \in K_n \text{ for all } t' \in T_n, X(t) \notin K_n\}$. Then $P(A_n(t)) = 0$; for if not,

$$P\{X(t') \in K_n \text{ for all } t' \in T_n\}$$
$$> P\{X(t') \in K_n \text{ for all } t' \in T_n, X(t) \in K_n\},$$

and we may replace T_n by $T_n \cup \{t\}$ to contradict the definition of λ_n.

By definition of $A_n(t)$, if $\omega \notin A_n(t)$, then

$$X(t', \omega) \in K_n \quad \text{for all } t' \in T_n \qquad \text{implies} \qquad X(t, \omega) \in K_n. \qquad (1)$$

Now set $A_t = \bigcup_{n=1}^{\infty} A_n(t)$, $T_0 = \bigcup_{n=1}^{\infty} T_n$. If K is any compact set, we claim that if $\omega \notin A_t$,

$$X(t', \omega) \in K \quad \text{for all } t' \in T_0 \qquad \text{implies} \qquad X(t, \omega) \in K. \qquad (2)$$

For K can be expressed as $\bigcap_{j=1}^{\infty} K_{n_j}$; thus if $\omega \notin A_t$, $X(t', \omega) \in K$ for all $t' \in T_0$ implies $X(t', \omega) \in K_{n_j}$ for all $t' \in T_0$, so $X(t, \omega) \in K_{n_j}$ by (1). Since this holds for any j, we have $X(t, \omega) \in K$.

Finally, if $\omega \notin A_t$, set $K = \overline{\{X(t', \omega) : t' \in T_0\}}$. Then $X(t', \omega) \in K$ for all $t' \in T_0$, and hence by (2), $X(t, \omega) \in K$.

(b)   The process $\{X(t), t \in T \subset R\}$ is separable iff there is a set A of probability 0 and a countable dense set $T_0 \subset T$ such that if $\omega \notin A$,

$$\{(t, X(t, \omega)) : t \in T \cap I\} \subset \overline{\{(t', X(t', \omega)) : t' \in T_0 \cap I\}}$$

for all open intervals $I \subset R$. (If T is a subset of the separable metric space M, I is replaced by an open subset of M.)

Assume this condition is satisfied, and let K be a compact subset of S, I an open interval of R. Assume that $\omega \notin A$ and $X(t', \omega) \in K$ for all $t' \in T_0 \cap I$. Then

$$\{(t, X(t, \omega)) : t \in T \cap I\} \subset \overline{\{(t', X(t', \omega)) : t' \in T_0 \cap I\}}$$

$$\subset \overline{I \times K} = \overline{I} \times \overline{K} = \overline{I} \times K.$$

Therefore $X(t, \omega) \in K$ for all $t \in T \cap I$. By 4.1.3, the process is separable.

Conversely, assume the process separable, with separating set T_0 and negligible set A. Let I be an open interval of R, and let $t \in T \cap I$, $t \notin T_0$ (if $t \in T_0$, there is nothing to prove). If $\omega \notin A$, we can find a sequence of points $t_n \in T_0 \cap I$ with $t_n \to t$ and $X(t_n, \omega) \to X(t, \omega)$. Therefore

$$(t_n, X(t_n, \omega)) \to (t, X(t, \omega)).$$

Consequently, $(t, X(t, \omega)) \in \overline{\{(t', X(t', \omega)) : t' \in T_0 \cap I\}}$.

(c) PROOF OF THE SEPARABILITY THEOREM We must produce a separable standard modification of $\{X(t), t \in T\}$. Let J be an open interval with rational endpoints. By (a) there is a countable set $T(J) \subset T \cap J$ and sets $A_t(J)$ of probability 0 $(t \in T \cap J)$ such that if $\omega \notin A_t(J)$, then $X(t, \omega)$ belongs to the closure of $\{X(t', \omega) : t' \in T(J)\}$. (If T is a subset of the separable metric space M, let D be a countable dense subset of M, and take the sets J to be open balls with rational radii and centers in D.) Define

$$A_t = \bigcup_J A_t(J), \qquad T_0' = \bigcup_J T(J),$$

and let T_0 be the union of T_0' and a fixed countable dense subset of T; thus T_0 is also a countable dense subset of T.

Now if $t \in T \cap J$ and $\omega \notin A_t$, then $\omega \notin A_t(J)$, so

$$X(t, \omega) \in \overline{\{X(t', \omega) : t' \in T(J)\}}$$

$$\subset \overline{\{X(t', \omega) : t' \in T_0 \cap J\}} = K(J, \omega). \tag{3}$$

Thus if $t \in T$ and $\omega \notin A_t$, we have $X(t, \omega) \in K(J, \omega)$ for all $J \supset \{t\}$, and hence

$$X(t, \omega) \in \bigcap_{J \supset \{t\}} K(J, \omega) = K(t, \omega),$$

a nonempty compact set.

We define a new process $\{Y(t), t \in T\}$ as follows. If $\omega \notin A_t$, take $Y(t, \omega) = X(t, \omega)$. If $\omega \in A_t$, let $Y(t, \omega)$ be any point of $K(t, \omega)$. Take $A = \bigcup_{t \in T_0} A_t$, where T_0 is as above. Since $P(A_t) = 0$ for all t, $\{Y(t)\}$ is a standard modification of $\{X(t)\}$. To show $\{Y(t)\}$ separable, we use (b). Let $\omega \notin A$, and let I be an open interval with $t \in T \cap I$. Then $t \in T \cap J$ for some rational interval $J \subset I$. If $\omega \notin A_t$, then

$$Y(t, \omega) = X(t, \omega)$$

$$\in K(J, \omega) = \overline{\{X(t', \omega) : t' \in T_0 \cap J\}} \qquad \text{by (3)}$$

$$\subset \overline{\{X(t', \omega) : t' \in T_0 \cap I\}}$$

$$= \overline{\{Y(t', \omega) : t' \in T_0 \cap I\}}$$

since $\omega \notin A_{t'}$ for $t' \in T_0$. If $\omega \in A_t$, then

$$Y(t, \omega) \in K(t, \omega) \subset K(J, \omega) \subset \overline{\{Y(t', \omega) : t' \in T_0 \cap I\}},$$

as above. If we note that the interval J can be chosen arbitrarily small, we see by (b) that $\{Y(t)\}$ is separable. $\|$

If the original process $\{X(t)\}$ is real, we may take S to be the compact space $\bar{R}$. Thus $Y(t)$ may take on the values $+\infty$ or $-\infty$; however, since for each t, $Y(t) = X(t)$ a.e., $Y(t)$ is almost surely finite for a fixed t.

Problems

1.   If $\{X(t), t \in T\}$ is a separable process, show that $\{\omega : X(\cdot, \omega)$ is continuous at $t_0\}$, $t_0 \in T$, and $\{\omega : X(\cdot, \omega)$ is uniformly continuous on $T\}$ are measurable sets, that is, belong to the completion of the σ-field $\mathscr{F}$ relative to the probability measure P.

2.   If $\{X(t), t \in T\}$ is an extended real separable process, show that for almost every ω, $\sup \{X(t, \omega) : t \in T\}$ (and similarly $\inf \{X(t, \omega) : t \in T\}$) is Borel measurable relative to the completion of $(\Omega, \mathscr{F}, P)$. Also, if t_0 is not an isolated point of T, show that for almost every ω, $\lim \sup_{t \to t_0} X(t, \omega)$ (and similarly $\lim \inf_{t \to t_0} X(t, \omega)$) is Borel measurable.

COMMENT   In Problems 1 and 2, if we redefine the process so that $X(t, \omega) \equiv 0$ for ω in the negligible set A, we need not complete the probability space $(\Omega, \mathscr{F}, P)$.

3.   Give an example of a separable process $\{X(t), t \geq 0\}$ with both of the following properties:

(a)   For each *fixed* t, $X(s) \to X(t)$ a.e. as $s \to t$, that is, for each t there is a set A_t of probability 0 such that if $\omega \notin A_t$, $\lim_{s \to t} X(s, \omega) = X(t, \omega)$.

(b) $P\{\omega : X(\cdot, \omega)$ is continuous on $[0, \infty)\} = 0$.

4.2 Measurability

As we have seen, separability allows us to handle events and random variables that involve uncountably many values of the parameter t of a stochastic process. However, if we wish to integrate the sample functions, a new concept is needed. Suppose that $\{X(t), t \in I\}$, I an interval of R, is a real stochastic process, and we ask whether almost every sample function is Lebesgue integrable on I. The problem can be attacked using Fubini's theorem:

$$\int_\Omega \int_I |X(t, \omega)| \, dt \, dP(\omega) = \int_I \int_\Omega |X(t, \omega)| \, dP(\omega) \, dt$$

$$= \int_I E[\,|X(t)|\,] \, dt.$$

Thus if $\int_I E[\,|X(t)|\,] \, dt < \infty$, then $\int_I |X(t, \omega)| \, dt$ will be finite for almost every ω, as desired.

The difficulty with this argument is that in order to apply Fubini's theorem, $X(t, \omega)$ should be measurable in *both* variables t and ω; thus the problem is to construct a measurable standard modification, retaining separability if possible.

In investigating the measurability problem, we shall for simplicity consider extended real processes, with T taken as $[0, \infty)$. This will cover all specific examples to be treated later in the chapter. Also, with little extra effort we can develop the stronger property of progressive measurability, which is useful in the general theory of Markov processes.

4.2.1 Definitions and Comments

Let $\{X(t), t \geq 0\}$ be a stochastic process; assume that the process is *adapted* to the family of σ-fields $\mathcal{F}(t)$, $t \geq 0$, that is, $s < t$ implies $\mathcal{F}(s) \subset \mathcal{F}(t)$, and $X(t)$ is $\mathcal{F}(t)$-measurable for each t. (All $\mathcal{F}(t)$ are assumed to be sub σ-fields of $\mathcal{F}$.) If no specific $\mathcal{F}(t)$ is mentioned, we take $\mathcal{F}(t)$ to be $\mathcal{F}(X(s), s \leq t)$, the smallest σ-field of subsets of Ω that makes each $X(s), s \leq t$, Borel measurable. The process is said to be *progressively measurable* iff for all $t > 0$ the map $(s, \omega) \to X(s, \omega)$ of $[0, t] \times \Omega \to \bar{R}$ is measurable relative to $\mathcal{B}[0, t] \times \mathcal{F}(t)$ and $\mathcal{B}(\bar{R})$.

A progressively measurable process is *measurable*, that is, the map $(s, \omega) \to X(s, \omega)$ is measurable: $([0, \infty) \times \Omega, \mathcal{B}[0, \infty) \times \mathcal{F}) \to (\bar{R}, \mathcal{B}(\bar{R}))$; to see this, observe that if $B \in \mathcal{B}(\bar{R})$, we have

$$\{(s, \omega) : X(s, \omega) \in B\} = \bigcup_{n=0}^{\infty} \{(s, \omega) : 0 \leq s \leq n, X(s, \omega) \in B\}$$

and

$$\{(s, \omega) : 0 \le s \le n, X(s, \omega) \in B\} \in \mathcal{B}[0, n] \times \mathcal{F}(n)$$

by definition of progressive measurability. Since $\mathcal{B}[0, n] \times \mathcal{F}(n) \subset \mathcal{B}[0, \infty) \times \mathcal{F}$, the result follows.

The sample functions of a measurable process are Borel measurable functions. Furthermore, if $\int_I E[\,|X(t)|\,]\,dt < \infty$, then almost all sample functions are Lebesgue integrable on I; the above argument using Fubini's theorem is now valid.

If $\{X(t), t \ge 0\}$ has state space $(S, \mathcal{S})$, the same definitions of progressive measurability and measurability may be used, with $(S, \mathcal{S})$ replacing $(\bar{R}, \mathcal{B}(\bar{R}))$.

Unlike the separability theorem, the theorem on progressively measurable standard modifications has a hypothesis of continuity in probability (see 4.1.6). This is, however, a relatively mild condition; it will be satisfied, for example, by any L^2 process with a continuous covariance.

4.2.2 Progressive Measurability Theorem

Let $\{X(t), t \ge 0\}$ be a real stochastic process adapted to the σ-fields $\mathcal{F}(t)$. If the process is continuous in probability, there is a standard modification $\{Y(t), t \ge 0\}$, also adapted to the $\mathcal{F}(t)$, that is progressively measurable and separable.

PROOF   Let M be the space of random variables on $(\Omega, \mathcal{F}, P)$, and let $d(X, Y) = E[\min(\,|X - Y|, 1)]$, $X, Y \in M$. Then (if functions that agree a.e. are identified), d is a metric on M and d-convergence is equivalent to convergence in probability (see Section 1.1, Problem 8).

By the continuity in probability hypothesis, the map $t \to X(t)$ of $[0, \infty)$ into M is continuous (relative to d on M). For each positive integer n, the map is uniformly continuous on $[0, n]$; hence for some $\delta_n > 0$ we have, for t, $t' \in [0, n]$, $d(X(t), X(t')) \le 2^{-n}$ whenever $|t - t'| \le \delta_n$. We may assume that the δ_n decrease and approach 0. Construct partitions $0 = t_0^{(n)} < t_1^{(n)} < \cdots < t_{a_n}^{(n)} = n$ such that

$$\max_{0 \le j \le a_n - 1} |t_{j+1}^{(n)} - t_j^{(n)}| \le \delta_n,$$

with the $t_j^{(n+1)}$ refining the $t_j^{(n)}$. Define

$$X_n(t) = X(t_{j-1}^{(n)}) \quad \text{if} \quad t_{j-1}^{(n)} \le t < t_j^{(n)}, \quad j = 1, 2, \ldots, a_n,$$

$$X_n(t) = X(n), \quad t \ge n.$$

We divide the remainder of the proof into several parts.

(a) $d(X_n(t), X_{n+1}(t)) \leq 2^{-n}$ for all $t < n$.

If $t < n$, then for some j and k we have

$$t^{(n)}_{j-1} \leq t < t^{(n)}_j, \qquad t^{(n+1)}_{k-1} \leq t < t^{(n+1)}_k.$$

Necessarily $\left| t^{(n)}_{j-1} - t^{(n+1)}_{k-1} \right| \leq \delta_n$, and the result follows.

(b) For each t, $X_n(t)$ converges a.e.

To prove this, write

$$P\{ |X_n(t) - X_{n+1}(t)| \geq n^{-2} \}$$
$$= P\{ |X_n(t) - X_{n+1}(t)| \wedge 1 \geq n^{-2} \}$$
$$\leq n^2 \, d(X_n(t), X_{n+1}(t)) \qquad \text{by Chebyshev's inequality}$$
$$\leq n^2 2^{-n}$$

if $n > t$, by (a). By the Borel–Cantelli lemma, with probability one we have $|X_n(t) - X_{n+1}(t)| < 1/n^2$ for sufficiently large n, and the result follows.

(c) The map $(s, \omega) \to X_n(s, \omega)$ of $[0, t] \times \Omega \to \bar{R}$ is $\mathscr{B}[0, t] \times \mathscr{F}(t)$-measurable for $n > t$.

We may express $X_n(s, \omega)$ as follows:

$$X_n(s, \omega) = \sum_{j=1}^{a_n} X(t^{(n)}_{j-1}, \omega) I_{[t_{j-1}^{(n)}, t_j) \times \Omega}(s, \omega)$$
$$+ X(n, \omega) I_{[n, \infty) \times \Omega}(s, \omega)$$

If we restrict s to $[0, t]$, the series is truncated, the last term being

$$X(t^{(n)}_{j-1}, \omega) I_{[t_{j-1}^{(n)}, t] \times \Omega}(s, \omega) \qquad \text{for} \quad t^{(n)}_{j-1} \leq t < t^{(n)}_j.$$

Since $X(t^{(n)}_{j-1}, \omega)$, as a function of ω, is $\mathscr{F}(t^{(n)}_{j-1}) \subset \mathscr{F}(t)$-measurable, the result follows.

(d) Define $Y(t, \omega) = \limsup_{n \to \infty} X_n(t, \omega)$, $\omega \in \Omega$, $t \geq 0$. Then $\{Y(t)\}$ is a standard modification of $\{X(t)\}$.

If $t^{(n)}_{j-1} \leq t < t^{(n)}_j$, we have $t^{(n)}_{j-1} \to t$, hence by hypothesis, $X_n(t) = X(t^{(n)}_{j-1}) \to X(t)$ in probability. By (b), $X_n(t) \to Y(t)$ a.e. But then $Y(t) = X(t)$ a.e.

Note that as in the separability theorem, the $Y(t)$ may take on the values $\pm \infty$; but since $Y(t) = X(t)$ a.e., $Y(t)$ is almost surely finite for a fixed t.

(e) $\{Y(t), t \geq 0\}$ is progressively measurable.

By (c) and the definition of $\{Y(t)\}$, the map $(s, \omega) \to Y(s, \omega)$ of $[0, t] \times \Omega \to \bar{R}$ is a $\limsup$ of measurable mappings relative to $\mathscr{B}[0, t] \times \mathscr{F}(t)$.

(f) $\{Y(t), t \geq 0\}$ is separable.

Let T_0 consist of all $t_j^{(n)}$, $n = 1, 2, \ldots, j = 1, 2, \ldots, a_n$. Then $Y(t, \omega) = \lim \sup_{n \to \infty} X_n(t, \omega)$ and (for t fixed, n sufficiently large), $X_n(t, \omega) = X(t_{j-1}^{(n)}, \omega)$ for some $j = j(n)$, where $t_{j-1}^{(n)} \leq t < t_j^{(n)}$. By definition of $Y(t, \omega)$, we can find a subsequence with

$$X(t_{j-1}^{(n_k)}, \omega) \to Y(t, \omega) \qquad (j = j(n_k)).$$

But $X(t_{j-1}^{(n_k)}, \omega) = Y(t_{j-1}^{(n_k)}, \omega)$ because if s is one of the partitioning points, then $X_n(s) = X(s)$ for large n, hence $Y(s) = X(s)$. Since $t_{j-1}^{(n_k)} \to t$, the separability condition is satisfied (with the negligible set A actually empty). The theorem is proved. ‖

The concept of a stopping time has already been introduced in connection with martingale theory (see RAP, 7.7, p. 302ff.). At that time we were concerned only with discrete parameter processes, and it will be convenient now to reformulate the definition in the continuous parameter case, and discuss a connection with progressive measurability.

4.2.3 Definitions

Let $\{\mathscr{F}(t), t \geq 0\}$ be an increasing family of sub σ-fields of $\mathscr{F}$, that is, $\mathscr{F}(s) \subset \mathscr{F}(t)$ for $s \leq t$. A *stopping time* for the $\mathscr{F}(t)$ is a map $T: \Omega \to [0, \infty]$ such that $\{T \leq t\} \in \mathscr{F}(t)$ for each $t \geq 0$. A stopping time for a stochastic process $\{X(t), t \geq 0\}$ is a stopping time for the σ-fields $\mathscr{F}(t) = \mathscr{F}(X(s), s \leq t)$.

If $A \in \mathscr{F}$, we say that A is *prior to* T iff $A \cap \{T \leq t\} \in \mathscr{F}(t)$ for all $t \geq 0$. If T is a stopping time for a stochastic process, then, intuitively, if $T \leq t$ we can tell by examination of the $X(s)$, $s \leq t$, whether or not A has occurred. The collection of all sets prior to T will be denoted by $\mathscr{F}(T)$; it follows quickly that $\mathscr{F}(T)$ is a σ-field.

The following properties may be developed from the definitions; all stopping times are relative to a fixed increasing family of σ-fields.

4.2.4 Theorem

(a) If S and T are stopping times, so are $S \wedge T = \min (S, T)$ and $S \vee T = \max (S, T)$. In particular, if $t \geq 0$ and T is a stopping time, so is $T \wedge t$.

(b) If T is a stopping time, then $T: (\Omega, \mathscr{F}(T)) \to ([0, \infty], \mathscr{B}[0, \infty])$; for short, T is $\mathscr{F}(T)$-measurable.

(c)   Let T be a stopping time, and S a nonnegative random variable, with $S \geq T$. If S is $\mathscr{F}(T)$-measurable, then S is a stopping time.

(d)   If S and T are stopping times and $A \in \mathscr{F}(S)$, then

$$A \cap \{S \leq T\} \in \mathscr{F}(T).$$

(e)   If S and T are stopping times and $S \leq T$, then $\mathscr{F}(S) \subset \mathscr{F}(T)$.

PROOF   (a)   If $t \geq 0$, $\{S \wedge T \leq t\} = \{S \leq t\} \cup \{T \leq t\} \in \mathscr{F}(t)$,

$$\{S \vee T \leq t\} = \{S \leq t\} \cap \{T \leq t\} \in \mathscr{F}(t).$$

(b)   If r is a real number, then $\{T \leq r\} \cap \{T \leq t\} = \{T \leq r \wedge t\} \in \mathscr{F}(r \wedge t) \subset \mathscr{F}(t)$. Thus $\{T \leq r\} \in \mathscr{F}(T)$.

(c)   If $t \geq 0$ we have $\{S \leq t\} = \{S \leq t\} \cap \{T \leq t\}$. Since S is $\mathscr{F}(T)$-measurable, $\{S \leq t\} \in \mathscr{F}(T)$, hence

$$\{S \leq t\} \cap \{T \leq t\} \in \mathscr{F}(t).$$

(d)   We have

$$A \cap \{S \leq T\} \cap \{T \leq t\} = A \cap \{S \leq t\} \cap \{T \leq t\} \cap \{S \wedge t \leq T \wedge t\}.$$

But $A \cap \{S \leq t\} \in \mathscr{F}(t)$ since $A \in \mathscr{F}(S)$, and $\{T \leq t\} \in \mathscr{F}(t)$. Also,

$$\{T \wedge t \leq r\} = \{T \wedge t \leq r \wedge t\} \in \mathscr{F}(r \wedge t) \subset \mathscr{F}(t),$$

and similarly $\{S \wedge t \leq r\} \in \mathscr{F}(t)$; hence $S \wedge t$ and $T \wedge t$ are $\mathscr{F}(t)$-measurable. It follows that $\{S \wedge t \leq T \wedge t\} \in \mathscr{F}(t)$. We conclude that

$$A \cap \{S \leq T\} \in \mathscr{F}(T).$$

(e)   If $A \in \mathscr{F}(S)$, then $A = A \cap \Omega = A \cap \{S \leq T\} \in \mathscr{F}(T)$ by (d). $\parallel$

If $\{X(t), t \geq 0\}$ is a stochastic process adapted to $\{\mathscr{F}(t), t \geq 0\}$, and T is a finite stopping time for the $\mathscr{F}(t)$, it is natural to consider the value $X(T)$ of the process when stopping occurs;

if $T(\omega) = t$, define $X(T)(\omega) = X(t, \omega)$.

It would be desirable that $X(T)$ be a random variable; for a progressively measurable process, more is true.

4.2.5   Theorem

Let $\{X(t), t \geq 0\}$ be a progressively measurable process adapted to the σ-fields $\mathscr{F}(t)$, $t \geq 0$. If T is a finite stopping time for the $\mathscr{F}(t)$, then $X(T)$ is $\mathscr{F}(T)$-measurable, that is, $X(T)$: $(\Omega, \mathscr{F}(T)) \to (\bar{R}, \mathscr{B}(\bar{R}))$.

PROOF   If $B \in \mathscr{B}(\bar{R})$, we must show that $\{X(T) \in B\} \in \mathscr{F}(T)$, that is, $\{X(T) \in B\} \cap \{T \le t\} \in \mathscr{F}(t)$ for each $t \ge 0$. Since

$$\{X(T) \in B\} \cap \{T \le t\} = \{X(T \wedge t) \in B\} \cap \{T \le t\},$$

it suffices to show that $\{X(T \wedge t) \in B\} \in \mathscr{F}(t)$ for each t, in other words, $X(T \wedge t)$ is $\mathscr{F}(t)$-measurable. But $X(T \wedge t)$ is the composition of the map $\omega \to ((T \wedge t)(\omega), \ \omega)$, which is measurable: $(\Omega, \mathscr{F}(t)) \to ([0, t] \times \Omega, \mathscr{B}[0, t] \times \mathscr{F}(t))$, and the map $(s, \omega) \to X(s, \omega)$, which is measurable: $([0, t] \times \Omega, \mathscr{B}[0, t] \times \mathscr{F}(t)) \to (\bar{R}, \mathscr{B}(\bar{R}))$ by the progressive measurability hypothesis. The result follows.   ‖

If progressive measurability is replaced by the weaker hypothesis of measurability, we can prove only that $X(T)$ is a random variable. To do this, note that $X(T)$ is the composition of the map $\omega \to (T(\omega), \omega)$, which is measurable: $(\Omega, \mathscr{F}) \to ([0, \infty) \times \Omega, \mathscr{B}[0, \infty) \times \mathscr{F})$, and the map $(s, \omega) \to X(s, \omega)$, which is measurable: $([0, \infty) \times \Omega, \mathscr{B}[0, \infty) \times \mathscr{F}) \to (\bar{R}, \mathscr{B}(\bar{R}))$.

Theorem 4.2.5 holds for a progressively measurable process with an arbitrary state space $(S, \mathscr{S})$; the proof is the same, with $(S, \mathscr{S})$ replacing $(\bar{R}, \mathscr{B}(\bar{R}))$.

Problems

1.   Give an example of a real stochastic process that is not measurable.

2.   If $\{X(t), t \ge 0\}$ is a process adapted to the σ-fields $\{\mathscr{F}(t), t \ge 0\}$, with right-continuous sample functions, show that the process is progressively measurable.

3.   Let S and T be stopping times for the σ-fields $\mathscr{F}(t)$, $t \ge 0$. Establish the following:

(a) $S + T$ is a stopping time.
(b) $\mathscr{F}(S \wedge T) = \mathscr{F}(S) \cap \mathscr{F}(T)$.
(c) $\{S > T\}, \{S < T\}$, and $\{S = T\}$ belong to $\mathscr{F}(S) \cap \mathscr{F}(T)$.

4.   Let $\{X(t), t \ge 0\}$ be a stochastic process adapted to the family of σ-fields $\mathscr{F}(t)$. Assume that the state space S is a metric space, and $\mathscr{S} = \mathscr{B}(S)$. If $B \in \mathscr{S}$, define the *hitting time* of B as

$$T(\omega) = \inf \{t \ge 0 : X(t, \omega) \in B\}$$

(if $X(t, \omega)$ is never in B, take $T(\omega) = \infty$).

If the sample functions are continuous and B is either open or closed, show that T is a stopping time for the σ-fields $\mathscr{F}(t^+) = \bigcap_{\delta > 0} \mathscr{F}(t + \delta)$.

4.3   One-Dimensional Brownian Motion

In this section, we analyze the one-dimensional Brownian motion process, that is, the process $\{B(t), t \geq 0\}$ that is Gaussian with zero mean and covariance function $K(s, t) = \sigma^2 \min (s, t)$. In 1.2.9, we saw that $B(t)$ can be regarded as the limiting case of a random walk as the size of each step and the time between steps approach 0. However, Brownian motion has many unusual properties not exhibited by a random walk.

By the separability theorem, there is always a separable process having a specified consistent set of finite-dimensional distributions; in particular, there is a separable version of Brownian motion. We shall restrict ourselves to this case; from now on $\{B(t), t \geq 0\}$ *will always denote separable Brownian motion.*

We first prove that almost all sample functions are continuous.

4.3.1   Theorem

For almost every ω, $B(\cdot, \omega)$ is continuous on $[0, \infty)$.

PROOF   We use 4.1.8; since $B(t + h) - B(t)$ is normally distributed with mean 0 and variance $\sigma^2 h$, we have

$$E[\,|B(t + h) - B(t)|^r] = E\left[\left|\frac{B(t + h) - B(t)}{\sigma\sqrt{h}}\right|^r\right]\sigma^r h^{r/2} = ch^{r/2},$$

where $c = \sigma^r E(\,|Z|^r)$, Z normal $(0, 1)$. By 4.1.8 (with r taken as any number greater than 2), for almost all ω, the sample function $B(\cdot, \omega)$ is continuous on $[0, n]$, where n is an arbitrary positive integer. Thus for almost all ω, $B(\cdot, \omega)$ is continuous on $[0, \infty)$. $\|$

The next result is the key to many properties of the Brownian sample functions; it may be regarded as an application of a continuous version of the reflection principle for random walks (see Ash, 1970, p. 188).

4.3.2   Theorem

If $a > 0$,

$$P\left\{\max_{0 \leq s \leq t} B(s) > a\right\} = 2P\{B(t) > a\}.$$

PROOF   We first argue intuitively. We have

$$P\left\{\max_{0 \le s \le t} B(s) > a,\, B(t) > a\right\} = P\left\{\max_{0 \le s \le t} B(s) > a,\, B(t) < a\right\} \tag{1}$$

since if $B(s) = a$ for some $s < t$, reflection about the horizontal line at a gives a one-to-one correspondence between paths with $B(t) > a$ and paths with $B(t) < a$ (see Figure 4.1). This argument is not easy to formalize; what is

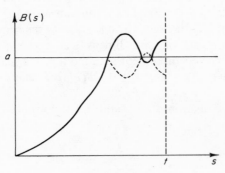

FIGURE 4.1

really involved here is the strong Markov property. The development of this concept takes considerable time and will be postponed until Section 4.8. (A direct proof of (1) is given in Problems 1–3.)

Now the sum of the two terms of (1) is $P\{\max_{0 \le s \le t} B(s) > a\}$ since $\{B(t) = a\}$ has probability 0. Thus

$$P\left\{\max_{0 \le s \le t} B(s) > a,\, B(t) > a\right\} = \tfrac{1}{2} P\left\{\max_{0 \le s \le t} B(s) > a\right\}.$$

But $\{\max_{0 \le s \le t} B(s) > a,\, B(t) > a\} = \{B(t) > a\}$, and the result follows.   ‖

The next two results yield properties of the Brownian sample functions for very large t and for very small t.

4.3.3   Theorem

$$P\left\{\sup_{t \ge 0} B(t) = +\infty\right\} = P\left\{\inf_{t \ge 0} B(t) = -\infty\right\} = 1.$$

Consequently, for almost every ω, $B(\cdot, \omega)$ is unbounded and has a zero in $[M, \infty)$ for every $M > 0$.

PROOF   If $a > 0$,

$$P\left\{\sup_{t \geq 0} B(t) > a\right\} \geq P\left\{\max_{0 \leq s \leq t} B(s) > a\right\}$$

$$= 2P\{B(t) > a\} \qquad \text{by 4.3.2}$$

$$= 2P\left\{\frac{B(t)}{\sigma\sqrt{t}} > \frac{a}{\sigma\sqrt{t}}\right\}$$

$$= 2[1 - F^*(a/\sigma\sqrt{t})]$$

where F^* is the normal $(0, 1)$ distribution function

$$\rightarrow 2[1 - F^*(0)] = 1 \qquad \text{as } t \rightarrow \infty.$$

Thus

$$P\left\{\sup_{t \geq 0} B(t) = \infty\right\} = P\left[\bigcap_{a=1}^{\infty} \left\{\sup_{t \geq 0} B(t) > a\right\}\right] = 1.$$

Now

$$P\left\{\inf_{t \geq 0} B(t) = -\infty\right\} = P\left\{\sup_{t \geq 0} [-B(t)] = \infty\right\} = 1$$

since $\{-B(t), t \geq 0\}$ is, by 4.1.4 or 4.1.5, also separable Brownian motion.   ‖

4.3.4   Theorem

If $h > 0$,

$$P\left\{\max_{0 \leq s \leq h} B(s) > 0\right\} = P\left\{\min_{0 \leq s \leq h} B(s) < 0\right\} = 1.$$

Consequently, for almost every ω, $B(\cdot, \omega)$ has a zero in $(0, h]$ for every $h > 0$.

PROOF   If $a > 0$,

$$P\left\{\max_{0 \leq s \leq h} B(s) > 0\right\} \geq P\left\{\max_{0 \leq s \leq h} B(s) > a\right\}$$

$$= 2[1 - F^*(a/\sigma\sqrt{h})] \qquad \text{as in 4.3.3}$$

$$\rightarrow 1 \qquad \text{as } a \rightarrow 0.$$

Thus $P\{\max_{0 \leq s \leq h} B(s) > 0\} = 1$; also, as in 4.3.3,

$$P\left\{\min_{0 \leq s \leq h} B(s) < 0\right\} = P\left\{\max_{0 \leq s \leq h} [-B(s)] > 0\right\} = 1.$$

The last statement of the theorem is proved by observing that

$$P\left\{ \max_{0 \le s \le h} B(s) > 0 \text{ for every } h > 0 \right\}$$

$$= P\left[\bigcap_{n=1}^{\infty} \left\{ \max_{0 \le s \le 1/n} B(s) > 0 \right\} \right] = 1,$$

and similarly,

$$P\left\{ \min_{0 \le s \le h} B(s) < 0 \text{ for every } h > 0 \right\} = 1. \quad \|$$

The following result indicates that the sample functions are perhaps not as well behaved as we might like.

4.3.5 Theorem

Almost every sample function of $\{B(t),\ t \ge 0\}$ is nowhere differentiable. Specifically, let

$$D = \{\omega : B(t, \omega) \text{ is differentiable for at least one } t \in [0, \infty)\}.$$

Then D is a subset of a set of probability 0. (Note that D need not belong to $\mathscr{F}$, but the theorem implies that D belongs to the completion of $\mathscr{F}$ relative to P.)

PROOF   Fix $k > 0$, and let

$$A = A(k) = \left\{ \omega : \limsup_{h \to 0^+} \frac{|B(t + h, \omega) - B(t, \omega)|}{h} < k \right.$$

$$\left. \text{for at least one } \quad t \in [0, 1) \right\}.$$

If $\omega \in A$, the sample function $B(\cdot, \omega)$ will lie inside the "fan" of slope k emanating from $B(t, \omega)$ (see Figure 4.2). We may take the positive integer m so large that $B(s, \omega)$ will be inside the fan for $t \le s \le (j + 3)/m$, where $(j - 1)/m \le t < j/m$. Thus if $\omega \in A$, then for sufficiently large m we have, for some integer $j = 1, 2, \ldots, m$,

$$\left| B\!\left(\frac{j + 1}{m}, \omega\right) - B\!\left(\frac{j}{m}, \omega\right) \right| \le \frac{3k}{m} \tag{1}$$

$$\left| B\!\left(\frac{j + 2}{m}, \omega\right) - B\!\left(\frac{j + 1}{m}, \omega\right) \right| \le \frac{5k}{m} \tag{2}$$

$$\left| B\!\left(\frac{j + 3}{m}, \omega\right) - B\!\left(\frac{j + 2}{m}, \omega\right) \right| \le \frac{7k}{m}. \tag{3}$$

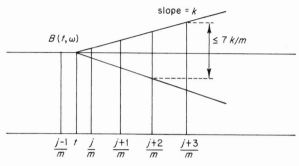

slope = k

$B(t, \omega)$

$\leq 7\,k/m$

$\dfrac{j-1}{m}\quad t \quad \dfrac{j}{m}\qquad \dfrac{j+1}{m}\qquad \dfrac{j+2}{m}\qquad \dfrac{j+3}{m}$

FIGURE 4.2

Now

$$P\{\,|B(t+h)-B(t)|\,<a\} = P\left\{\frac{|B(t+h)-B(t)|}{\sigma\sqrt{h}} < \frac{a}{\sigma\sqrt{h}}\right\}$$

$$= \int_{-a/\sigma\sqrt{h}}^{a/\sigma\sqrt{h}} \frac{1}{\sqrt{2\pi}} e^{-x^2/2}\, dx$$

$$\leq \frac{2a}{\sigma\sqrt{2\pi h}}.$$

Thus if A_{mj} is the event that inequalities (1), (2), and (3) hold, then by independence of increments,

$$P(A_{mj}) \leq \frac{8}{\sigma^3(2\pi/m)^{3/2}}\left(\frac{3k}{m}\right)\left(\frac{5k}{m}\right)\left(\frac{7k}{m}\right) = cm^{-3/2}.$$

If $A_m = \bigcup_{j=1}^{m} A_{mj}$, then $P(A_m) \leq \sum_{j=1}^{m} P(A_{mj}) \leq cm^{-1/2}$. Take $m = n^4$; by the Borel–Cantelli lemma, $\lim\sup_n A_{n^4}$ has probability 0. But

$$A \subset \lim_{m}\inf A_m \subset \lim_{n}\inf A_{n^4} \subset \lim_{n}\sup A_{n^4},$$

so if $D_0 = \{\omega : B(\cdot, \omega)$ is differentiable for at least one $t \in [0, 1)\}$, then $D_0 \subset \bigcup_{k=1}^{\infty} A(k)$, and hence D_0 is included in a set of probability 0. Since $B(t_0 + t) - B(t_0)$ is also separable Brownian motion, $D_n = \{\omega : B(\cdot, \omega)$ is differentiable for at least one $t \in [n, n+1)\}$ is included in a set of probability 0. But $D = \bigcup_{n=0}^{\infty} D_n$ and the result follows. ‖

4.3.6 Corollary

If $L > 0$, almost every sample function has infinite variation on $[0, L]$, and hence with probability one, the graph of $B(t)$, $0 \leq t \leq L$, is not rectifiable. Furthermore, for almost every ω, there is no interval (a, b), $a < b$, on which $B(\cdot, \omega)$ is monotone.

PROOF If $B(\cdot, \omega)$ were of bounded variation on $[0, L]$, it would be differentiable almost everywhere [Lebesgue measure]. Also, if $B(\cdot, \omega)$ were monotone on the interval (c, d), with c, d rational, it would be differentiable a.e. on (c, d). The result now follows from 4.3.5. ‖

The next section will be devoted to a proof of a basic sample function property, the "law of the iterated logarithm," that is, for almost every ω,

$$\limsup_{t \to \infty} \frac{B(t, \omega)}{(2\sigma^2 t \ln \ln t)^{1/2}} = 1$$

and

$$\liminf_{t \to \infty} \frac{B(t, \omega)}{(2\sigma^2 t \ln \ln t)^{1/2}} = -1.$$

Problems

1. Let $X_n = \sum_{k=1}^{n} Y_k$, where the Y_k are independent, symmetric random variables. ("Symmetric" means that $P\{Y_k \in B\} = P\{-Y_k \in B\}$ for all Borel sets B. Equivalently, by RAP, 1.5(d), p. 325, the characteristic function of Y_k is real valued.) Show that for any real number a,

$$P\left\{ \max_{1 \le i \le n} X_i > a \right\} \le 2P\{X_n > a\}.$$

(Let $A_i = \{X_1 \le a, \ldots, X_{i-1} \le a, X_i > a\}$, $1 \le i \le n$, $B = \{X_n > a\}$; show that $P(A_i \cap B) \ge \frac{1}{2}P(A_i)$.)

2. As in Problem 1, let $Y_1, Y_2, \ldots, Y_n$ be independent, symmetric random variables, with $X_n = \sum_{k=1}^{n} Y_k$. If $a > 0$, let $T = \min \{i : X_i > a\}$, $T = \infty$ if there is no such i. Establish the following, for any $\varepsilon > 0$:

(a) $P\{T \le n - 1, X_n - X_T \le -\varepsilon\} \le P\{T \le n - 1, X_n \le a\}$

$$+ \sum_{j=1}^{n-1} P\{Y_j > \varepsilon\}.$$

(b) $P\{T \le n - 1, X_n > a + 2\varepsilon\} \le P\{T \le n - 1, X_n - X_T > \varepsilon\}$

$$+ \sum_{j=1}^{n-1} P\{Y_j > \varepsilon\}.$$

(c) $P\{X_n > a + 2\varepsilon\} \le P\{T \le n - 1, X_n > a + 2\varepsilon\} + P\{Y_n > 2\varepsilon\}.$

(d) $P\left\{\max_{1 \le i \le n} X_i > a,\ X_n \le a\right\} \ge P\{X_n > a + 2\varepsilon\}$

$$- P\{Y_n > 2\varepsilon\} - 2\sum_{j=1}^{n-1} P\{Y_j > \varepsilon\}.$$

(e) $P\left\{\max_{1 \le i \le n} X_i > a\right\} \ge 2P\{X_n > a + 2\varepsilon\} - 2\sum_{j=1}^{n} P\{Y_j > \varepsilon\}.$

3.   Use Problems 1 and 2(e) to establish Eq. (1) of 4.3.2. (The following result will also be needed. Let X_n be normally distributed with mean 0 and variance $b = b(n)$. If $a = a(n)$ and $a(n)[b(n)]^{-1/2} \to \infty$, then

$$P\{X_n > a\} \sim \frac{\sqrt{b}}{\sqrt{2\pi}\,a} e^{-a^2/2b}$$

in the sense that the ratio of the two sides approaches 1 as $n \to \infty$.

To prove this, note that $X_n/\sqrt{b}$ is normal $(0, 1)$, and apply a standard property of the normal density:

$$\int_x^{\infty} \frac{1}{\sqrt{2\pi}} e^{-t^2/2}\, dt \sim \frac{1}{\sqrt{2\pi}\,x} e^{-x^2/2}$$

(see Ash, 1970, p. 176).)

4.   (a)   Show that $\{c^{-1/2}B(ct),\ t \ge 0\}$ is separable Brownian motion.

(b)   If $Z(t) = tB(1/t),\ t > 0;\ Z(0) = 0$, show that $\{Z(t),\ t \ge 0\}$ is separable Brownian motion.

5.   Use Problem 4(b) to show that $t^{-1}B(t) \to 0$ a.e. as $t \to \infty$.

6.   If c is a real number, let $S_c(\omega) = \{t \ge 0 : B(t, \omega) = c\}$. For almost every ω, $S_c(\omega)$ is closed (by 4.3.1) and unbounded (by 4.3.3). Show that for almost every ω, $S_c(\omega)$ has Lebesgue measure 0.

4.4   Law of the Iterated Logarithm

If $Y_1, Y_2, \ldots$ are independent, identically distributed random variables with 0 mean, the strong law of large number states that $X_n/n \to 0$ a.e., where $X_n = Y_1 + \cdots + Y_n$. Thus for any $k > 0$, $|X_n|$ will be less than kn for sufficiently large n. We might say that X_n oscillates with an amplitude less than kn. We ask for more precise information about the oscillation; for example, is it true that $|X_n|$ will be less than $k\sqrt{n}$ eventually? It is this type of question that we investigate here, for the special case of normally distributed random variables; we then apply the results of Brownian motion.

The remarkable fact is that the oscillation can be measured precisely by $f(n) = (2\sigma^2 n \ln \ln n)^{1/2}$, where σ^2 is the variance of the Y_i. In other words, if $\varepsilon > 0$, then $|X_n| < (1 + \varepsilon)f(n)$ eventually; also, $X_n > (1 - \varepsilon)f(n)$ for infinitely many n, and $X_n < -(1 - \varepsilon)f(n)$ for infinitely many n.

We first consider normal random variables with mean 0 and variance 1.

4.4.1   Lemma

Let $X_n = \sum_{k=1}^{n} Y_k$, $n = 1, 2, \ldots$, where the Y_k are independent and each Y_k is normal $(0, 1)$. Then for almost every ω,

$$\limsup_{n \to \infty} \frac{X_n(\omega)}{(2n \ln \ln n)^{1/2}} \le 1.$$

PROOF   Fix $\lambda > 1$, and let $n_k = \lambda^k$, $k = r, r + 1, r + 2, \ldots$ where r is the smallest positive integer such that $\lambda^r \ge 3$ (so $(\ln \ln n_k)^{1/2}$ is well defined for $k \ge r$). Let

$$A_k = \{X_n > \lambda(2n \ln \ln n)^{1/2} \quad \text{for some} \quad n \in (n_k, n_{k+1}]\},$$

and take $a(n) = \lambda(2n \ln \ln n)^{1/2}$. Then

$$P(A_k) \le P\{X_n > a(n_k) \text{ for some } n \le n_{k+1}\}$$

$$\le 2P\{X_{n_{k+1}} > a(n_k)\} \qquad \text{by Problem 1, Section 4.3}$$

$$\sim \frac{2\sqrt{n_{k+1}}}{\sqrt{2\pi}\, a(n_k)} \exp\left[-a^2(n_k)/2n_{k+1}\right]$$

by Problem 3, Section 4.3

$$\le c \exp\left[-\lambda \ln \ln \lambda^k\right]$$

$$\le c' \exp\left[-\lambda \ln k\right] = c' k^{-\lambda}.$$

But $\sum_k k^{-\lambda} < \infty$, so by the Borel–Centalli lemma, with probability one only finitely many A_k occur. But then, almost surely, $X_n \le \lambda(2n \ln \ln n)^{1/2}$ for sufficiently large n. Since this holds for $\lambda = 1 + 1/m$, $m = 1, 2, \ldots$, we conclude that $P\{\text{for every } \lambda > 1, X_n \le \lambda (2n \ln \ln n)^{1/2} \text{ eventually}\} = 1$. ∥

4.4.2   Lemma

In 4.4.1,

$$\limsup_{n \to \infty} \frac{X_n}{(2n \ln \ln n)^{1/2}}$$

actually equals 1 a.e.

PROOF   If $\lambda < 1$, we must show that with probability 1,

$$\frac{X_n}{(2n \ln \ln n)^{1/2}} > \lambda$$

for infinitely many n. We may apply 4.4.1 to $\{-X_n\}$ to obtain, almost surely, $-X_n \leq 2(2n \ln \ln n)^{1/2}$ eventually. Therefore if $m_k = M^k$ $(M > 1)$, we have

$$X_{m_{k-1}} \geq -2(2m_{k-1} \ln \ln m_{k-1})^{1/2} \qquad \text{for large } k.$$

Thus if $Z_k = X_{m_k} - X_{m_{k-1}}$, then

$$X_{m_k} \geq Z_k - 2(2m_{k-1} \ln \ln m_{k-1})^{1/2} \qquad \text{for large } k.$$

Therefore it suffices to show that

$$Z_k > \lambda(2m_k \ln \ln m_k)^{1/2} + 2(2m_{k-1} \ln \ln m_{k-1})^{1/2}$$

$$\text{for infinitely many } k.$$

Pick $\lambda' \in (\lambda, 1)$. Then for some M we have

$$\lambda'[2(M^k - M^{k-1}) \ln \ln M^k]^{1/2}$$

$$> \lambda(2M^k \ln \ln M^k)^{1/2} + 2(2M^{k-1} \ln \ln M^{k-1})^{1/2} \qquad \text{for all } k$$

(the ratio of right to left sides is less than

$$\frac{\lambda}{\lambda'}\left(1 - \frac{1}{M}\right)^{-1/2} + \frac{2}{\lambda'}(M - 1)^{-1/2}\right).$$

Thus it suffices to show that

$$Z_k > \lambda'(2(M^k - M^{k-1}) \ln \ln M^k)^{1/2} \qquad \text{for infinitely many } k.$$

Now Z_k is normal $(0, M^k - M^{k-1})$, so by Problem 3, Section 4.3,

$$P(Z_k > \lambda'[2(M^k - M^{k-1}) \ln \ln M^k]^{1/2})$$

$$\sim \frac{1}{\sqrt{2\pi}\,\lambda'(2 \ln \ln M^k)^{1/2}} \exp\left[-\lambda'^2 \ln \ln M^k\right]$$

$$\geq \frac{c}{(\ln k)^{1/2}} k^{-\lambda'^2}$$

$$\geq \frac{c}{k \ln k} \qquad \text{since } \lambda' < 1.$$

But $\sum 1/k \ln k = \infty$, and the result follows from the second Borel–Cantelli lemma. ‖

We now have the desired result for normal random variables.

4.4.3   Theorem

Let $X_n = \sum_{k=1}^{n} Y_k$, $n = 1, 2, \ldots$, where the Y_k are independent and each Y_k is normal $(0, \sigma^2)$. Then for almost every ω,

$$\limsup_{n \to \infty} \frac{X_n(\omega)}{(2\sigma^2 n \ln \ln n)^{1/2}} = 1$$

and

$$\liminf_{n \to \infty} \frac{X_n(\omega)}{(2\sigma^2 n \ln \ln n)^{1/2}} = -1.$$

PROOF   For the "lim sup" statement, apply 4.4.2 to $\{X_n/\sigma\}$. For the "lim inf" statement, apply 4.4.2 to $\{-X_n/\sigma\}$. ‖

For Brownian motion, we may establish a continuous parameter analog of 4.4.3.

4.4.4   Law of the Iterated Logarithm for Brownian Motion

For almost every ω,

$$\limsup_{t \to \infty} \frac{B(t, \omega)}{(2\sigma^2 t \ln \ln t)^{1/2}} = 1$$

and

$$\liminf_{t \to \infty} \frac{B(t, \omega)}{(2\sigma^2 t \ln \ln t)^{1/2}} = -1.$$

PROOF   We may assume $\sigma^2 = 1$ (if $\sigma^2 \neq 1$ consider $\{B(t)/\sigma\}$). The "lim inf" statement follows from the "lim sup" statement by considering $\{-B(t)\}$, so it suffices to consider the lim sup.

Now $B(n) = B(1) + (B(2) - B(1)) + \cdots + (B(n) - B(n-1))$. The summands are independent and normal $(0, 1)$, and hence by 4.4.3,

$$\limsup_{n \to \infty} \frac{B(n)}{(2n \ln \ln n)^{1/2}} = 1 \quad \text{a.e.}$$

Therefore

$$\limsup_{t \to \infty} \frac{B(t)}{(2t \ln \ln t)^{1/2}} \geq 1 \quad \text{a.e.}$$

Now

$$\max_{n \le t \le n+1} B(t) = B(n) + \max_{n \le t \le n+1} [B(t) - B(n)]$$

and

$$P\left\{ \max_{n \le t \le n+1} [B(t) - B(n)] > a \right\}$$

$$= P\left\{ \max_{0 \le t \le 1} B(t) > a \right\}$$

$$= 2P\{B(1) > a\} \qquad \text{by 4.3.2}$$

$$\sim \frac{2}{\sqrt{2\pi}\,a} e^{-a^2/2} \qquad \text{by Problem 3, Section 4.3.}$$

Let $a = n^{1/4}$; then $e^{-a^2/2} = e^{-\sqrt{n}/2}$. Since $\sum_n e^{-\sqrt{n}/2} < \infty$, the Borel–Cantelli lemma shows that with probability one

$$\max_{n \le t \le n+1} [B(t) - B(n)] \le n^{1/4} \qquad \text{for sufficiently large } n.$$

Thus if $\lambda' > 1$, $\varepsilon > 0$ we have, for large enough n,

$$\max_{n \le t \le n+1} B(t) < \lambda'(2n \ln \ln n)^{1/2} + n^{1/4}$$

$$< (\lambda' + \varepsilon)(2n \ln \ln n)^{1/2}$$

$$\le (\lambda' + \varepsilon)(2t \ln \ln t)^{1/2} \qquad \text{if } n \le t \le n+1.$$

Thus if $\lambda > 1$ we have, for sufficiently large t,

$$B(t) < \lambda(2t \ln \ln t)^{1/2}.$$

Hence

$$\limsup_{t \to \infty} \frac{B(t)}{(2t \ln \ln t)^{1/2}} \le 1 \quad \text{a.e.} \quad \|$$

Problem

1. Prove the *Local Law of the Iterated Logarithm* for Brownian motion, that is, for each $t_0 \ge 0$,

$$\limsup_{h \to 0^+} \frac{B(t_0 + h) - B(t_0)}{(2\sigma^2 h \ln \ln h^{-1})^{1/2}} = 1 \qquad \text{a.e.}$$

and

$$\liminf_{h \to 0^+} \frac{B(t_0 + h) - B(t_0)}{(2\sigma^2 h \ln \ln h^{-1})^{1/2}} = -1 \qquad \text{a.e.}$$

4.5   Markov Processes

The Markov property is familiar to us in the case of a discrete parameter process with a countable state space, that is, a Markov chain. In this section we discuss the general concept of a Markov process, and show how such processes can be constructed from a specified transition probability.

4.5.1   Definition

Let I be an arbitrary totally ordered index set, $\{X(t), t \in I\}$ a stochastic process on $(\Omega, \mathscr{F}, P)$ with state space $(S, \mathscr{S})$. Assume $\{X(t)\}$ is adapted to the family of σ-fields $\mathscr{F}(t)$, $t \in I$, that is, for all $s, t \in I$, $s < t$ implies $\mathscr{F}(s) \subset \mathscr{F}(t)$, and $X(t)$ is $\mathscr{F}(t)$-measurable.

We say that $\{X(t)\}$ is a *Markov process* relative to the $\mathscr{F}(t)$, or that $\{X(t), \mathscr{F}(t)\}$ is Markov, iff for all $B \in \mathscr{S}$ and $s, t \in I$, $s < t$,

$$P\{X(t) \in B \mid \mathscr{F}(s)\} = P\{X(t) \in B \mid X(s)\} \quad \text{a.e.} \tag{1}$$

This is called the *Markov property*.

Equivalently, for all $s, t \in I$, $s < t$, and all $g: (S, \mathscr{S}) \to (R, \mathscr{B}(R))$ such that $E[g(X(t))]$ is finite,

$$E[g(X(t)) \mid \mathscr{F}(s)] = E[g(X(t)) \mid X(s)] \quad \text{a.e.} \tag{2}$$

To show that (2) implies (1), let $g = I_B$; to prove that (1) implies (2), start with indicators and proceed in the usual way.

The "a.e." is implicit in any statement involving conditional probability or expectation, and will usually be suppressed.

If we say that $\{X(t)\}$ is a Markov process without mentioning the $\mathscr{F}(t)$, we always assume $\mathscr{F}(t) = \mathscr{F}_0(t) = \mathscr{F}(X(r), r \leq t)$. In this case, (1) becomes

$$P\{X(t) \in B \mid X(r), r \leq s\} = P\{X(t) \in B \mid X(s)\}.$$

Note that if $\{X(t), \mathscr{F}(t)\}$ is a Markov process, so is $\{X(t), \mathscr{F}_0(t)\}$. For $\mathscr{F}(X(s)) \subset \mathscr{F}_0(s) \subset \mathscr{F}(s)$, so if we condition both sides of (2) with respect to $\mathscr{F}_0(s)$, we obtain

$$E[g(X(t)) \mid \mathscr{F}_0(s)] = E[g(X(t)) \mid X(s)].$$

4.5.2 Comments

(a) If $\{X(t), t \in I\}$ is a Markov process, then so is $\{X(t), t \in J\}$ for any $J \subset I$.

It is immediate from the definition that if $\{X(t), \mathscr{F}(t), t \in I\}$ is Markov, so is $\{X(t), \mathscr{F}(t), t \in J\}$, but this is not quite what we have here since $\mathscr{F}(X(r), r \leq t, r \in I)$ need not equal $\mathscr{F}(X(r), r \leq t, r \in J)$. We know, however, that for $s < t$, $s, t \in J$,

$$E[g(X(t)) \mid X(r), r \leq s, r \in I] = E[g(X(t)) \mid X(s)];$$

condition both sides with respect to $\mathscr{F}(X(r), r \leq s, r \in J)$ to obtain the desired result.

(b) If $\{X(t), t \in I_0\}$ is Markov for all finite subsets I_0 of I, that is,

$$P\{X(t_n) \in B \mid X(t_1), \ldots, X(t_{n-1})\} = P\{X(t_n) \in B \mid X(t_{n-1})\}$$

for all $t_1 < t_2 < \cdots < t_n$, $n = 1, 2, \ldots$ and all $B \in \mathscr{S}$, then $\{X(t), t \in I\}$ is Markov.

To prove this, let $A \in \mathscr{F}(X(r_1), \ldots, X(r_n))$, where $r_1, \ldots, r_n \leq s < t$; since $A \in \mathscr{F}(X(r_1), \ldots, X(r_n), X(u))$ for any $u \in I$, we may assume $r_n = s$. Then if $B \in \mathscr{S}$, we have

$$P[A \cap \{X(t) \in B\}]$$

$$= \int_A P\{X(t) \in B \mid X(r_1), \ldots, X(r_n)\} \, dP$$

by definition of conditional probability

$$= \int_A P\{X(t) \in B \mid X(s)\} \, dP \qquad \text{by hypothesis.}$$

An application of the monotone class theorem shows that this holds for all $A \in \mathscr{F}(X(r), r \leq s)$, and it follows that

$$P\{X(t) \in B \mid X(r), r \leq s\} = P\{X(t) \in B \mid X(s)\}.$$

(c) If $s < t$, $g, h: (S, \mathscr{S}) \to (R, \mathscr{B}(R))$, and $E[g(X(t)) \mid \mathscr{F}(s)] = h(X(s))$, then $h(X(s)) = E[g(X(t)) \mid X(s)]$, so 4.5.1, Eq. (2) holds.

For $E[E(g(X(t)) \mid \mathscr{F}(s)) \mid X(s)] = E[g(X(t)) \mid X(s)]$, and

$$E[h(X(s)) \mid X(s)] = h(X(s)).$$

(d) If $\{X(t), \mathscr{F}(t)\}$ is Markov, and $A \in \mathscr{F}(X(r), r \geq t)$, then $P(A \mid \mathscr{F}(t)) = P(A \mid X(t))$.

To prove this, let $B_1, B_2, \ldots, B_n \in \mathscr{S}, t \le t_1 < t_2 < \cdots < t_n$; then

$$P\{X(t_1) \in B_1, \ldots, X(t_n) \in B_n \,|\, \mathscr{F}(t)\} = E[I_{B_1}(X(t_1)) \cdots I_{B_n}(X(t_n)) \,|\, \mathscr{F}(t)]$$

$$= E[E(I_{B_1}(X(t_1)) \cdots I_{B_n}(X(t_n)) \,|\, \mathscr{F}_{t_{n-1}}) \,|\, \mathscr{F}(t)] \quad \text{since } \mathscr{F}(t) \subset \mathscr{F}(t_{n-1})$$

$$= E[I_{B_1}(X(t_1)) \cdots I_{B_{n-1}}(X(t_{n-1})) E(I_{B_n}(X(t_n)) \,|\, \mathscr{F}(t_{n-1})) \,|\, \mathscr{F}(t)]$$

$$\text{since } I_{B_1}(X(t_1)) \cdots I_{B_{n-1}}(X(t_{n-1})) \text{ is } \mathscr{F}(t_{n-1})\text{-measurable}$$

$$= E[I_{B_1}(X(t_1)) \cdots I_{B_{n-1}}(X(t_{n-1})) E(I_{B_n}(X(t_n)) \,|\, X(t_{n-1})) \,|\, \mathscr{F}(t)]$$

$$\text{by the Markov property.}$$

This is of the form

$$E[I_{B_1}(X(t_1)) \cdots I_{B_{n-2}}(X(t_{n-2})) g(X(t_{n-1})) \,|\, \mathscr{F}(t)]$$

$$= E[E(I_{B_1}(X(t_1)) \cdots I_{B_{n-2}}(X(t_{n-2})) g(X(t_{n-1})) \,|\, \mathscr{F}(t_{n-2})) \,|\, \mathscr{F}(t)]$$

$$= E[I_{B_1}(X(t_1)) \cdots I_{B_{n-2}}(X(t_{n-2})) E(g(X(t_{n-1})) \,|\, X(t_{n-2})) \,|\, \mathscr{F}(t)]$$

$$= E[I_{B_1}(X(t_1)) \cdots I_{B_{n-3}}(X(t_{n-3})) h(X(t_{n-2})) \,|\, \mathscr{F}(t)].$$

Proceeding inductively, we obtain an expression of the form $E[g_1(X(t_1)) \,|\, \mathscr{F}(t)]$, which equals $E[g_1(X(t_1)) \,|\, X(t)]$ by the Markov property. Thus $P\{X(t_1) \in B_1, \ldots, X(t_n) \in B_n \,|\, \mathscr{F}(t)\}$ is a function of $X(t)$, hence equals $P\{X(t_1) \in B_1, \ldots, X(t_n) \in B_n \,|\, X(t)\}$ as in (c). Therefore the desired result holds when $A = \{X \in B\}$, $X = (X(r), r \ge t)$, B a measurable rectangle in $\prod_{r \ge t} \mathscr{S}$. A passage to finite disjoint unions of measurable rectangles and an application of the monotone class theorem completes the proof.

We may use 4.5.2(d) to formulate the definition of a Markov process in a symmetrical way. First, a preliminary result on conditional probability.

4.5.3   Theorem

Let $(\Omega, \mathscr{F}_0, P)$ be a probability space, and let $\mathscr{F}, \mathscr{G}$, and $\mathscr{H}$ be sub σ-fields of $\mathscr{F}_0$. The following statements are equivalent.

(a) $P(A \cap B \,|\, \mathscr{H}) = P(A \,|\, \mathscr{H}) P(B \,|\, \mathscr{H})$ for all $A \in \mathscr{F}, B \in \mathscr{G}$.

(b) $P(A \,|\, \mathscr{G} \vee \mathscr{H}) = P(A \,|\, \mathscr{H})$ for all $A \in \mathscr{F}$, where $\mathscr{G} \vee \mathscr{H}$ is the minimal σ-field over $\mathscr{G} \cup \mathscr{H}$.

If either (and hence both) of these conditions are satisfied, we say that $\mathscr{F}$ and $\mathscr{G}$ are *conditionally independent given* $\mathscr{H}$.

PROOF Assume that (b) holds. Then

$$P(A \mid \mathcal{H})P(B \mid \mathcal{H}) = E(I_A \mid \mathcal{H})E(I_B \mid \mathcal{H})$$
$$= E[I_B E(I_A \mid \mathcal{H}) \mid \mathcal{H}]$$

$$\text{since } E(I_A \mid \mathcal{H}) \text{ is } \mathcal{H}\text{-measurable}$$

$$= E[I_B E(I_A \mid \mathcal{G} \vee \mathcal{H}) \mid \mathcal{H}] \qquad \text{by (b)}$$
$$= E[E(I_A I_B \mid \mathcal{G} \vee \mathcal{H}) \mid \mathcal{H}]$$

$$\text{since } I_B \text{ is } \mathcal{G} \vee \mathcal{H}\text{-measurable}$$

$$= E(I_A I_B \mid \mathcal{H}) \qquad \text{since } \mathcal{H} \subset \mathcal{G} \vee \mathcal{H}$$
$$= P(A \cap B \mid \mathcal{H}).$$

Now assume (a). We must show that if $D \in \mathcal{G} \vee \mathcal{H}$,

$$P(A \cap D) = \int_D P(A \mid \mathcal{H}) \, dP.$$

Since $\mathcal{G} \vee \mathcal{H}$ is the minimal σ-field containing the field of finite disjoint unions of sets $B \cap C, B \in \mathcal{G}, C \in \mathcal{H}$ (see RAP, p. 12, Problem 9), the monotone class theorem shows that it suffices to consider only sets D of the form $B \cap C, B \in \mathcal{G}, C \in \mathcal{H}$. Now

$$P(A \cap D) = P(A \cap B \cap C)$$

$$= \int_C P(A \cap B \mid \mathcal{H}) \, dP$$

$$= \int_C E(I_A \mid \mathcal{H})E(I_B \mid \mathcal{H}) \, dP \qquad \text{by (a)}$$

$$= \int_C E[I_B E(I_A \mid \mathcal{H}) \mid \mathcal{H}] \, dP$$

$$\text{since } E(I_A \mid \mathcal{H}) \text{ is } \mathcal{H}\text{-measurable}$$

$$= \int_C I_B E(I_A \mid \mathcal{H}) \, dP$$

$$\text{by definition of conditional expectation}$$

$$= \int_{B \cap C} E(I_A \mid \mathcal{H}) \, dP$$

$$= \int_D P(A \mid \mathcal{H}) \, dP. \quad \|$$

4.5.4   Theorem

The process $\{X(t), \mathscr{F}(t), t \in I\}$ is Markov iff for each $t \in I, \mathscr{F}(X(r), r \geq t)$ and $\mathscr{F}(t)$ are conditionally independent given $X(t)$ (that is, given $\mathscr{F}(X(t))$).

PROOF   The "only if" part follows from 4.5.2(d) and 4.5.3(b) if we note that if $\mathscr{G} = \mathscr{F}(t), \mathscr{H} = \mathscr{F}(X(t))$, then $\mathscr{G} \vee \mathscr{H} = \mathscr{G}$. For the "if" part, let $B \in \mathscr{S}$, $t' > t$. Then

$$P\{X(t') \in B \mid \mathscr{F}(t)\}$$
$$= P\{X(t') \in B \mid \mathscr{F}(t) \vee \mathscr{F}(X(t))\}$$
$$= P\{X(t') \in B \mid X(t)\} \qquad \text{by conditional independence.} \quad \|$$

By 4.5.4, $\{X(t)\}$ is Markov iff $\mathscr{F}(X(r), r \geq t)$ and $\mathscr{F}(X(r), r \leq t)$ are conditionally independent given $X(t)$. But by 4.5.3(a), $\mathscr{F}$ and $\mathscr{G}$ are conditionally independent given $\mathscr{H}$ iff $\mathscr{G}$ and $\mathscr{F}$ are conditionally independent given $\mathscr{H}$. Thus *a Markov process reversed in time is still Markov*, that is, if a new ordering $<'$ is introduced on I, with $s <' t$ iff $s > t$, $\{X(t)\}$ is still a Markov process.

We now describe the standard method for constructing a Markov process.

4.5.5   The Basic Construction

Suppose we are given a *transition probability*, in other words, a function $p(x, B)$, $x \in S$, $B \in \mathscr{S}$, such that for each fixed x, $p(x, \cdot)$ is a probability measure on $\mathscr{S}$, and for each fixed B, $p(\cdot, B)$ is Borel measurable, that is, $p(\cdot, B): (S, \mathscr{S}) \to (R, \mathscr{B}(R))$. Let P_0 be an arbitrary probability measure on $\mathscr{S}$ (the *initial distribution*). We may construct a probability space $(\Omega, \mathscr{F}, P)$ and a sequence of measurable functions $X_n: (\Omega, \mathscr{F}) \to (S, \mathscr{S})$ such that for any $B \in \mathscr{S}^{n+1}$, $n = 0, 1, \ldots$ we have

$$P\{(X_0, X_1, \ldots, X_n) \in B\}$$
$$= \int_S P_0(dx_0) \int_S p(x_0, dx_1) \cdots \int_S I_B(x_0, \ldots, x_n) p(x_{n-1}, dx_n); \qquad (1)$$

also, if $g: (S^{n+1}, \mathscr{S}^{n+1}) \to (R, \mathscr{B}(R))$ and g is nonnegative or bounded, then

$$E[g(X_0, \ldots, X_n)] = \int_S P_0(dx_0) \int_S p(x_0, dx_1) \cdots$$
$$\int_S g(x_0, \ldots, x_n) p(x_{n-1}, dx_n). \qquad (2)$$

(To establish (1) and (2), use RAP, 2.6.7, 2.6.8, and 2.7.2, pp. 104–109, with $P_{x_0 \cdots x_j}(B) = p(x_j, B)$.)

We are going to prove that $\{X_n, n \ge 0\}$ is a Markov process. The key step is to verify that if $B \in \mathcal{S}$, then

$$P\{X_{j+1} \in B \mid X_0, \ldots, X_j\} = p(X_j, B), \qquad j = 0, 1, \ldots. \tag{3}$$

To prove this, let $C \in \mathcal{S}^{j+1}$. Then by (1),

$$P[\{X_{j+1} \in B\} \cap \{(X_0, \ldots, X_j) \in C\}]$$

$$= \int_S P_0(dx_0) \int_S p(x_0, dx_1) \cdots \int_S I_B(x_{j+1})$$

$$\times I_C(x_0, \ldots, x_j) p(x_j, dx_{j+1})$$

$$= \int_S P_0(dx_0) \int_S p(x_0, dx_1) \cdots \int_S p(x_j, B)$$

$$\times I_C(x_0, \ldots, x_j) p(x_{j-1}, dx_j)$$

$$= \int_{\{(X_0, \ldots, X_j) \in C\}} p(X_j, B) \, dP \qquad \text{by (2), proving (3)}.$$

By 4.5.2(c),

$$P\{X_{j+1} \in B \mid X_0, \ldots, X_j\} = P\{X_{j+1} \in B \mid X_j\}.$$

But

$$P\{X_{j+k} \in B \mid X_0, \ldots, X_j\}$$

$$= E[E(I_B(X_{j+k}) \mid X_0, \ldots, X_{j+k-1}) \mid X_0, \ldots, X_j]$$

$$= E[E(I_B(X_{j+k}) \mid X_{j+k-1}) \mid X_0, \ldots, X_j]$$

$$\text{(see 4.5.1, Eq. (2))}$$

$$= E[g(X_{j+k-1}) \mid X_0, \ldots, X_j].$$

Thus an induction argument shows that

$$P\{X_{j+k} \in B \mid X_0, \ldots, X_j\} = P\{X_{j+k} \in B \mid X_j\},$$

establishing the Markov property.

We call $\{X_n, n \ge 0\}$ the Markov process generated by the initial distribution P_0 and the transition probability p. The same construction works if we have "nonstationary transition probabilities," that is, a sequence of transition probabilities p_{01}, p_{12}, ... We can construct a Markov process $\{X_n, n \geq 0\}$ satisfying

$$P\{(X_0, \ldots, X_n) \in B\} = \int_S P_0(dx_0) \int_S p_{01}(x_0, dx_1) \cdots$$

$$\int_S I_B(x_0, \ldots, x_n)p_{n-1, n}(x_{n-1}, dx_n),$$

$$B \in \mathscr{S}^{n+1}, \quad n = 0, 1, \ldots.$$

The argument is the same as before, with minor notational changes. In particular, (3) becomes

$$P\{X_{j+1} \in B \mid X_0, \ldots, X_j\} = p_{j, j+1}(X_j, B). \tag{4}$$

A Markov process generated by an initial distribution and a single transition probability is said to be *time-homogeneous* or to have *stationary transition probabilities*.

Now let us generalize the proof of (3). First we define the *composition* of two transition probabilities p and q by

$$(p * q)(x, B) = \int_S p(x, dy)q(y, B), \qquad x \in S, \quad B \in \mathscr{S};$$

$p * q$ is also a transition probability, and the composition operation is associative (Problem 2). To avoid cumbersome notation, we consider a specific example. Say $(X_0, X_1, X_2, X_3, X_4, X_5)$ is a Markov process generated by P_0 and the transition probabilities p_i, $i = 1, 2, 3, 4, 5$. Let g_0, g_2, g_5: $(S, \mathscr{S}) \to (R, \mathscr{B}(R))$, with all $g_i \geq 0$. Then by (2),

$$E[g_0(X_0)g_2(X_2)g_5(X_5)]$$

$$= \int_S P_0(dx_0) \int_S p_1(x_0, dx_1) \int_S p_2(x_1, dx_2) \int_S p_3(x_2, dx_3)$$

$$\times \int_S p_4(x_3, dx_4) \int_S p_5(x_4, dx_5)g_0(x_0)g_2(x_2)g_5(x_5).$$

But

$$\int_S p_3(x_2, dx_3) \int_S p_4(x_3, dx_4) \int_S p_5(x_4, dx_5)g_5(x_5)$$

$$= \int_S (p_3 * p_4 * p_5)(x_2, dx_5)g_5(x_5)$$

(prove this by starting with indicators). Proceed backward to obtain

$$E[g_0(X_0)g_2(X_2)g_5(X_5)] = \int_S P_0(dx_0) \int_S (p_1 * p_2)(x_0, dx_2)$$

$$\times \int_S (p_3 * p_4 * p_5)(x_2, dx_5)g_0(x_0)g_2(x_2)g_5(x_5).$$

Set $g_i = I_{B_i}$ to conclude that (X_0, X_2, X_5) is a Markov process generated by $P_0, p_1 * p_2$ and $p_3 * p_4 * p_5$. Then set $g_0 \equiv 1$ to conclude that (X_2, X_5) is a Markov process generated by the initial distribution $P_0 * p_1 * p_2$ and the transition probability $p_3 * p_4 * p_5$, where the composition of the probability measure P_0 and the transition probability p is defined by

$$(P_0 * p)(B) = \int_S P_0(dx)p(x, B).$$

(Note $(P_0 * p) * q = P_0 * (p * q)$; see Problem 3.)

The above reasoning establishes the following result.

4.5.6 Generation of Markov Subsequences

If $\{X_n, n = 0, 1, \ldots\}$ is a Markov process generated by the initial distribution P_0 and the transition probability p, and $0 < t_1 < t_2 < \cdots$, then X_0, $X_{t_1}, X_{t_2}, \ldots$ is a Markov process generated by P_0 and the transition probabilities $p_{t_1}, p_{t_2-t_1}, p_{t_3-t_2}, \ldots$, where p_k is the *k-step transition probability* $p * \cdots * p$ (*k* times). Also, $X_{t_1}, X_{t_2}, \ldots$ is a Markov process generated by $P_{t_1} = P_0 * p_{t_1}$ and the $p_{t_{i+1}-t_i}, i = 1, 2, \ldots$.

It follows from (4) that

$$P\{X_{t_{n+1}} \in B \mid X_{t_1}, \ldots, X_{t_n}\} = p_{t_{n+1}-t_n}(X_{t_n}, B). \tag{5}$$

It follows from (5) that if $g: (S, \mathscr{S}) \to (R, \mathscr{B}(R))$, then

$$E[g(X_{t_{n+1}}) \mid X_{t_1}, \ldots, X_{t_n}] = \int_S g(y)p_{t_{n+1}-t_n}(X_{t_n}, dy) \tag{6}$$

if g is nonnegative or bounded (start with indicators and apply (5)). A similar statement can be made in the case of nonstationary transition probabilities.

We now consider the continuous parameter case. For convenience, we take $I = [0, \infty)$. Suppose we are given an initial distribution P_0 and, for each $t > 0$, a transition probability p_t. We try to construct a corresponding Markov process. Again let us use particular numbers to simplify the notation. We specify, say, that $(X_0, X_{1.7}, X_\pi, X_{10}, X_{10.1}, X_{25}, X_{30})$ be a Markov process generated by $P_0, p_{1.7}, p_{\pi-1.7}, p_{10-\pi}, p_{.1}, p_{14.9}$, and p_5. It follows from 4.5.6 that (X_π, X_{10}, X_{30}) is a Markov process generated by $P_0 * p_{1.7} * p_{\pi-1.7}, p_{10-\pi}$ and $p_{.1} * p_{14.9} * p_5$. But the natural specification of the distribution of (X_π, X_{10}, X_{30}) (forgetting about $X_{1.7}, X_{10.1}$, etc.) is that of a Markov process generated by $P_0 * p_\pi, p_{10-\pi}$, and p_{20}. In order to have consistency, it appears that we need relations of the form $p_{s+t} = p_s * p_t$ (which then implies $p_{s_1 + \cdots + s_n} = p_{s_1} * p_{s_2} * \cdots * p_{s_n}$). This does not finish the story, for in order to apply the Kolmogorov extension theorem, we must assume that S is a complete separable metric space. We may then state the final result.

4.5.7 Generation of Continuous Parameter Markov Processes

Let S be a complete separable metric space, with $\mathscr{S} = \mathscr{B}(S)$. Let P_0 be an arbitrary probability measure on $\mathscr{S}$, and let $\{p_t, t > 0\}$ be a family of transition probabilities satisfying the *Chapman–Kolmogorov equation*

$$p_{s+t} = p_s * p_t,$$

that is,

$$p_{s+t}(x, B) = \int_S p_s(x, dy) p_t(y, B), \qquad x \in S, \quad B \in \mathscr{S}, \quad s, t > 0.$$

There is a Markov process $\{X_t, t \geq 0\}$ such that if $0 < t_1 < \cdots < t_n$, then $(X_0, X_{t_1}, \ldots, X_{t_n})$ is generated by $P_0, p_{t_1}, p_{t_2 - t_1}, \ldots, p_{t_n - t_{n-1}}$.

A similar construction can be carried out for nonstationary transition probabilities. If we are given a family of transition probabilities p_{rs}, $0 \leq r < s$, then a Markov process can be constructed so that $(X_0, X_{t_1}, \ldots, X_{t_n})$ is generated by $P_0, p_{0t_1}, \ldots, p_{t_{n-1}t_n}$, provided $p_{rt} = p_{rs} * p_{st}$ for $r < s < t$.

Note that in the discrete parameter case the Chapman–Kolmogorov equation $p_{rt} = p_{rs} * p_{st}$, $r < s < t$, is automatically satisfied. For example, $p_{13} * p_{36} = (p_{12} * p_{23}) * (p_{34} * p_{45} * p_{56}) = p_{16}$. Note also that in the case of a Markov chain, $p_{n, n+1}$ is given by a transition matrix, that is, $p_{n, n+1}(i, B) = \sum_{j \in B} {}_n p_{ij}$ for some stochastic matrix $[{}_n p_{ij}]$.

Problems

1. Show that $\{X(t), t \in T\}$ is a Markov process iff $\mathscr{F}(X(r), r < t)$ and $\mathscr{F}(X(r), r > t)$ are conditionally independent given $X(t)$.

2. Show that composition of transition probabilities is associative.

3. If p and q are transition probabilities on $(S, \mathscr{S})$ and P_0 a probability measure on $\mathscr{S}$, show that $(P_0 * p) * q = P_0 * (p * q)$.

4.6 Processes with Independent Increments

The processes to be studied in this section yield a convenient source of explicit examples of continuous parameter Markov processes.

4.6.1 Definitions and Comments

Let $\{X(t), t \geq 0\}$ be a real stochastic process. The process is said to have *independent increments* iff whenever $0 < t_1 < \cdots < t_n$, $X(0)$, $X(t_1) - X(0)$, $X(t_2) - X(t_1)$, ..., $X(t_n) - X(t_{n-1})$ are independent.

Note that if $Y(t) = X(t) - X(0)$, then $X(0)$, $Y(t_1)$, $Y(t_2) - Y(t_1)$, ..., $Y(t_n) - Y(t_{n-1})$ are independent, hence $X(0)$ and $(Y(t_1), Y(t_2) - Y(t_1), ..., Y(t_n) - Y(t_{n-1}))$ are independent. Since functions of independent random vectors are independent, $X(0)$ and $(Y(t_1), Y(t_2), ..., Y(t_n))$ are independent. Since $t_1, ..., t_n$ are arbitrary, the monotone class theorem implies that $X(0)$ and $\{Y(t), t \geq 0\}$ are independent. Also, $Y(0) \equiv 0$, and $Y(t) - Y(s) = X(t) - X(s)$, $s < t$, so that $\{Y(t), t \geq 0\}$ has independent increments.

Conversely, if $\{Y(t), t \geq 0\}$ has independent increments, $Y(0) \equiv 0$, and we define $X(t) = X(0) + Y(t)$, where $X(0)$ and $\{Y(t), t \geq 0\}$ are independent, then $\{X(t), t \geq 0\}$ has independent increments. Thus in studying processes with independent increments, no generality is lost if we subtract the initial random variable $X(0)$.

If $\{X(t), t \geq 0\}$ has independent increments and $X(t) - X(s)$ has the same distribution as $X(t + h) - X(s + h)$ for all s, t, $h \geq 0$, $s < t$, the process is said to have *stationary independent increments*. Such processes may be conveniently characterized, as follows.

4.6.2 Theorem

Let $\{X(t), t \geq 0\}$ have stationary independent increments, with $Y(t) = X(t) - X(0)$. Then for each $s < t$, $Y(t) - Y(s)$ is infinitely divisible. If h_t is the characteristic function of $Y(t)$, and $h_t(u)$ is continuous (or more generally, Borel measurable) in t for each fixed u, then

$$h_t(u) = [h_1(u)]^t = \exp\left[t \log h_1(u)\right],$$

where "log" means the unique continuous logarithm of h_1 such that $\log h_1(0) = 0$ (see Ash, 1971, pp. 49–51). Conversely, if h_1 is any infinitely divisible characteristic function, there is a process $\{Y(t), t \geq 0\}$ with stationary independent increments, such that for each t, $Y(t)$ has characteristic function h_1^t.

PROOF   If $\{X(t)\}$ has stationary independent increments, then

$$Y(t) - Y(s) = \sum_{k=1}^{n} \left[Y\left(s + \frac{k(t-s)}{n}\right) - Y\left(s + \frac{(k-1)(t-s)}{n}\right)\right],$$

so $Y(t) - Y(s)$ is infinitely divisible. Since $Y(s + t) = Y(s) + (Y(s + t) - Y(s))$, and, by stationarity of increments, $Y(s + t) - Y(s)$ has the same distribution as $Y(t)$, we have $h_{s+t}(u) = h_s(u)h_t(u)$. Since $h_t(u)$ is Borel measurable in t for u fixed, $h_t(u)$ is of the form $A(u) \exp [B(u)t]$.

Since $Y(0) \equiv 0$, we may set $t = 0$ to obtain $A(u) = 1$. Then set $t = 1$ to obtain $h_1(u) = e^{B(u)}$, so that $B(u)$ is a logarithm of $h_1(u)$. If B is discontinuous at u_0, then h_t will be discontinuous at u_0 for some t, a contradiction; thus B is continuous. Since B and $\log h_1$ are continuous logarithms of the same function h_1, we have $B(u) = \log h_1(u) + i2k\pi$ for some integer k. Therefore $h_t(u) = \exp [t \log h_1(u)]$, as desired.

Conversely, assume h_1 infinitely divisible. We first show that h_1^t is a characteristic function. If q is a positive integer, then $h^q = h_1$ for some characteristic function h. But $[\exp (q^{-1} \log h_1)]^q = h_1$, and hence $h = \exp [q^{-1} \log h_1]$ (see the discussion before RAP, 8.5.6, p. 353). Thus $h^p = \exp [pq^{-1} \log h_1]$ is a characteristic function for any positive integer p. Let $pq^{-1} \to t$ to show, by Lévy's theorem, that h_1^t is a characteristic function.

Now if $0 \le t_1 < t_2 < \cdots < t_n$, we specify the joint distribution of $Y(t_1)$, $\ldots$, $Y(t_n)$ by requiring that $Y(t_1)$, $Y(t_2) - Y(t_1)$, $\ldots$, $Y(t_n) - Y(t_{n-1})$ be independent, with $Y(t_k) - Y(t_{k-1})$ having characteristic function $(h_1)^{t_k - t_{k-1}}$. This specification satisfies the Kolmogorov consistency conditions, and the result follows. ∥

4.6.3   Examples

(a)   Let $h_1(u) = e^{-u^2\sigma^2/2}$, the characteristic function of a normal $(0, \sigma^2)$ random variable. Then $h_1^t(u) = e^{-u^2\sigma^2 t/2}$, so that $Y(s + t) - Y(s)$ is normal $(0, \sigma^2 t)$. Since $Y(t_1)$, $Y(t_2) - Y(t_1)$, $\ldots$, $Y(t_n) - Y(t_{n-1})$ are independent, normal random variables, $(Y(t_1), Y(t_2), \ldots, Y(t_n))$ is Gaussian, so that $\{Y(t), t \ge 0\}$ is a Gaussian process. The covariance function is given by

$$E[Y(s)Y(t)] = E[Y(s)(Y(t) - Y(s) + Y(s))] = E[Y^2(s)] = \sigma^2 s, \qquad s \le t,$$

hence $\{Y(t), t \ge 0\}$ is Brownian motion. The process $\{X(0) + Y(t), t \ge 0\}$, where $X(0)$ and $\{Y(t), t \ge 0\}$ are independent, is called *Brownian motion starting at $X(0)$*.

(b)   Let $h_1(u) = e^{-|u|}$, so that $h_1^t(u) = e^{-t|u|}$, and $Y(t)$ has the Cauchy density with parameter t:

$$f_t(y) = t/\pi(t^2 + y^2).$$

The process $\{Y(t), t \geq 0\}$ is called the *Cauchy process*.

(c) Let $h_1(u) = \exp[\lambda(e^{iu} - 1)]$, so that $h_1^t(u) = \exp[\lambda t(e^{iu} - 1)]$, where $\lambda > 0$. Thus $Y(t)$ has the Poisson distribution with parameter λt; moreover, if $0 \leq t_1 < t_2 < \cdots < t_n$, $Y(t_1)$, $Y(t_2) - Y(t_1)$, $\ldots$, $Y(t_n) - Y(t_{n-1})$ are independent, with $Y(t_k) - Y(t_{k-1})$ Poisson with parameter $\lambda(t_k - t_{k-1})$. Since the Poisson process (Section 1.1) also satisfies this property, which in turn determines the finite-dimensional distributions, it follows that $\{Y(t), t \geq 0\}$ is a Poisson process. Thus we have another way of constructing the Poisson process, faster but less intuitive than the approach of Section 1.1. Properties of this construction will be examined in Problem 2.

We are going to show that a process with independent increments is a Markov process. The basic point is the following.

4.6.4 Lemma

If $X_n = \sum_{k=1}^{n} Y_k$, $n = 1, 2, \ldots$ where the Y_k are independent random variables, then $\{X_n\}$ is a Markov process with respect to the σ-fields $\mathscr{F}(Y_1, \ldots, Y_n)$, and hence with respect to the σ-fields $\mathscr{F}(X_1, \ldots, X_n)$.

PROOF If $C, D \in \mathscr{B}(R)$, then

$$P\{X_{n-1} \in C, Y_n \in D \mid Y_1, \ldots, Y_{n-1}\} = P\{X_{n-1} \in C, Y_n \in D \mid X_{n-1}\}.$$

For

$$P\{X_{n-1} \in C, Y_n \in D \mid Y_1, \ldots, Y_{n-1}\}$$

$$= E[I_C(X_{n-1})I_D(Y_n) \mid Y_1, \ldots, Y_{n-1}]$$

$$= I_C(X_{n-1})E[I_D(Y_n)].$$

But

$$P\{X_{n-1} \in C, Y_n \in D \mid X_{n-1}\} = E[I_C(X_{n-1})I_D(Y_n) \mid X_{n-1}]$$

$$= I_C(X_{n-1})E[I_D(Y_n)].$$

It follows that

$$P\{(X_{n-1}, Y_n) \in A \mid Y_1, \ldots, Y_{n-1}\} = P\{(X_{n-1}, Y_n) \in A \mid X_{n-1}\}$$

for all $A \in \mathscr{B}(R^2)$. In particular, if $B \in \mathscr{B}(R)$, then

$$P\{X_{n-1} + Y_n \in B \mid Y_1, \ldots, Y_{n-1}\} = P\{X_{n-1} + Y_n \in B \mid X_{n-1}\}.$$

Since $X_{n-1} + Y_n = X_n$, the result follows. $\|$

4.6.5   Theorem

A process $\{X(t),\ t \geq 0\}$ with independent increments is a Markov process.

PROOF   If $0 \leq t_1 < t_2 < \cdots < t_n$, then

$$X(t_n) = \sum_{k=1}^{n} (X(t_k) - X(t_{k-1})) = \sum_{k=1}^{n} Y_k,$$

where the Y_k are independent. By 4.6.4,

$$P\{X(t_n) \in B \mid X(t_1), \ldots, X(t_{n-1})\} = P\{X(t_n) \in B \mid X(t_{n-1})\}.$$

By 4.5.2(b), $\{X(t)\}$ is a Markov process. $\|$

Now suppose that $\{X(t),\ t \geq 0\}$ is a Markov process generated by an initial distribution P_0 and transition probabilities $p_{rs},\ r < s$. We ask what properties the transition probabilities should have in order that the process have independent increments.

4.6.6   Theorem

If the transition probabilities p_{rs} are *spatially homogeneous*, that is, $p_{rs}(x, B) = p_{rs}(x + h,\ B + h)$ for all $r < s$, $x \in R$, $B \in \mathscr{B}(R)$, then $\{X(t),\ t \geq 0\}$ has independent increments.

PROOF   Let $X_j = X(t_j)$, $p_{j,j+1} = p_{t_j, t_{j+1}}$, $P_j = P_0 * p_{0t_j}$, $0 = t_0 < t_1 < \cdots < t_n$. Then

$$P\{X_0 \in A_0,\ X_1 - X_0 \in A_1, \ldots, X_n - X_{n-1} \in A_n\}$$

$$= \int_{x_0 \in A_0} P_0(dx_0) \int_{x_1 - x_0 \in A_1} p_{01}(x_0,\ dx_1) \cdots$$

$$\int_{x_n - x_{n-1} \in A_n} p_{n-1,n}(x_{n-1},\ dx_n).$$

Now

$$\int_{x_n - x_{n-1} \in A_n} p_{n-1,n}(x_{n-1},\ dx_n)$$

$$= p_{n-1,n}(x_{n-1},\ A_n + x_{n-1})$$

$$= p_{n-1,n}(0,\ A_n) \qquad \text{by spatial homogeneity.}$$

But

$$P\{X_n - X_{n-1} \in A_n\}$$

$$= \int_R P_{n-1}(dx_{n-1}) \int_{x_n - x_{n-1} \in A_n} p_{n-1,n}(x_{n-1}, dx_n)$$

$$= P_{n-1}(R)p_{n-1,n}(0, A_n)$$

$$= p_{n-1,n}(0, A_n).$$

Continuing in this fashion, we obtain

$$P\{X_0 \in A_0, X_1 - X_0 \in A_1, \ldots, X_n - X_{n-1} \in A_n\}$$

$$= P\{X_0 \in A_0\}P\{X_1 - X_0 \in A_1\} \cdots P\{X_n - X_{n-1} \in A_n\}. \quad \|$$

The above proof shows that

$$P\{X(s + t) - X(s) \in B\} = p_{s, s+t}(0, B).$$

Thus if $\{X(t)\}$ has stationary transition probabilities (so that $p_{s, s+t} = p_t$), the process has stationary independent increments. For example, to generate Brownian motion we may take

$$p_t(0, B) = P\{X_t \in B\} = \frac{1}{\sqrt{2\pi t}\,\sigma} \int_B e^{-y^2/2\sigma^2 t} \, dy,$$

hence

$$p_t(x, B) = p_t(0, B - x)$$

$$= \frac{1}{\sqrt{2\pi t}\,\sigma} \int_B e^{-(y-x)^2/2\sigma^2 t} \, dy.$$

To generate the Poisson process, let

$$p_t(j, j + k) = p_t(0, k) = \frac{e^{-\lambda t}(\lambda t)^k}{k!}, \qquad j, k = 0, 1, \ldots,$$

where, for example, $p_t(0, k)$ stands for $p_t(0, \{k\})$; $p_t(0, B)$ is given by $\sum_{k \in B} p_t(0, k)$.

In the case of a Markov chain with transition matrix $[p_{ij}]$, the spatial homogeneity condition becomes $p_{ij} = p_{0, j-i} = q_{j-i}$. In this case the chain corresponds to a sequence of sums of independent random variables, which is the discrete parameter analog of a process with independent increments.

Problems

1. Let $\{Y(t), t \geq 0\}$ be a process with stationary (not necessarily independent) increments, and let h_t be the characteristic function of $Y(t)$; assume $Y(0) \equiv 0$. If $h_t(u) \to 1$ as $t \to 0$ for each fixed u, show that $\{Y(t)\}$ is continuous in probability, and hence by 4.2.2, has a standard modification that is progressively measurable and separable.

2. Let $\{Y(t), t \geq 0\}$ be a Poisson process, that is, $\{Y(t)\}$ has stationary independent increments, with $Y(t) - Y(s)$ Poisson with parameter $\lambda(t - s)$, $s < t$. For the remainder of the problem, we assume $\{Y(t)\}$ separable.

(a) Show that almost every sample function is integer-valued and increasing.

(b) Show that, almost surely, the sample functions have jumps of height 1 and are constant between jumps.

4.7 Continuous Parameter Martingales

If it is known that a particular discrete parameter process forms an L^1 bounded martingale, significant information is obtained about the sample functions; they are almost surely convergent. In this section we extend the martingale concept to the continuous parameter case; and again we find that the sample functions have desirable properties.

4.7.1 Definitions and Comments

Let T be a totally ordered index set and let $\{X(t), t \in T\}$ be a stochastic process adapted to the family of σ-fields $\mathscr{F}(t)$, $t \in T$. We say that $\{X(t)\}$ is a *martingale* relative to the $\mathscr{F}(t)$ (or that $\{X(t), \mathscr{F}(t)\}$ is a martingale) iff each $X(t)$ is integrable, and, for $s < t$,

$$E[X(t) \,|\, \mathscr{F}(s)] = X(s) \quad \text{a.e.}$$

If $=$ is replaced by $\geq$, we obtain a submartingale; by $\leq$, a supermartingale.

If $\{X(t)\}$ is a martingale relative to the $\mathscr{F}(t)$, it is automatically a martingale relative to the σ-fields $\mathscr{F}_0(t) = \mathscr{F}(X(r), r \leq t)$. To see this, condition both sides of the defining relation with respect to $\mathscr{F}_0(s)$; similar statements hold for sub and supermartingales. If we do not mention the $\mathscr{F}(t)$ explicitly, we always mean $\mathscr{F}(t) = \mathscr{F}_0(t)$.

If $\{X(t), t \in I\}$ is a martingale for each finite set $I \subset T$, then $\{X(t), t \in T\}$ is a martingale. We prove this by the technique of 4.5.2(b). Let $r_1, \ldots,$

$r_n \le s < t$, with $r_n = s$. If $A \in \mathscr{F}(X(r_1), \ldots, X(r_n))$, then by definition of conditional expectation,

$$\int_A E[X(t) \mid X(r), r \le s] \, dP = \int_A X(t) \, dP$$

$$= \int_A E[X(t) \mid X(r_1), \ldots, X(r_n)] \, dP$$

$$= X(r_n) = X(s)$$

by hypothesis. By the monotone class theorem, this holds for all $A \in \mathscr{F}(X(r), r \le s)$, and the result follows.

Conversely, if $\{X(t), t \in T\}$ is a martingale, then $\{X(t), t \in I\}$ is a martingale for all $I \subset T$. This follows as in 4.5.2(a); similar statements hold for sub and supermartingales.

We now exhibit a large class of continuous parameter martingales.

4.7.2 Theorem

If $\{X(t), t \ge 0\}$ has independent increments and $E(\mid X(t) \mid) < \infty$ for all t, then $\{X(t) - E[X(t)], t \ge 0\}$ is a martingale.

PROOF Since $\{X(t) - E[X(t)]\}$ also has independent increments, we may assume without loss of generality that $E[X(t)] \equiv 0$. If $0 \le t_1 < \cdots < t_{n+1}$, then $X(0), X(t_1) - X(0), \ldots, X(t_{n+1}) - X(t_n)$ are independent, and therefore $X(t_{n+1}) - X(t_n)$ and $(X(0), X(t_1) - X(0), \ldots, X(t_n) - X(t_{n-1}))$ are independent. But then $X(t_{n+1}) - X(t_n)$ and $(X(t_1), \ldots, X(t_n))$ are independent, and hence

$$E[X(t_{n+1}) \mid X(t_1), \ldots, X(t_n)]$$

$$= X(t_n) + E[X(t_{n+1}) - X(t_n) \mid X(t_1), \ldots, X(t_n)]$$

$$= X(t_n) + E[X(t_{n+1}) - X(t_n)] \qquad \text{by independence}$$

$$= X(t_n).$$

Thus $\{X(t), t \in I\}$ is a martingale for each finite subset I of $[0, \infty)$, and the result follows from 4.7.1. ∥

We now examine the behavior of the sample functions of sub and supermartingales. All results will be stated for submartingales, but hold equally well for supermartingales upon replacing $X(t)$ by $-X(t)$.

4.7.3   Theorem

Let T be an interval of R, and let $\{X(t), t \in T\}$ be a separable submartingale. Then for almost every ω, $X(\cdot, \omega)$ is bounded on each bounded subinterval of T.

PROOF   Let T_0 be the separating set. If $t_1, \dots, t_n \in T_0, \lambda > 0$, then by the sub and supermartingale inequalities (RAP, p. 308, Problem 3),

$$P\left\{ \max_{1 \leq i \leq n} X(t_i) > \lambda \right\} \leq \frac{1}{\lambda} E[X(t_n)^+],$$

$$P\left\{ \min_{1 \leq i \leq n} X(t_i) < -\lambda \right\} = P\left\{ \max_{1 \leq i \leq n} (-X(t_i)) > \lambda \right\}$$

$$\leq \frac{1}{\lambda} (E[-X(t_1)] + E[(-X(t_n))^-])$$

$$= \frac{1}{\lambda} (-E[X(t_1)] + E[X(t_n)^+]).$$

If $t_1, \dots, t_n \in [c, d]$, then $E[X(c)] \leq E[X(t_1)]$, $E[X(t_n)^+] \leq E[X(d)^+]$; thus we obtain, upon letting $n \to \infty$,

$$P\left\{ \sup_{\substack{t \in T_0 \\ c \leq t \leq d}} X(t) > \lambda \right\} \leq \frac{1}{\lambda} E[X(d)^+],$$

$$P\left\{ \inf_{\substack{t \in T_0 \\ c \leq t \leq d}} X(t) < -\lambda \right\} \leq \frac{1}{\lambda} (-E[X(c)] + E[X(d)^+]).$$

By separability, we may replace T_0 by T; thus

$$P\left\{ \sup_{\substack{t \in T \\ c \leq t \leq d}} X(t) = \infty \right\} = \lim_{\lambda \to \infty} P\left\{ \sup_{\substack{t \in T \\ c \leq t \leq d}} X(t) > \lambda \right\}$$

$$\leq \lim_{\lambda \to \infty} \frac{1}{\lambda} E[X(d)^+]$$

$$= 0.$$

Similarly,

$$P\left\{ \inf_{\substack{t \in T \\ c \leq t \leq d}} X(t) = -\infty \right\} = 0.$$

Consequently,

$P\{\omega : X(\cdot, \omega)$ bounded on every bounded subinterval of $T\}$

$$= P\left[\bigcap_{n=1}^{\infty} \{\omega : X(\cdot, \omega) \text{ bounded on } [-n, n] \cap T\}\right]$$

$$= 1. \quad \|$$

4.7.4   Theorem

Under the hypothesis of 4.7.3, for almost every ω, $X(\cdot, \omega)$ has no oscillatory discontinuities, that is,

$$X(t^+, \omega) = \lim_{t' \downarrow t} X(t', \omega) \qquad \text{and} \qquad X(t^-, \omega) = \lim_{t' \uparrow t} X(t', \omega)$$

exist for all t.

PROOF   Let $t_1, \ldots, t_n \in [c, d] \cap T_0$, $t_1 < \cdots < t_n$. If $a < b$ and U_{ab} is the number of upcrossings of $[a, b]$, that is, the number of excursions from a value less than a to a value greater than b, by $X(t_1), \ldots, X(t_n)$, then

$$E(U_{ab}) \le \frac{1}{b - a} E[(X(t_n) - a)^+] \qquad \text{by RAP, 7.4.2, p. 291}$$

$$\le \frac{1}{b - a} E[(X(d) - a)^+]$$

since $\{(X(t) - a)^+\}$ is a submartingale

(RAP, 7.3.6, p. 288).

Let $n \to \infty$ to conclude that the average number of upcrossings of $[a, b]$ by $X(t)$, $t \in T_0 \cap [c, d]$, is finite, hence for almost every ω, $(X(t, \omega),$ $t \in T_0 \cap [c, d])$ has a finite number of upcrossings of $[a, b]$ for all rational a and b, $a < b$. By separability, the same is true for $(X(t, \omega), t \in T \cap [c, d])$. But if $f: T \to R$ and f has, say, no left limit at t, then we can find distinct $t_n \uparrow t$ such that $\lim \inf_{n \to \infty} f(t_n) = u < v = \lim \sup_{n \to \infty} f(t_n)$. Pick rational a and b with $u < a < b < v$. Then $f(t_n)$ will be less than a infinitely often and greater than b infinitely often, and hence f will have an infinite number of upcrossings of $[a, b]$. Take $f = X(\cdot, \omega)$ to conclude that almost every sample function has right and left limits on $T \cap [c, d]$. Since c and d are arbitrary, the result follows. $\|$

A remarkable fact about continuous parameter submartingales is that even without a separability hypothesis, strong conclusions can be made about sample function behavior.

4.7.5 Theorem

Let T be an open interval of R, and let $\{X(t), t \in T\}$ be a submartingale, not necessarily separable. If T_0 is an arbitrary countable dense result of T, then for almost every ω,

$$\lim_{\substack{t' \uparrow t \\ t' \in T_0}} X(t', \omega) \qquad \text{and} \qquad \lim_{\substack{t' \downarrow t \\ t' \in T_0}} X(t', \omega)$$

exist for all t.

PROOF Just as in 4.7.4, for almost every ω, the number of upcrossings of $[a, b]$ by $(X(t, \omega), t \in T_0 \cap [c, d])$ is finite. But if, say, $t \in [c, d]$ and

$$\lim_{\substack{t' \uparrow t \\ t' \in T_0 \cap [c, d]}} X(t', \omega)$$

fails to exist, the argument of 4.7.4 shows that $(X(t', \omega), t' \in T_0 \cap [c, d])$ will have an infinite number of upcrossings of $[a, b]$ for some rational a, b. Since c and d are arbitrary, the result follows. ‖

If T is allowed to be closed or semiclosed, Theorem 4.7.5 still holds, if we require that T_0 contain the endpoints of T. The same remark applies to Problems 1 and 2.

Problems

1. Let T_0 be a countable dense subset of the open interval $T \subset R$. Assume that $f: T \to R$, and

$$g(t) = \lim_{\substack{t' \downarrow t \\ t' \in T_0}} f(t'), \qquad h(t) = \lim_{\substack{t' \uparrow t \\ t' \in T_0}} f(t')$$

exist for all $t \in T$. Show that g is right-continuous with left limits.

2. Let T_0 be a countable dense subset of the open interval $T \subset R$. Let $\{X(t), t \in T\}$ be a stochastic process that is continuous in probability from the right, in other words, $X(t) \xrightarrow{P} X(t_0)$ as $t \downarrow t_0$. Assume that almost every sample function has right and left limits on sequences in T_0, that is, for almost every ω,

$$Y(t, \omega) = \lim_{\substack{t' \downarrow t \\ t' \in T_0}} X(t', \omega), \qquad Z(t, \omega) = \lim_{\substack{t' \uparrow t \\ t' \in T_0}} X(t', \omega)$$

exist for all t. (Thus by Problem 1, for almost every ω, $Y(\cdot, \omega)$ is right-continuous with left limits.) Show that for each t, $X(t) = Y(t)$ a.e.

Thus by 4.7.5, a continuous parameter submartingale that is continuous in probability has a standard modification whose sample functions are right-continuous with left limits.

3. Let T be an interval of R, and let $\{X(t), t \in T\}$ be a separable submartingale which is continuous in probability. Fix t, and show that for almost every ω, either $X(t^+, \omega) = X(t, \omega)$ or $X(t^-, \omega) = X(t, \omega)$.

4. Let T be an interval of R, and let $\{X(t), t \in T\}$ be continuous in probability, with sample functions that are right-continuous with left limits. Show that for each fixed t, $X(t) = X(t^-)$ a.e.

5. Let $\{X(t), t \geq 0\}$ be a martingale with right-continuous sample functions, and let T be a finite stopping time for $\{X(t)\}$. Assume that

(a) $E\left[\sup_{0 \leq s \leq t} |X(s)| \right] < \infty$ for each t.

(b) $E(|X(T)|) < \infty$.

(c) $\liminf_{t \to \infty} \int_{\{T > t\}} |X(t)|\ dP = 0$.

Show that $\{X(0), X(T)\}$ is a martingale (and hence $E[X(0)] = E[X(T)]$).

4.8   The Strong Markov Property

In this section we consider the strong Markov property for general Markov processes. Intuitively, if $\{X(t), t \geq 0\}$ is Markov and T is a stopping time, we would like to show that the process starts from scratch at time T, in other words, $\{X(T + t), t \geq 0\}$ has the same character as the original process. We start with the discrete parameter case.

4.8.1   Theorem

Let $\{X_n, \mathscr{F}_n, n = 0, 1, \ldots\}$ be a Markov process with state space $(S_0, \mathscr{S}_0)$, generated by an initial distribution P_0 and transition probabilities p_{mn}, $m, n = 0, 1, \ldots, m < n$. For convenience of notation we write $p(m, n, x, B)$ instead of $p_{mn}(x, B)$.

Let S be a stopping time relative to the σ-fields $\mathscr{F}_n$, and let T be an $\mathscr{F}(S)$-measurable map of Ω to $\{0, 1, \ldots, \infty\}$; assume $T \geq S$, so that by 4.2.4(c), T is also a stopping time. If $B \in \mathscr{S}_0$, then

$$P\{X(T) \in B, T < \infty \mid \mathscr{F}(S)\}$$
$$= p(S, T, X(S), B) \quad \text{a.e.} \qquad \text{on } \{T < \infty\}.$$

PROOF   Let $A \in \mathcal{F}(S)$; then

$$P[\{X(T) \in B, \, T < \infty\} \cap A]$$

$$= \sum_{n=0}^{\infty} \sum_{m=0}^{n} P[\{S = m, \, T = n, \, X_n \in B\} \cap A].$$

Since T is $\mathcal{F}(S)$-measurable, $\{T = n\} \in \mathcal{F}(S)$, hence $A \cap \{T = n\} \in \mathcal{F}(S)$, and therefore $A \cap \{T = n\} \cap \{S = m\} \in \mathcal{F}_m$. Consequently,

$$P[A \cap \{S = m, \, T = n\} \cap \{X_n \in B\}]$$

$$= \int_{A \cap \{S=m, \, T=n\}} P\{X_n \in B \,|\, \mathcal{F}_m\} \, dP.$$

Since $P\{X_n \in B \,|\, \mathcal{F}_m\} = P\{X_n \in B \,|\, X_m\}$, this becomes, by Eq. (4) of 4.5.5,

$$\int_{A \cap \{S=m, \, T=n\}} p(m, n, X_m, B) \, dP$$

$$= \int_{A \cap \{S=m, \, T=n\}} p(S, T, X(S), B) \, dP.$$

Sum over m and n to obtain

$$P[\{X(T) \in B, \, T < \infty\} \cap A]$$

$$= \int_{A \cap \{T<\infty\}} p(S, T, X(S), B) \, dP.$$

Since S, T, and $X(S)$ are $\mathcal{F}(S)$-measurable, so is $p(S, T, X(S), B)$, and the result follows. $\|$

If we replace S by $S + n$, T by $S + n + 1$ and assume S finite, then

$$P\{X(S + n + 1) \in B \,|\, \mathcal{F}(S + n)\} = p(S + n, S + n + 1, X(S + n), B)$$

which implies that $\{X(S + n), \, n = 0, 1, \ldots\}$ is a Markov process whose conditional probabilities are calculated by the transition probabilities of the original process.

We turn now to the continuous parameter case. If $p = p(x, B)$, $x \in S_0$, $B \in \mathcal{S}_0$, is a transition probability, the *transition operator* Q associated with p is defined by

$$(Qf)(x) = \int_{S_0} f(y)p(x, dy)$$

where f belongs to the class $B(S_0)$ of bounded Borel measurable mappings from S_0 to R; thus Q maps $B(S_0)$ into itself. If a Markov process $\{X(t), t \geq 0\}$

has transition probabilities p_{rs}, the Chapman–Kolmogorov equation may be expressed as $Q_{st} \circ Q_{rs} = Q_{rt}$, $r < s < t$. (This may be proved by starting with indicators.) Note also that

$$(Q_{st} f)(x) = E[f(X(t)) \mid X(s) = x]$$

(see Eq. (6) of 4.5.6).

4.8.2   Theorem

Let $\{X(t), t \geq 0\}$ be a Markov process with state space $(S_0, \mathscr{S}_0)$, where S_0 is a compact metric space and $\mathscr{S}_0 = \mathscr{B}(S_0)$. Let the process be generated by an initial distribution P_0 and transition probabilities $p(r, s)$, $r < s$. Assume that for all ω, $X(\cdot, \omega)$ is right-continuous. Let S be a stopping time relative to the σ-fields $\mathscr{F}(t^+)$, where $\mathscr{F}(t) = \mathscr{F}(X(s), s \leq t)$ (see Problem 4, Section 4.2). Let T be an $\mathscr{F}(S)$-measurable mapping of Ω into $[0, \infty]$, with $T \geq S$; by 4.2.4(c), T is also a stopping time relative to the $\mathscr{F}(t^+)$.

Let $C(S_0)$ be the class of all continuous real-valued functions on S_0. Assume that for each $f \in C(S_0)$, $(Q_{st} f)(x)$ is jointly continuous in s, t, and x (in particular, $Q_{st} f \in C(S_0)$). If $B \in \mathscr{S}_0$, then

$$P\{X(T) \in B, \, T < \infty \mid \mathscr{F}(S)\}$$
$$= p(S, T, X(S), B) \qquad \text{a.e.} \qquad \text{on } \{T < \infty\}.$$

PROOF   First note that right-continuity implies progressive measurability, which implies that $X(T)$ is $\mathscr{F}(T)$-measurable; see 4.2.5 and Problem 2, Section 4.2.

Let $T^{(n)} = j2^{-n}$ if $(j - 1)2^{-n} \leq T < j2^{-n}$, $j, n = 1, 2, \ldots$; set $T^{(n)} = \infty$ if $T = \infty$, and define $S^{(n)}$ similarly. Then for each n, $\{X(j2^{-n}), \mathscr{F}(j2^{-n}), j = 1, 2, \ldots\}$ is Markov, and $S^{(n)}$ and $T^{(n)}$ are stopping times relative to the σ-fields $\mathscr{F}(j2^{-n})$, $j = 1, 2, \ldots$, since

$$\{S^{(n)} = j2^{-n}\} = \{(j - 1)2^{-n} \leq S < j2^{-n}\} \in \mathscr{F}(j2^{-n}).$$

Now $\mathscr{F}(S) \subset \mathscr{F}(S^{(n)})$ by 4.2.4(e), so that T is $\mathscr{F}(S^{(n)})$-measurable, and therefore so is $T^{(n)}$. Also, $S^{(n)} \leq T^{(n)}$, $S^{(n)} \downarrow S$, $T^{(n)} \downarrow T$. By 4.8.1, if $B \in \mathscr{S}_0$,

$$P\{X(T^{(n)}) \in B, \, T^{(n)} < \infty \mid \mathscr{F}(S^{(n)})\}$$
$$= p(S^{(n)}, T^{(n)}, X(S^{(n)}), B) \quad \text{a.e.}$$
$$\text{on } \{T^{(n)} < \infty\} = \{T < \infty\}.$$

But this says that

$$E[f(X(T^{(n)}))I_{\{T < \infty\}} \mid \mathscr{F}(S^{(n)})]$$
$$= I_{\{T < \infty\}}(Q_{S^{(n)}T^{(n)}} f)(X(S^{(n)})) \quad \text{a.e.} \tag{1}$$

for f an indicator, and hence by the usual passage for f an arbitrary bounded Borel measurable function.

If $A \in \mathscr{F}(S) \subset \mathscr{F}(S^{(n)})$, then by (1),

$$\int_{A \cap \{T < \infty\}} f(X(T^{(n)})) \, dP$$

$$= \int_{A \cap \{T < \infty\}} (Q_{S^{(n)}T^{(n)}} f)(X(S^{(n)})) \, dP. \tag{2}$$

Assuming f continuous, the right-continuity of the process, joint continuity hypothesis and the dominated convergence theorem yield

$$\int_{A \cap \{T < \infty\}} f(X(T)) \, dP = \int_{A \cap \{T < \infty\}} (Q_{ST} f)(X(S)) \, dP. \tag{3}$$

If C is an open subset of S_0, then I_C is lower semicontinuous, and hence there is an increasing sequence of continuous functions converging to I_C. Since (3) holds for all continuous functions, it holds for I_C, by the monotone convergence theorem.

Let $\mathscr{H}$ be the collection of functions $f \colon S_0 \to R$ such that (3) holds for f. Then $\mathscr{H}$ is a vector space, $I_C \in \mathscr{H}$ for every open set C, and $\mathscr{H}$ is closed under pointwise limits of bounded sequences of functions. By RAP, 4.1.4, p. 169, $I_B \in \mathscr{H}$ for every Borel set B (in fact every bounded Borel measurable function belongs to $\mathscr{H}$). A similar argument shows that if f is bounded Borel measurable, $(Q_{st} f)(x)$ is jointly measurable in s, t, and x; consequently, if $B \in \mathscr{B}(S_0)$, then $(Q_{ST} I_B)(X(S)) = p(S, T, X(S), B)$ is $\mathscr{F}(S)$-measurable. Since (3) holds for I_B, we obtain

$$P\{X(T) \in B, \, T < \infty \mid \mathscr{F}(S)\} = I_{\{T < \infty\}} p(S, T, X(S), B). \quad \parallel$$

Just as in the discussion after 4.8.1, if T is finite then $\{X(T + t), t \geq 0\}$ is a Markov process whose conditional probabilities are calculated using the original transition probabilities.

We now use the strong Markov property to give a proof of Theorem 4.3.2. Set

$$S = \inf \{t' : B(t') > a\} \qquad (S = \infty \text{ if } B(t') \text{ never goes above } a)$$

$$T = t \quad \text{if } S < t, \qquad T = \infty \quad \text{if } S \geq t.$$

It is not difficult to check that the transition operators for Brownian motion (see the end of Section 4.6) satisfy the joint continuity hypothesis of 4.8.2 (the state space is taken to be $\bar{R}$), and thus the strong Markov property applies. We have, with $B = (a, \infty)$,

$$P\{B(T) \in B, \, T < \infty \mid \mathscr{F}(S)\} = p(S, T, B(S), B) I_{\{T < \infty\}}.$$

But if $T < \infty$, then $B(S) \equiv a$, and $p(S, T, B(S), B) = \frac{1}{2}$. Integrate both sides of the above equation over Ω to obtain

$$P\{B(t) > a, S < t\} = \tfrac{1}{2}P\{S < t\}.$$

An identical argument with $B = (-\infty, a)$ proves

$$P\{B(t) < a, S < t\} = \tfrac{1}{2}P\{S < t\}.$$

Therefore

$$P\{B(t) > a, S < t\} = P\{B(t) < a, S < t\}$$

which is Eq. (1) of 4.3.2.

4.9 Notes

An alternative construction of one-dimensional Brownian motion, and many additional properties of the process, may be found in Freedman (1971). This book also considers diffusion processes, a natural generalization of the one-dimensional Brownian motion. Diffusion processes are also discussed by Breiman (1968).

The general theory of Markov processes makes constant use of the concepts of separability and progressive measurability; discrete and continuous parameter martingales are also important. Comprehensive developments are given by Blumenthal and Getoor (1968) and Dynkin (1965).

A further discussion of sample function continuity of stochastic processes is given by Gikhman and Skorokhod (1969).

Chapter 5

The Itô Integral and Stochastic Differential Equations

5.1 Definition of the Itô Integral

In Chapter 2 we considered stochastic integrals and stochastic differential equations in the L^2 sense. Now that we have available many results on the behavior of the sample functions of continuous parameter processes, we are in a position to study a large class of differential equations with random coefficients, and discuss the sample function behavior of the solutions. The basic tool will be the Itô stochastic integral, which we proceed to define.

5.1.1 Definitions and Comments

Let $\{B(s),\ a \le s \le t\}$ be Brownian motion on a probability space $(\Omega, \mathscr{F}, P)$; for convenience we take $\sigma^2 = 1$. We are going to define the stochastic integral

$$\int_a^b f(t, \omega)\, dB(t, \omega)$$

or for short

$$\int_a^b f(t)\, dB(t)$$

210

for an appropriate class of integrands f; namely, for the class $\mathscr{C}$ of real-valued functions on $[a, b] \times \Omega$ satisfying:

(a) f is $\mathscr{B}[a, b] \times \mathscr{F}$-measurable.

(b)   For each $t \in [a, b], f(t) = f(t, \cdot)$ is $\mathscr{F}(t)$-measurable, where $\mathscr{F}(t) = \mathscr{F}(B(s), a \le s \le t)$.

(c)   For each $t \in [a, b], f(t) \in L^2(\Omega, \mathscr{F}, P)$, and $\int_a^b E(|f(t)|^2)\, dt < \infty$.

Intuitively, (b) says that f is nonanticipating with respect to $\{B(t)\}$; in other words, the value of $f(t)$ depends only on past and present values of the Brownian notion. Note also that in contrast with the stochastic integrals of Chapter 2, both the integrand f and the "differential" dB are random.

We first define the integral for a simple type of integrand. Call f a *step function* if $f \in \mathscr{C}$ and there is a partition $a = t_0 < t_1 < \cdots < t_n = b$ with associated random variables $f_0, f_1, \ldots, f_{n-1}$ such that

$$f(t, \omega) = \sum_{i=0}^{n-1} f_i(\omega) I_{[t_i, t_{i+1})}(t). \tag{1}$$

(To avoid complicating the notation, we adopt the convention that $[t_{n-1}, t_n) = [t_{n-1}, b]$.) We define the integral $\Phi(f) = \int_a^b f(t)\, dB(t)$ by

$$\Phi(f)(\omega) = \sum_{i=0}^{n-1} f_i(\omega)[B(t_{i+1}, \omega) - B(t_i, \omega)]. \tag{2}$$

If f and g are step functions, they can be represented (as in (1)) using the same partition $a = t_0 < t_1 < \cdots < t_n = b$, but with possibly different random variables f_i and g_i. This observation allows us to establish the following basic properties of the integral (2).

5.1.2   Lemma

If f and g are step functions, then

$$E\left[\int_a^b f(t)\, dB(t)\right] = 0$$

and

$$E\left[\int_a^b f(t)\, dB(t) \int_a^b g(t)\, dB(t)\right] = \int_a^b E[f(t)g(t)]\, dt.$$

PROOF By definition of the integral, we have

$$E\left[\int_a^b f(t)\, dB(t)\right] = E\left[\sum_{i=0}^{n-1} f_i(B(t_{i+1}) - B(t_i))\right]$$

$$= E\left[\sum_{i=0}^{n-1} E\{f_i(B(t_{i+1}) - B(t_i)) \mid \mathscr{F}(t_i)\}\right]$$

$$= E\left[\sum_{i=0}^{n-1} f_i E\{B(t_{i+1}) - B(t_i) \mid \mathscr{F}(t_i)\}\right]$$

since $f_i = f(t_i)$ is $\mathscr{F}(t_i)$-measurable

$$= E\left[\sum_{i=0}^{n-1} f_i E\{B(t_{i+1}) - B(t_i)\}\right] = 0$$

since $\{B(t)\}$ has independent increments. Also,

$$E\left[\int_a^b f(t)\, dB(t) \int_a^b g(t)\, dB(t)\right]$$

$$= E\left[\sum_{i,\, j=0}^{n-1} f_i g_j (B(t_{i+1}) - B(t_i))(B(t_{j+1}) - B(t_j))\right]$$

$$= E\left[\sum_{i,\, j=0}^{n-1} E\{f_i g_j (B(t_{i+1})\right.$$

$$\left. - B(t_i))(B(t_{j+1}) - B(t_j)) \mid \mathscr{F}(t_i \vee t_j)\}\right].$$

If $i < j$, then $t_i \vee t_j = t_j$, and since f_i, g_j, and $B(t_{i+1}) - B(t_i)$ are $\mathscr{F}(t_j)$-measurable, the inner expectation is 0 as above. If $i = j$, the inner expectation is

$$f_i g_i E[\,|B(t_{i+1}) - B(t_i)|^2 \mid \mathscr{F}(t_i)]$$

$$= f_i g_i E[\,|B(t_{i+1}) - B(t_i)|^2]$$

since $\{B(t)\}$ has independent increments

$$= f(t_i)g(t_i)(t_{i+1} - t_i).$$

Thus

$$E\left[\int_a^b f(t)\, dB(t) \int_a^b g(t)\, dB(t)\right] = \sum_{i=0}^{n-1} E[f(t_i)g(t_i)](t_{i+1} - t_i)$$

$$= \int_a^b E[f(t)g(t)]\, dt. \quad \|$$

Now $\mathscr{C}$ is a closed subspace of the Hilbert space $L^2([a, b] \times \Omega, \mathscr{B}[a, b] \times \mathscr{F}, m \times P)$ (where m is Lebesgue measure, and functions that agree a.e. $[m \times P]$ are identified), and is therefore itself a Hilbert space. The inner product is given by

$$\langle f, g \rangle = \int_a^b \int_\Omega f(t, \omega)g(t, \omega) \, dP(\omega) \, dt$$

$$= \int_a^b E[f(t)g(t)] \, dt.$$

By Lemma 5.1.2 and the definition of the integral, the map Φ from the space S of step functions to $L^2(\Omega, \mathscr{F}, P)$ is linear and preserves inner products. Therefore Φ has a unique continuous linear extension (also denoted by Φ) to the closure of S in $\mathscr{C}$. But, as the following lemma shows, the closure of S is $\mathscr{C}$ itself, so that we may extend the definition of the integral to all of $\mathscr{C}$.

5.1.3 Lemma

Step functions are dense in $\mathscr{C}$.

PROOF   We show successively that the closure $\bar{S}$ of S in $\mathscr{C}$ contains

(a)   all L^2-continuous $f \in \mathscr{C}$, that is, all $f \in \mathscr{C}$ such that the map $t \to f(t, \cdot)$ of $[a, b]$ to $L^2(\Omega, \mathscr{F}, P)$ is continuous (hence uniformly continuous),

(b)   all bounded $f \in \mathscr{C}$, and

(c)   all $f \in \mathscr{C}$.

(a)   If f is an L^2-continuous function in $\mathscr{C}$ and $a = t_0 < t_1 < \cdots < t_n = b$, define

$$g(t, \omega) = \sum_{i=0}^{n-1} f(t_i, \omega)I_{[t_i, t_{i+1})}(t).$$

Then g is a step function, and if $t_i \leq t < t_{i+1}$, we have $g(t, \omega) = f(t_i, \omega)$. It follows by L^2-continuity that $E(|g(t) - f(t)|^2)$ can be made arbitrarily small, uniformly in t. Therefore $\int_a^b E(|g(t) - f(t)|^2) \, dt$ can be made arbitrarily small, so that $f \in \bar{S}$.

(b)   If f is a bounded function in $\mathscr{C}$, define

$$f_n(t, \omega) = \int_0^\infty e^{-x} f\left(t - \frac{x}{n}, \omega\right) dx$$

where we take $f(t, \omega) = 0$ for $t \notin [a, b]$. Then $f_n \in \mathscr{C}$ (see Problem 1), and f_n is L^2-continuous:

$$E[\,|\,f_n(t + s) - f_n(t)\,|^2\,] = E\left[\left|\int_0^\infty e^{-x}\left[f\left(t + s - \frac{x}{n}\right) - f\left(t - \frac{x}{n}\right)\right] dx\right|^2\right]$$

$$\leq E\left[\int_0^\infty e^{-x}\left|f\left(t + s - \frac{x}{n}\right) - f\left(t - \frac{x}{n}\right)\right|^2 dx\right]$$

by Jensen's inequality

$$\leq nE\left[\int_0^\infty |f(t + s - y) - f(t - y)|^2\right] dy$$

$$\to 0 \qquad \text{as} \quad s \to 0 \text{ by Appendix 2, Lemma A2.3.}$$

Thus by (a), $f_n \in \bar{S}$. We now show that $f_n \to f$ in the norm of $\mathscr{C}$, so that $f \in \bar{S}$. We have

$$\|f_n - f\|^2 = E\left[\int_a^b \left|f(t) - \int_0^\infty e^{-x}f\left(t - \frac{x}{n}\right) dx\right|^2 dt\right]$$

$$= E\left[\int_a^b \left|\int_0^\infty e^{-x}\left[f(t) - f\left(t - \frac{x}{n}\right)\right] dx\right|^2 dt\right]$$

$$\leq E\left[\int_a^b \int_0^\infty e^{-x}\left|f(t) - f\left(t - \frac{x}{n}\right)\right|^2 dx\, dt\right]$$

by Jensen's inequality

$$= E\left[\int_0^\infty e^{-x}\left(\int_a^b \left|f(t) - f\left(t - \frac{x}{n}\right)\right|^2 dt\right) dx\right]$$

$$\to 0 \qquad \text{as} \quad n \to \infty \text{ by A2.3 and the}$$

dominated convergence theorem.

(c) If $f \in \mathscr{C}$, define $f_n = f$ if $|f| \leq n$; $f_n = 0$ if $|f| > n$. By (b), $f_n \in \bar{S}$. But

$$\|f_n - f\|^2 = \int_a^b E[\,|\,f_n(t) - f(t)\,|^2\,]\, dt$$

$$= \int_a^b \int_{\{\omega\,:\,|f(t,\,\omega)| > n\}} |f(t, \omega)|^2\, dP(\omega)\, dt$$

$$\to 0 \qquad \text{as} \quad n \to \infty \text{ by the dominated convergence theorem.}$$

Thus $f \in \bar{S}$. $\|$

5.1.4 Theorem

The integral $\Phi(f) = \int_a^b f(t)\, dB(t), f \in S$, has a unique continuous linear extension to $\mathscr{C}$. If $f, g \in \mathscr{C}$, we have

$$E\left[\int_a^b f(t)\, dB(t)\right] = 0$$

$$E\left[\int_a^b f(t)\, dB(t) \int_a^b g(t)\, dB(t)\right] = \int_a^b E[f(t)g(t)]\, dt.$$

PROOF This is immediate from 5.1.2, 5.1.3, and the continuity of the inner product. ‖

We now consider the stochastic integral as a function of its upper limit. Fix the function $f \in \mathscr{C}$, and define

$$X(t) = \int_a^t f(s)\, dB(s), \qquad a \le t \le b.$$

The following result is basic.

5.1.5 Theorem

The process $\{X(t), a \le t \le b\}$ is an L^2-continuous martingale.

PROOF To establish the martingale property it suffices to show that $\{X(t), \mathscr{F}(t), a \le t \le b\}$ is a martingale (see 4.7.1). First assume f is a step function; if $s < t$,

$$X(t) - X(s) = \int_s^t f(u)\, dB(u)$$

$$= \sum_{i=0}^{n-1} f(t_i)(B(t_{i+1}) - B(t_i))$$

where $s = t_0 < t_1 < \cdots < t_n = t$. Set $s = a$ to observe that $X(t)$ is $\mathscr{F}(t)$-measurable. Now we must show that $E[X(t) - X(s)\,|\,\mathscr{F}(s)] = 0$; this is a consequence of the following computation:

$$E[f(t_i)(B(t_{i+1}) - B(t_i))\,|\,\mathscr{F}(s)]$$
$$= E\{E[f(t_i)(B(t_{i+1}) - B(t_i))\,|\,\mathscr{F}(t_i)]\,|\,\mathscr{F}(s)\}$$

and

$$E[f(t_i)(B(t_{i+1}) - B(t_i))\,|\,\mathscr{F}(t_i)]$$
$$= f(t_i)E[B(t_{i+1}) - B(t_i)\,|\,\mathscr{F}(t_i)]$$
$$= 0.$$

If f is an arbitrary function in $\mathscr{C}$, let $\{f_n\}$ be a sequence of step functions converging to f in $L^2(m \times P)$. Define

$$X_n(t) = \int_a^t f_n(u) \, dB(u).$$

Then $\{X_n(t), \mathscr{F}(t), a \le t \le b\}$ is a martingale, and for $s < t$ we have

$$E[X(t) - X(s) \,|\, \mathscr{F}(s)] = E[X(t) - X_n(t) \,|\, \mathscr{F}(s)]$$
$$+ E[X_n(t) - X_n(s) \,|\, \mathscr{F}(s)] + E[X_n(s) - X(s) \,|\, \mathscr{F}(s)].$$

The second term on the right-hand side is 0 by the martingale property, and the first and third terms converge to 0 in $L^2(\Omega, \mathscr{F}, P)$ as $n \to \infty$; to see this, observe that for any t we have

$$E[\,|\, E(X(t) - X_n(t) \,|\, \mathscr{F}(s))\,|^2]$$
$$\le E[E(\,|\, X(t) - X_n(t)\,|^2) \,|\, \mathscr{F}(s))]$$

by Jensen's inequality

$$= E[\,|\, X(t) - X_n(t)\,|^2]$$

$$= E\left[\left|\int_a^t [f(u) - f_n(u)] \, dB(u)\right|^2\right]$$

$$= \int_a^t E(\,|\, f(u) - f_n(u)\,|^2) \, du \qquad \text{by 5.1.4}$$

$$\to 0 \qquad \text{as} \quad n \to \infty \text{ by the choice of } \{f_n\}.$$

This computation also shows that $X_n(t) \to X(t)$ in $L^2(\Omega, \mathscr{F}, P)$ as $n \to \infty$, so that $X(t)$ is $\mathscr{F}(t)$-measurable. (We should complete the measure P to avoid any problems here.)

We have now shown that $\{X(t)\}$ is a martingale; L^2-continuity follows from

$$E[\,|\, X(t) - X(s)\,|^2] = E\left[\left|\int_s^t f(u) \, dB(u)\right|^2\right]$$

$$= \int_s^t E(\,|\, f(u)\,|^2) \, du$$

$$\to 0 \qquad \text{as} \quad t \to s \text{ by condition (c) of 5.1.1}$$

and the dominated convergence theorem. $\quad \|$

It follows from 4.2.2 that $\{X(t), a \leq t \leq b\}$ has a standard modification that is progressively measurable and separable; from now on, we assume that we are working with such a version of $\{X(t)\}$, as well as with a separable version of $\{B(t)\}$. Furthermore, by 4.1.6, any countable dense subset of $[a, b]$ can be used as the separating set. We can then prove that every sample function of $\{X(t)\}$ is continuous, with the aid of the following submartingale inequality.

5.1.6 Theorem

If $\{Y(t), a \leq t \leq b\}$ is a separable submartingale and $\lambda > 0$, then

$$P\left\{ \sup_{a \leq t \leq b} Y(t) > \lambda \right\} \leq \frac{1}{\lambda} E[Y(b)^+].$$

PROOF   See the proof of 4.7.3.   ‖

5.1.7 Theorem

Almost every sample function of $\{X(t), a \leq t \leq b\}$ is continuous on $[a, b]$.

PROOF   First let $X(t) = \int_a^t f(u)\, dB(u)$ where f is a step function. Then if $a = t_0 < t_1 < \cdots < t_n = b$ is the partition associated with f, and $t_m \leq t \leq t_{m+1}$ $(m = m(t))$, we have

$$X(t, \omega) = \sum_{i=0}^{m-1} f(t_i, \omega)[B(t_{i+1}, \omega) - B(t_i, \omega)]$$
$$+ f(t_m, \omega)[B(t, \omega) - B(t_m, \omega)],$$

and the result follows from the sample function continuity of Brownian motion (4.3.1).

If f is an arbitrary function in $\mathscr{C}$, choose step functions f_n such that $\|f - f_n\|^2 \leq 1/n^4$, and let $\{X_n(t), a \leq t \leq b\}$, be a progressively measurable and separable version of

$$\left\{ \int_a^t f_n(u)\, dB(u), a \leq t \leq b \right\}.$$

By 5.1.5, 4.1.4, and RAP, 7.3.6(b), p. 288,

$$\{|X(t) - X_n(t)|^2, a \leq t \leq b\}$$

is a progressively measurable and separable submartingale. By 5.1.6,

$$P\left\{ \sup_{a \le t \le b} |X(t) - X_n(t)|^2 > 1/n^2 \right\}$$

$$\le n^2 E[|X(b) - X_n(b)|^2]$$

$$= n^2 E\left[\left| \int_a^b [f(u) - f_n(u)] \, dB(u) \right|^2 \right]$$

$$= n^2 \int_a^b E[|f(u) - f_n(u)|^2] \, du \qquad \text{by 5.1.4}$$

$$= n^2 \| f - f_n \|^2$$

$$\le 1/n^2.$$

By the Borel–Cantelli lemma, for almost every ω we have, for sufficiently large n (depending on ω),

$$|X(t, \omega) - X_n(t, \omega)| \le 1/n \qquad \text{for all} \quad t \in [a, b].$$

Therefore $X_n(t, \omega) \to X(t, \omega)$ as $n \to \infty$, uniformly in t. But then $X(\cdot, \omega)$ is a uniform limit of continuous functions, and hence is continuous. ‖

Problems

1. If f is a bounded function in $\mathscr{C}$ and

$$f_n(t, \omega) = \int_0^\infty e^{-x} f\left(t - \frac{x}{n}, \omega \right) dx,$$

show that $f_n \in \mathscr{C}$.

2. (*Quadratic Variation of Brownian Motion*) Let $\{B(t), a \le t \le b\}$ be Brownian motion on $[a, b]$, and let $P: a = t_0 < t_1 < \cdots < t_n = b$ be a partition of $[a, b]$. Form the quadratic variation

$$V(P) = \sum_{i=0}^{n-1} [B(t_{i+1}) - B(t_i)]^2.$$

(a) If $\{P_m\}$ is any sequence of partitions such that

$$|P_m| = \max_i |t_{i+1} - t_i| \to 0 \text{ as } m \to \infty,$$

show that $V(P_m) \to \sigma^2(b - a)$ in L^2.

(b) In (a), show that if $\sum_{m=1}^\infty |P_m| < \infty$, then $V(P_m)$ converges to $\sigma^2(b - a)$ a.e.

(c) For the partition P define

$$S(P) = \sum_{i=0}^{n-1} \left| (B(t_{i+1}) - B(t_i))^2 - (t_{i+1} - t_i) \right|.$$

If $\{P_m\}$ is a sequence of partitions such that $\sum_{m=1}^{\infty} |P_m| < \infty$, show that $S(P_m) \to 0$ a.e.

3. Suppose $f \in \mathscr{C}$, but f is a function of t alone (that is, $f(t, \omega) = f(t)$), then $\int_a^b f(t)\, dB(t)$ may be defined as in Chapter 2 (see 2.1.6) or as in this chapter. Show that the two definitions agree in this case.

5.2 Existence and Uniqueness Theorems for Stochastic Differential Equations

In this section we consider the stochastic differential equation

$$X'(t) = m(X(t), t) + \sigma(X(t), t)B'(t) \tag{1}$$

where $\{B(t)\}$ is Brownian notion. Before proceeding to the formal details, let us discuss a physical model that leads to a differential equation of this type. Suppose that $X(t)$ is the position at time t of a microscopic particle suspended in a liquid. If the liquid is homogeneous and macroscopically motionless (in other words, no external forces are acting on it), then $\{X(t)\}$ is Brownian motion. We consider the situation in which the liquid is not homogeneous and not motionless. For simplicity, we assume one-dimensional motion (think of the liquid as enclosed in a long thin tube).

Let $m(x, t)$ be the macroscopic velocity of a small volume V of fluid located at position x at time t. A microscopic particle within V will execute Brownian motion (relative to V) with variance parameter $\sigma^2(x, t)$. Now the change in position of the particle in the small time interval $[t, t + dt]$ arises from two sources: the macroscopic motion of the liquid, which contributes $m(X(t), t)\, dt$, and the molecular bombardment, which contributes $\sigma(X(t), t) \times [B(t + dt) - B(t)]$ where $\{B(t)\}$ is Brownian motion with variance parameter 1. Thus

$$dX(t) = m(X(t), t)\, dt + \sigma(X(t), t)\, dB(t)$$

which is equivalent to (1).

We now give an appropriate mathematical model for (1). Let $\{B(t), a \le t \le b\}$ be a separable Brownian notion on $(\Omega, \mathscr{F}, P)$. The stochastic differential equation (1) is interpreted as

$$X(t) - X(a) = \int_a^t m(X(s), s)\, ds + \int_a^t \sigma(X(s), s)\, dB(s). \tag{2}$$

The functions m and σ are Borel measurable maps of $R \times [a, b]$ into R, and are assumed to satisfy (for some $k > 0$) a Lipschitz condition

$$\left| m(x, t) - m(y, t) \right| \le k \left| x - y \right|, \qquad \left| \sigma(x, t) - \sigma(y, t) \right| \le k \left| x - y \right| \tag{3}$$

for all x, y, t; also, we assume

$$\left| m(x, t) \right| \le k(1 + x^2)^{1/2}, \qquad \left| \sigma(x, t) \right| \le k(1 + x^2)^{1/2}. \tag{4}$$

(Note that if (3) is assumed, (4) is equivalent to boundedness of the maps $t \to m(0, t)$ and $t \to \sigma(0, t)$; to see this, write $m(x, t) = m(x, t) - m(0, t) + m(0, t)$.)

Let X be a given random variable in $L^2(\Omega, \mathscr{F}, P)$ assumed to be $\mathscr{F}(a)$-measurable, where $\mathscr{F}(t) = \mathscr{F}\{B(s), a \le s \le t\}$. By a *solution* of (1) on $[a, b]$ with initial condition $X(a) = X$ we mean a stochastic process $\{X(t), a \le t \le b\}$ such that a.s., $X(a) = X$; and for each $t \in [a, b]$, $X(t)$ satisfies (2). The main result is as follows.

5.2.1 Theorem

There is a separable, progressively measurable, L^2-continuous process $\{X(t), a \le t \le b\}$ with the following properties:

(a) For each $t \in [a, b]$, $X(t)$ is $\mathscr{F}(t)$-measurable.

(b) $\int_a^b E[\,|X(t)|^2]\,dt < \infty$.

(c) $\{X(t), a \le t \le b\}$ satisfies (2) with $X(a) = X$.

(d) Almost every sample function of $\{X(t)\}$ is continuous on $[a, b]$.

(e) $\{X(t)\}$ is unique in the sense that if $\{Y(t), a \le t \le b\}$ is any separable process satisfying (2) and the initial condition $Y(a) = X$, then

$$P\{X(t) = Y(t) \text{ for all } t \in [a, b]\} = 1.$$

(f) $\{X(t), a \le t \le b\}$ is a Markov process.

PROOF We define a sequence of approximations to the solution:

$$X_0(t) \equiv X$$

$$X_{n+1}(t) = X + \int_a^t m(X_n(s), s)\,ds + \int_a^t \sigma(X_n(s), s)\,dB(s). \tag{5}$$

We show inductively that the processes $\{X_n(t), a \le t \le b\}$ have the following properties.

(i) $\{X_n(t), a \le t \le b\}$ is well defined and L^2-continuous, and may be taken to be separable and progressively measurable and to have continuous sample functions.

(ii) For almost every ω, $s \to m(X_n(s, \omega), s)$ is a bounded Borel measurable function.

(iii) For each t, the maps

$$(s, \omega) \to m(X_n(s, \omega), s) \quad \text{and} \quad (s, \omega) \to \sigma(X_n(s, \omega), s) \quad (a \le s \le t)$$

are $\mathscr{B}[a, t] \times \mathscr{F}(t)$-measurable.

(iv) For each s, $X_n(s)$ is $\mathscr{F}(s)$-measurable, hence $\omega \to \sigma(X_n(s, \omega), s)$ is $\mathscr{F}(s)$-measurable.

(v) $\int_a^b E[\,|\sigma(X_n(s), s)|^2]\, ds < \infty$.

It is easy to check that $\{X_0(t)\}$ satisfies (i)–(v), so we assume that $\{X_n(t)\}$ has these properties and consider the corresponding results for $\{X_{n+1}(t)\}$ (labeled with primes).

(i)′ By (ii), $\int_a^t m(X_n(s), s)\, ds$ is well defined for almost every ω, and by (iii), (iv), and (v), $\int_a^t \sigma(X_n(s), s)\, dB(s)$ is well defined; therefore, $\{X_{n+1}(t)\}$ is well defined. To establish L^2-continuity, we compute, for $s < t$,

$$E[\,|X_{n+1}(t) - X_{n+1}(s)|^2]$$

$$= E\left[\left|\int_s^t m(X_n(u), u)\, du + \int_s^t \sigma(X_n(u), u)\, dB(u)\right|^2\right]$$

$$\le 2\left\{E\left[\left|\int_s^t m(X_n(u), u)\, du\right|^2\right]\right.$$

$$\left. + E\left[\left|\int_s^t \sigma(X_n(u), u)\, dB(u)\right|^2\right]\right\}$$

$$\le 2\left\{(t - s)\int_s^t E[\,|m(X_n(u), u)|^2]\, du\right.$$

$$\left. + \int_s^t E[\,|\sigma(X_n(u), u)|^2]\, du\right\}$$

by the Cauchy–Schwarz inequality and 5.1.4

$$\le 2[1 + (t - s)]k^2 \int_s^t [1 + E(\,|X_n(u)|^2)]\, du \qquad \text{by (4)}$$

$$\to 0 \qquad \text{as } t \to s \text{ since } \{X_n(t)\} \text{ is } L^2\text{-continuous,}$$

and hence L^2-bounded, that is, the map $u \to E(\,|X_n(u)|^2)$ is bounded.

By 5.1.7 and the discussion preceding 5.1.6, we may take the right-hand side of (5) to be separable and progressively measurable and to have continuous sample functions; this proves (i)′.

(ii)′ This follows from (4) and the fact that for almost every ω, $X_n(\cdot, \omega)$ is continuous, and therefore bounded.

(iii)′ Since the process $\{X_n(t)\}$ is progressively measurable, the maps in question are compositions of measurable functions and hence measurable.

(iv)′ By (iii) and Fubini's theorem, $\int_a^t m(X_n(s), s) \, ds$ is $\mathscr{F}(t)$-measurable, and by 5.1.5, $\int_a^t \sigma(X_n(s), s) \, dB(s)$ is $\mathscr{F}(t)$-measurable; the result follows.

(v)′ By (4),

$$\int_a^b E[\,|\sigma(X_{n+1}(s), s)|^2] \, ds \le k^2 \int_a^b [1 + E(|X_{n+1}(s)|^2)] \, ds$$

$$< \infty$$

since $\{X_{n+1}(t)\}$ is L^2-bounded.

We next show that for each $t \in [a, b]$, the sequence $\{X_n(t), n = 0, 1, \ldots\}$ converges in $L^2(\Omega, \mathscr{F}, P)$ to a random variable $X(t)$. Define

$$\Delta_0(t) = X$$

$$\Delta_n(t) = X_n(t) - X_{n-1}(t), \qquad n = 1, 2, \ldots. \tag{6}$$

Then

$$E[\,|\Delta_{n+1}(t)|^2] = E[\,|X_{n+1}(t) - X_n(t)|^2]$$

$$= E\left\{\left|\int_a^t [m(X_n(s), s) - m(X_{n-1}(s), s)] \, ds \right.\right.$$

$$\left.\left. + \int_a^t [\sigma(X_n(s), s) - \sigma(X_{n-1}(s), s)] \, dB(s)\right|^2\right\}.$$

Exactly as in the proof of (i)′, we obtain

$$E[\,|\Delta_{n+1}(t)|^2]$$

$$\le 2\left\{(t - a)\int_a^t E[\,|m(X_n(s), s) - m(X_{n-1}(s), s)|^2] \, ds\right.$$

$$\left. + \int_a^t E[\,|\sigma(X_n(s), s) - \sigma(X_{n-1}(s), s)|^2] \, ds\right\}$$

$$\le 2[1 + (t - a)]k^2 \int_a^t E[\,|X_n(s) - X_{n-1}(s)|^2] \, ds \qquad \text{by (3)}$$

$$\le A \int_a^t E[\,|\Delta_n(s)|^2] \, ds \tag{7}$$

where $A = 2k^2[1 + (b - a)]$.

Using (7) we obtain

$$E[\,|\Delta_1(t)|^2] \le AE(\,|X|^2)(t-a)$$

and, inductively,

$$E[\,|\Delta_n(t)|^2] \le A^n E(\,|X|^2)(t-a)^n/n!. \qquad (8)$$

Now $X_n(t) - X_m(t) = \sum_{j=m+1}^{n} \Delta_j(t)$ for $n \ge m$, so

$$|X_n(t) - X_m(t)|^2 = \left| \sum_{j=m+1}^{n} 2^{-j/2} 2^{j/2} \Delta_j(t) \right|^2$$

$$\le \sum_{j=m+1}^{n} 2^{-j} \sum_{j=m+1}^{n} 2^j |\Delta_j(t)|^2$$

by the Cauchy–Schwarz inequality. Thus for all $t \in [a, b]$,

$$E[\,|X_n(t) - X_m(t)|^2] \le \sum_{j=m+1}^{n} 2^j E[\,|\Delta_j(t)|^2]$$

$$\le \sum_{j=m+1}^{n} 2^j A^j E(\,|X|^2) \frac{(t-a)^j}{j!} \qquad \text{by (8)}$$

$$\le E(\,|X|^2) \sum_{j=m+1}^{n} \frac{[2A(b-a)]^j}{j!}$$

$$\to 0 \qquad \text{as} \quad n, m \to \infty \qquad (9)$$

as desired. To prove L^2-continuity of $\{X(t)\}$, note that by (9), the maps $t \to X_n(t)$ of $[a, b]$ into $L^2(\Omega, \mathscr{F}, P)$ converge uniformly to the map $t \to X(t)$.

We now show that $\{X(t), a \le t \le b\}$ satisfies (a)–(f).

(a)   This is immediate from (i) and the L^2-convergence of $X_n(t)$ to $X(t)$.

(b)   This follows from the continuity, hence boundedness, of the map $t \to X(t)$.

(c)   Clearly $X(a) = X$; to show that (2) is satisfied, consider the difference

$$D(t) = X(t) - X - \int_a^t m(X(s), s)\, ds - \int_a^t \sigma(X(s), s)\, dB(s)$$

$$= X(t) - X_{n+1}(t) - \int_a^t [m(X(s), s) - m(X_n(s), s)]\, ds$$

$$- \int_a^t [\sigma(X(s), s) - \sigma(X_n(s), s)]\, dB(s).$$

Estimates just like those used in the proofs of (i)' and (v)', and the fact that $X_n(t) \to X(t)$ in L^2, uniformly in t, imply that for each t, $D(t) = 0$ a.s.

In fact more is true. We may choose separable versions of all processes involved in the definition of $D(t)$, with a common separating set (by 4.1.6). Then $\{D(t), a \le t \le b\}$ is separable, so that

$$P\{D(t) = 0 \text{ for all } t \in [a, b]\} = 1.$$

(d)   This is immediate from (2) and 5.1.7.

(e)   Since $\{X(t)\}$ and $\{Y(t)\}$ satisfy (2),

$$X(t) - Y(t) = \int_a^t [m(X(s), s) - m(Y(s), s)] \, ds$$

$$+ \int_a^t [\sigma(X(s), s) - \sigma(Y(s), s)] \, dB(s).$$

As in the proof of (7),

$$E[\,|X(t) - Y(t)|^2] \le A \int_a^t E[\,|X(s) - Y(s)|^2] \, ds = AF(t).$$

This inequality is of the form $dF(t)/dt \le AF(t)$, or $d[e^{-At}F(t)]/dt \le 0$. Integrate from a to t to obtain $e^{-At}F(t) - e^{-Aa}F(a) \le 0$. But $F(a) = 0$, so that $F(t) \le 0$. Since $F(t) \ge 0$ by definition, we must have $F(t) \equiv 0$. The desired result now follows by separability.

(f)   It suffices to show (see 4.5.1 and 4.5.3) that if $s < t$, $X(t)$ and $\mathscr{F}(s)$ are conditionally independent given $X(s)$. But by (2),

$$X(t) = X(s) + \int_s^t m(X(u), u) \, du + \int_s^t \sigma(X(u), u) \, dB(u). \tag{10}$$

Since the right-hand side of (10) is the unique solution of the stochastic differential equation on $[s, t]$, (a) implies that given $X(s) = x$, $X(t)$ is the sum of x and a random variable Y that is measurable relative to $\mathscr{F}\{B(u), s \le u \le t\}$. Thus given $X(s) = x$ we have $X(t) = x + Y$ where (by independence of increments of Brownian notion), Y and $\mathscr{F}(s)$ are independent. It follows that $X(t)$ and $\mathscr{F}(s)$ are conditionally independent given $X(s)$. ‖

Problems

1.   (*Kolmogorov's Forward Equation, Also Called the Fokker–Planck Equation*)   Let $\{X(t), a \le t \le b\}$ be a Markov process with a transition density $p(x, t \,|\, x_0, t_0)$, in other words, for each Borel set A,

$$P\{X(t) \in A \,|\, X(t_0) = x_0\} = \int_A p(x, t \,|\, x_0, t_0) \, dx.$$

Let $f: R \rightarrow R$ be infinitely differentiable with compact support, and consider the integral

$$I = I(t \mid x_0, t_0) = \int_{-\infty}^{\infty} f(x)p(x, t \mid x_0, t_0)\, dx$$

$$= E[f(X(t)) \mid X(t_0) = x_0].$$

Let $M_n(z, t, \Delta t) = (\Delta t)^{-1} E[(X(t + \Delta t) - X(t))^n \mid X(t) = z]$ and assume that as $\Delta t \rightarrow 0$,

$$M_1(z, t, \Delta t) \rightarrow m(z, t) = M_1(z, t)$$
$$M_2(z, t, \Delta t) \rightarrow \sigma^2(z, t) = M_2(z, t)$$
$$M_3(z, t, \Delta t) \rightarrow \quad 0 \quad = M_3(z, t)$$

and that each $|M_i(z, t, \Delta t)|$ is bounded by a finite constant K_i.

(a) Use the Chapman–Kolmogorov equation

$$p(x, t + \Delta t \mid x_0, t_0) = \int_{-\infty}^{\infty} p(x, t + \Delta t \mid z, t)p(z, t \mid x_0, t_0)\, dz$$

(this follows directly from the corresponding statement for transition probabilities) and a Taylor expansion of f to show that

$$\frac{\partial I}{\partial t} = \sum_{k=1}^{2} \frac{1}{k!} \int_{-\infty}^{\infty} M_k(x, t)p(x, t \mid x_0, t_0)f^{(k)}(x)\, dx.$$

(b) Assume the partial derivatives $\partial p/\partial t$, $\partial p/\partial x$, and $\partial^2 p/\partial x^2$ are continuous, and also bounded in (x, t) for each fixed (x_0, t_0). Assume also that for fixed (x_0, t_0), $|\partial p/\partial t|$ is bounded by a Lebesgue integrable $g(x)$. Integrate the result of (a) by parts to show that

$$\int_{-\infty}^{\infty} f(x)\left| -\frac{\partial p}{\partial t} + \sum_{k=1}^{2} \frac{(-1)^k}{k!} \frac{\partial^k}{\partial x^k} (M_k(x, t)p(x, t \mid x_0, t_0)) \right| dx = 0$$

for all infinitely differentiable f with compact support. Thus the term in the absolute value signs must be 0, so that

$$\frac{\partial p}{\partial t} = -\frac{\partial}{\partial x}[m(x, t)p(x, t \mid x_0, t_0)] + \frac{1}{2} \frac{\partial^2}{\partial x^2}[\sigma^2(x, t)p(x, t \mid x_0, t_0)]$$

which is Kolmogorov's forward equation. The word "forward" is used because time derivatives are computed at the front end of the interval $[t_0, t]$. The "backward" equation will be considered in the next problem.

(c) Set up the forward equation for the Brownian motion process, and verify that the transition density is in fact a solution.

2. (*Kolmogorov's Backward Equation*) Let $\{X(t)\}$, I, and M_n be as defined in Problem 1, and assume that as $\Delta t \to 0$,

$$M_1(x_0, t_0, \Delta t) \to m(x_0, t_0) = M_1(x_0, t_0)$$

$$M_2(x_0, t_0, \Delta t) \to \sigma^2(x_0, t_0) = M_2(x_0, t_0)$$

$$M_3(x_0, t_0, \Delta t) \to \quad 0 \quad = M_3(x_0, t_0).$$

Assume $|M_i| \le K_i < \infty$ as before, and that $\partial p/\partial t_0$, $\partial p/\partial x_0$, and $\partial^2 p/\partial x_0^2$ are continuous, and also bounded in (x_0, t_0) for fixed (x, t). Use the Chapman–Kolmogorov equation

$$p(x, t \mid x_0, t_0) = \int_{-\infty}^{\infty} p(x, t \mid z, t_0 + \Delta t) p(z, t_0 + \Delta t \mid x_0, t_0)\, dz$$

and Taylor expansion of $p(x, t \mid z, t_0 + \Delta t)$ to obtain the backward equation

$$-\frac{\partial p}{\partial t_0} = m(x_0, t_0)\frac{\partial p}{\partial x_0} + \frac{1}{2}\sigma^2(x_0, t_0)\frac{\partial^2 p}{\partial x_0^2}.$$

5.3 Stochastic Differentials: A Chain Rule

In discussing stochastic differentials it is convenient to define the stochastic integral for a larger class of integrands; this new class $\mathscr{C}_1$ is obtained from $\mathscr{C}$ (see 5.1.1) by weakening the requirement that $\int_a^b E(|f(t)|^2)\, dt < \infty$ to $\int_a^b |f(t)|^2\, dt < \infty$ a.s. The details of this extension of the definition of the stochastic integral are explored in Problem 1.

Suppose $\alpha, \beta \in \mathscr{C}_1$; actually, for α we need only require

$$\int_a^b |\alpha(t)|\, dt < \infty \qquad \text{a.s.}$$

rather than

$$\int_a^b |\alpha(t)|^2\, dt < \infty \quad \text{a.s.}$$

If for all s, t such that $a \le s \le t \le b$, we have

$$X(t) - X(s) = \int_s^t \alpha(u)\, du + \int_s^t \beta(u)\, dB(u) \quad \text{a.s.}$$

we call $\alpha(t)\, dt + \beta(t)\, dB(t)$ the *stochastic differential* of $\{X(t)\}$ and we write

$$dX(t) = \alpha(t)\, dt + \beta(t)\, dB(t).$$

Note that if $\{X(t)\}$ has a stochastic differential, then almost every sample function of $\{X(t)\}$ is continuous, and for each t, $X(t)$ is $\mathscr{F}(t) = \mathscr{F}\{B(s),\ a \leq s \leq t\}$-measurable.

We now examine (heuristically first) the stochastic differential of a process obtained from $\{X(t)\}$ by

$$Y(t) = f(X(t), t).$$

According to the chain rule of ordinary calculus, we would expect that

$$
\begin{aligned}
dY(t) &= f_x(X(t), t)\, dX(t) + f_t(X(t), t)\, dt \\
&= f_x(X(t), t)[\alpha(t)\, dt + \beta(t)\, dB(t)] + f_t(X(t), t)\, dt \\
&= [\alpha(t)f_x(X(t), t) + f_t(X(t), t)]\, dt \\
&\quad + \beta(t)f_x X(t), t)\, dB(t)
\end{aligned}
$$

where $f_x = \partial f/\partial x$, $f_t = \partial f/\partial t$.

This result is not what we obtain, however; there is another term in the coefficient of dt that arises in the following way. We expand $dY(t)$ by Taylor's theorem and collect terms of order dt or larger, remembering that $[dB(t)]^2 \sim dt$:

$$
\begin{aligned}
dY(t) &= Y(t + dt) - Y(t) \\
&= f(X(t + dt), t + dt) - f(X(t), t) \\
&= f_x\, dX(t) + f_t\, dt \\
&\quad + \tfrac{1}{2}[f_{xx}(dX(t))^2 + 2f_{xt}\, dX(t)\, dt + f_{tt}(dt)^2] + \cdots \\
&= f_x(\alpha\, dt + \beta\, dB(t)) + f_t\, dt \\
&\quad + \tfrac{1}{2}[f_{xx}(\alpha^2(dt)^2 + 2\alpha\beta\, dB(t)\, dt + \beta^2(dB(t))^2) \\
&\quad + 2f_{xt}(\alpha\, dt + \beta\, dB(t))\, dt + f_{tt}(dt)^2] + \cdots \\
&= [\alpha f_x + f_t + \tfrac{1}{2}\beta^2 f_{xx}]\, dt + \beta f_x\, dB(t).
\end{aligned}
$$

The formal statement is as follows.

5.3.1   Itô's Differentiation Formula

Let $f: R \times [a, b] \to R$, and assume that f, f_x, f_{xx}, f_t are all continuous. Set $Y(t) = f(X(t), t)$. If $\{X(t),\ a \leq t \leq b\}$ has stochastic differential $dX(t) = \alpha(t)\, dt + \beta(t)\, dB(t)$, then $\{Y(t),\ a \leq t \leq b\}$ has stochastic differential

$$
\begin{aligned}
dY(t) &= [\alpha(t)f_x(X(t), t) + f_t(X(t), t) + \tfrac{1}{2}\beta^2(t)f_{xx}(X(t), t)]\, dt \\
&\quad + \beta(t)f_x(X(t), t)\, dB(t).
\end{aligned}
$$

PROOF Let P: $s = t_0 < t_1 < \cdots < t_n = t$ be a partition of $[s, t]$. Then

$$Y(t) - Y(s) = \sum_{i=0}^{n-1} [Y(t_{i+1}) - Y(t_i)]$$

$$= \sum_{i=0}^{n-1} [f(X(t_{i+1}), t_{i+1}) - f(X(t_i), t_i)]. \tag{1}$$

By Taylor's formula,

$$f(X(t_{i+1}), t_{i+1}) - f(X(t_i), t_i)$$
$$= f(X(t_{i+1}), t_{i+1}) - f(X(t_i), t_{i+1})$$
$$+ f(X(t_i), t_{i+1}) - f(X(t_i), t_i)$$
$$= f_x(X(t_i), t_{i+1})(X(t_{i+1}) - X(t_i))$$
$$+ \tfrac{1}{2} f_{xx}(\xi_i, t_{i+1})(X(t_{i+1}) - X(t_i))^2$$
$$+ f_t(X(t_i), \tau_i)(t_{i+1} - t_i) \tag{2}$$

where $t_i < \tau_i < t_{i+1}$ and $\xi_i = \xi_i(\omega)$ is between $X(t_i, \omega)$ and $X(t_{i+1}, \omega)$. Since almost every sample function of $\{X(t)\}$ is continuous, we may write (a.s.) $\xi_i(\omega) = X(\sigma_i(\omega))$ where $t_i < \sigma_i(\omega) < t_{i+1}$.

We first assume that α and β do not depend on t (but may depend on ω) so that $dX(t) = \alpha\, dt + \beta\, dB(t)$ becomes

$$X(t_{i+1}) - X(t_i) = \alpha(t_{i+1} - t_i) + \beta[B(t_{i+1}) - B(t_i)].$$

We now analyze the three components (see (2)) of the summation (1) as $|P| = \max |t_{i+1} - t_i| \to 0$.

First, we note that

$$\sum_{i=0}^{n-1} f_t(X(t_i), \tau_i)(t_{i+1} - t_i) \xrightarrow{\text{a.s.}} \int_s^t f_t(X(u), u)\, du.$$

This follows because for almost every ω, $u \to X(u, \omega)$ is continuous, so the integrand $u \to f_t(X(u, \omega), u)$ is continuous. Now

$$\sum_{i=0}^{n-1} f_x(X(t_i), t_{i+1})(X(t_{i+1}) - X(t_i))$$

$$= \sum_{i=0}^{n-1} f_x(X(t_i), t_{i+1})[\alpha(t_{i+1} - t_i) + \beta[B(t_{i+1}) - B(t_i)]].$$

As above, we have

$$\sum_{i=0}^{n-1} f_x(X(t_i), t_{i+1})\alpha(t_{i+1} - t_i) \xrightarrow{\text{a.s.}} \alpha \int_s^t f_x(X(u), u)\, du.$$

Also,

$$\sum_{i=0}^{n-1} f_x(X(t_i),\, t_{i+1})\beta[B(t_{i+1}) - B(t_i)] \xrightarrow{P} \beta \int_s^t f_x(X(u),\, u)\, dB(u).$$

This follows from Problem 1 and the fact that the left side is the integral of a step function approximation to $\beta f_x(X(u),\, u)$. Finally,

$$\sum_{i=0}^{n-1} f_{xx}(\xi_i,\, t_{i+1})(X(t_{i+1}) - X(t_i))^2$$

$$= \sum_{i=0}^{n-1} f_{xx}(\xi_i,\, t_{i+1})[\alpha^2(t_{i+1} - t_i)^2$$

$$+ 2\alpha\beta(B(t_{i+1}) - B(t_i))(t_{i+1} - t_i) + \beta^2(B(t_{i+1}) - B(t_i))^2] \quad (3)$$

where

$$\alpha^2 \sum_{i=0}^{n-1} f_{xx}(\xi_i,\, t_{i+1})(t_{i+1} - t_i)^2 \xrightarrow{\text{a.s.}} 0$$

since for almost every $\omega f_{xx}(\xi_i,\, t_{i+1})$ is bounded and

$$\sum_{i=0}^{n-1} (t_{i+1} - t_i)^2 \le (b - a)|P| \to 0 \text{ as } |P| \to 0.$$

The second term of (3)

$$2\alpha\beta \sum_{i=0}^{n-1} f_{xx}(\xi_i,\, t_{i+1})(B(t_{i+1}) - B(t_i))(t_{i+1} - t_i) \xrightarrow{\text{a.s.}} 0$$

since for almost every ω f_{xx} is bounded and $t \to B(t,\, \omega)$ is uniformly continuous on $[a,\, b]$; thus $|B(t_{i+1}) - B(t_i)| < \varepsilon$ provided $|P|$ is small enough (depending on ω). The third term of (3)

$$\beta^2 \sum_{i=0}^{n-1} f_{xx}(\xi_i,\, t_{i+1})(B(t_{i+1}) - B(t_i))^2 \to \beta^2 \int_s^t f_{xx}(X(u),\, u)\, du$$

in probability. To see this, replace $(B(t_{i+1}) - B(t_i))^2$ by $(t_{i+1} - t_i)$ to get

$$\beta^2 \sum_{i=0}^{n-1} f_{xx}(\xi_i,\, t_{i+1})(t_{i+1} - t_i) \xrightarrow{\text{a.s.}} \beta^2 \int_s^t f_{xx}(X(u),\, u)\, du$$

since for almost every ω, $u \to f_{xx}(X(u),\, u)$ is continuous. The error committed in making this replacement is

$$e(P) = \beta^2 \sum_{i=0}^{n-1} f_{xx}(\xi_i,\, t_{i+1})\, \Delta_i$$

where $\Delta_i = (B(t_{i+1}) - B(t_i))^2 - (t_{i+1} - t_i)$; we show that this error converges to zero in probability. If $\{P_n\}$ is any sequence of partitions with $|P_n| \to 0$, we may extract a subsequence $\{Q_m\}$ such that $\sum_{m=1}^{\infty} |Q_m| < \infty$. By Problem 2(c) of 5.1 $\sum_{i=0}^{n-1} |\Delta_i| \xrightarrow{\text{a.s.}} 0$ and since $f_{xx}(\xi_i, t_{i+1})$ is bounded for almost every ω, the error $e(Q_m) \to 0$ as $m \to \infty$. Thus $e(P) \to 0$ in probability as $|P| \to 0$. This proves the theorem for α, β functions only of ω. If α and β are step functions, then the theorem follows by applying what we have just proved to each of the intervals on which both α and β do not depend on t. The general case follows from Problem 2.   ‖

Problems

1. In this problem we extend the definition of the stochastic integral Φ defined by $\Phi(f) = \int_a^b f(t) \, dB(t)$ for all f in the class $\mathscr{C}$ (see 5.1.1). Let $\mathscr{C}_1$ be the class obtained from $\mathscr{C}$ by weakening the requirement (i) $\int_a^b E(f^2(t)) \, dt < \infty$ to (i)' $\int_a^b |f(t)|^2 \, dt < \infty$ a.s. Define a metric d_1 on the vector space $\mathscr{C}_1$ by

$$d_1(f, g) = N\left(\left[\int_a^b |f(t) - g(t)|^2 \, dt\right]^{1/2}\right)$$

where $N(X) = E(|X|/(1 + |X|))$ (see 1.1, Problem 8). Then $f_n \to f$ in the metric d_1 iff $\int_a^b |f_n(t) - f(t)|^2 \, dt \to 0$ in probability.

(a)   For $f \in \mathscr{C}$, $\varepsilon > 0$, and $\eta \geq 0$ show that

$$P(|\Phi(f)| > \eta) \leq \varepsilon^2/\eta^2 + ((1 + \varepsilon)/\varepsilon)N\left(\int_a^b |f(t)|^2 \, dt\right).$$

(*Hint:* Consider the two cases $\int_a^b |f(t)|^2 \, dt \leq \varepsilon^2$ or $> \varepsilon^2$ and show that

$$P\left(\int_a^b |f(t)|^2 \, dt > \varepsilon^2\right) \leq ((1 + \varepsilon)/\varepsilon)N\left(\int_a^b |f(t)|^2 \, dt\right)$$

and

$$P\left(|\Phi(f)| > \eta \text{ and } \int_a^b |f(t)|^2 \, dt \leq \varepsilon^2\right) \leq \varepsilon^2.$$

(b)   Use (a) with appropriate ε and η to establish the inequality

$$N(\Phi(f)) \leq 4\left\{N\left(\left[\int_a^b |f(t)|^2 \, dt\right]^{1/2}\right)\right\}^{1/3}$$

for all $f \in \mathscr{C}$. (*Hint:* $N(X) \leq \eta + P(|X| > \eta)$ for all $\eta \geq 0$.)

(c)   Let M be the space of random variables on $(\Omega, \mathscr{F}, P)$ with the metric d_2 of convergence in probability (take $d_2(X, Y) = N(X - Y)$) and consider $\mathscr{C}$ and $\mathscr{C}_1$ with the metric d_1. Show that Φ is a uniformly continuous mapping from $(\mathscr{C}, d_1)$ to (M, d_2) and show that $(\mathscr{C}, d_1)$ is dense in $(\mathscr{C}_1, d_1)$. Since $(\mathscr{C}_1, d_1)$ and (M, d_2) are topological vector spaces (that is, addition and scalar multiplication are continuous), it follows that Φ extends uniquely to a uniformly continuous linear map Φ^* from $(\mathscr{C}_1, d_1)$ to (M, d_2)—thus we have extended the stochastic integral from $\mathscr{C}$ to $\mathscr{C}_1$.

2.   Assume the hypotheses of 5.3.1.

(a)   Show that there exist step functions α_n, β_n such that

(i) $\displaystyle \int_a^b |\alpha_n(u) - \alpha(u)|^2 \, du \to 0$ a.s.

(ii) $\displaystyle \int_a^b |\beta_n(u) - \beta(u)|^2 \, du \to 0$ a.s.

(iii) $\displaystyle \sup_{a \le u \le b} |X_n(u) - X(u)| \to 0$ a.s.

where

$$X_n(t) = X(a) + \int_a^t \alpha_n(u) \, du + \int_a^t \beta_n(u) \, dB(u).$$

(b)   Define $Y_n(t) = f(X_n(t), t)$ and $Y(t) = f(X(t), t)$. For $a \le s < t \le b$ find $Y_n(t) - Y_n(s)$ in terms of α_n and β_n; then let n approach infinity to find $Y(t) - Y(s)$ in terms of α and β. Conclude that 5.3.1 holds for all $\alpha, \beta \in \mathscr{C}_1$.

3.   We want to attach a meaning to the white-noise differential equation

$$\frac{d}{dt} X(t) = m(X(t), t) + \sigma(X(t), t)W(t) \tag{1}$$

where $\{W(t)\}$ is a white-noise process. We shall call $\{X(t), a \le t \le b\}$ a solution to (1) on $[a, b]$ if $\{X(t)\}$ satisfies a corresponding stochastic differential equation; to find the corresponding stochastic differential equation we proceed heuristically.

Let $\{W_n(t)\}$ be an approximation to white noise with continuous sample functions. If m and σ are nice, then for each ω the initial value problem

$$dX_n(t, \omega)/dt = m(X_n(t, \omega), t) + \sigma(X_n(t, \omega), t)W_n(t, \omega), \tag{2}$$

with $X_n(a, \omega) = X(\omega)$ given, has a solution $\{X_n(t)\}$. If the processes $\{X_n(t), a \le t \le b\}$ converge in some sense to a process $\{X(t), a \le t \le b\}$ as $\{W_n(t)\}$

converges to $\{W(t)\}$, then we would want to call $\{X(t), a \le t \le b\}$ a solution to (1) on $[a, b]$. Write (2) as

$$dX_n(t) = m(X_n(t), t)\, dt + \sigma(X_n(t), t)\, dB_n(t) \tag{3}$$

where $B_n(t, \omega) = \int_a^t W_n(s, \omega)\, ds$; then

$$B_n(t) - B_n(a) \to B(t) - B(a) \quad \text{a.e.}$$

and $\{B(t)\}$ is a Brownian motion process since $\{W_n(t)\}$ converges to white noise $\{W(t)\}$. Put

$$\psi(x, t) = \int_0^x \frac{dy}{\sigma(y, t)}.$$

Then for each sample function we have

$$d\psi(X_n(t), t) = \psi_t(X_n(t), t)\, dt + \psi_x(X_n(t), t)\, dX_n(t)$$

$$= \psi_t(X_n(t), t)dt + \frac{m(X_n(t), t)}{\sigma(X_n(t), t)}\, dt + dB_n(t)$$

or in integral form

$$\psi(X_n(t), t) - \psi(X_n(a), a)$$

$$= \int_a^t \left[\psi_t(X_n(s), s) + \frac{m(X_n(s), s)}{\sigma(X_n(s), s)} \right] ds + B_n(t) - B_n(a).$$

Let $n \to \infty$ to get

$$\psi(X(t), t) - \psi(X(a), a)$$

$$= \int_a^t \left[\psi_t(X(s), s) + \frac{m(X(s), s)}{\sigma(X(s), s)} \right] ds + B(t) - B(a). \tag{4}$$

Hence, heuristically at least, we want $\{X(t)\}$ to satisfy (4). Assume that $\{X(t)\}$ satisfies some stochastic differential equation $dX(t) = \alpha(t)\, dt + \beta(t)\, dB(s)$. Show that for (4) to hold we should have $\alpha(s) = m(X(s), s) + \frac{1}{2}\sigma(X(s), s)\sigma_x(X(s), s)$ and $\beta(s) = \sigma(X(s), s)$. Hence we call $\{X(t)\}$ a solution to the white-noise differential equation

$$dX(t)/dt = m(X(t), t) + \sigma(X(t), t)W(t)$$

if it is a solution to the stochastic differential equation

$$dX(t) = [m(X(t), t) + \tfrac{1}{2}\sigma(X(t), t)\sigma_x(X(t), t)]\, dt$$

$$+ \sigma(X(t), t)\, dB(t).$$

Note that the solution is a Markov process with continuous sample functions by 5.2.1.

4. Let $dX(t) = X(t)[m(t)\,dt + \sigma(t)\,dB(t)]$ where m and σ are in $\mathscr{C}_1$ but do not depend on ω. Show that for $t \geq 0$

$$X(t) = X(0)\exp\left\{\int_0^t \left[m(s) - \frac{\sigma^2(s)}{2}\right]ds + \int_0^t \sigma(s)\,dB(s)\right\}.$$

Compute $E(X^n(t) \mid X(0) = x)$.

5. Suppose a system is governed by the white-noise differential equation

$$dX(t)/dt = c(t)(M + \sigma W(t)), \qquad t > 0,$$

where $\{W(t)\}$ is white noise, $X(t)$ is the state of the system at time t, and $c(t)$ is the control applied at time t. A control function of the form $c(t) = f(X(0), t)$ is called open-loop control and one of the form $c(t) = \alpha(t)X(t)$ is called linear closed-loop control. In each of these cases find the control function that minimizes $E(X^2(b) \mid X(0) = x)$.

5.4 Notes

A nice introduction to stochastic integrals and stochastic differential equations may be found in Wong (1971). For a thorough treatment of the topics introduced in this chapter see the books by McKean (1969) and Itô and McKean (1965) as well as the notes by Doleans–Dade and Meyer (1969) and the chapter on diffusion processes in Gikhman and Skorokhod (1969).

Appendix 1

Some Results from Complex Analysis

In this appendix we prove several theorems of complex analysis that are needed in the solution of the general prediction problem.

A1.1 Definitions and Comments

Define

$$Q_z(t) = \frac{e^{it} + z}{e^{it} - z}, \qquad |z| < 1, \quad t \text{ real.}$$

If $z = re^{i\theta}, 0 \le r < 1$, then

$$\text{Re } Q_z(t) = P_r(\theta - t),$$

where $P_r(x)$, the *Poisson kernel*, is given by

$$P_r(x) = \frac{1 - r^2}{1 - 2r \cos x + r^2}$$

(see Ash, 1971, p. 105 ff). The Poisson kernel can be expressed as

$$P_r(x) = \sum_{k=-\infty}^{\infty} r^{|k|} e^{ikx};$$

this follows upon expressing the above infinite series in terms of two geometric series. Also,

$$\frac{1}{2\pi} \int_{-\pi}^{\pi} P_r(x) \, dx = 1$$

(see Ash, 1971, p. 108).

Now assume $f \in L^1[-\pi, \pi]$, and define $\mu(E) = \int_E f(t)\, dt$, $E \in \mathscr{B}[-\pi, \pi]$. If

$$G(z) = \frac{1}{2\pi} \int_{-\pi}^{\pi} Q_z(t) f(t)\, dt = \frac{1}{2\pi} \int_{-\pi}^{\pi} Q_z(t)\, d\mu(t),$$

then G is analytic on $D = \{z : |z| < 1\}$. To see this, note that

$$Q_z(t) = (1 + ze^{-it}) \sum_{n=0}^{\infty} (ze^{-it})^n$$

$$= \sum_{n=0}^{\infty} [e^{-int}z^n + e^{-i(n+1)t}z^{n+1}],$$

where the series converges uniformly in t. We may therefore integrate term by term to obtain

$$G(z) = \sum_{n=0}^{\infty} [a_n z^n + a_{n+1} z^{n+1}],$$

where

$$a_n = \frac{1}{2\pi} \int_{-\pi}^{\pi} e^{-int} f(t)\, dt.$$

Thus G is given by a convergent power series, and hence is analytic on D. Since the real part of an analytic function is harmonic, we have the following conclusion:

If f is a real-valued function in $L^1[-\pi, \pi]$, and we define

$$F(re^{i\theta}) = \frac{1}{2\pi} \int_{-\pi}^{\pi} P_r(\theta - t) f(t)\, dt, \qquad 0 \leq r < 1, \quad -\pi \leq \theta \leq \pi, \quad (1)$$

then F is harmonic on the open unit disk D.

The following result is preparatory to the next theorem.

A1.2   Lemma

Let μ be a finite signed measure on $\mathscr{B}[-\pi, \pi]$ and let $I(\theta; s)$ be the interval $(\theta - s, \theta + s)$ modulo 2π (in other words, the points x and $x + 2\pi n$ are identified for all integers n). Suppose that for some positive real number A and $\delta \in (0, \pi)$, we have

$$\mu(I(\theta; s)) < 2sA \qquad \text{for} \quad 0 < s \leq \delta.$$

If

$$F(re^{i\theta}) = \frac{1}{2\pi} \int_{-\pi}^{\pi} P_r(\theta - t) \, d\mu(t), \qquad 0 \le r < 1,$$

then

$$F(re^{i\theta}) \le A + (P_r(\delta)/2\pi) |\mu| [-\pi, \pi].$$

PROOF Break up the integral defining F into two parts, corresponding to $\delta \le |\theta - t| \le \pi$ and $|\theta - t| < \delta$. In the first integral, we have $P_r(\theta - t) \le P_r(\delta)$ by definition of P_r, so the first integral is bounded above by $(2\pi)^{-1} P_r(\delta) |\mu| [-\pi, \pi]$. To estimate the second integral, we integrate $(dP_r(s)/ds) \, ds \, d\mu(t)$ over the triangular region in the s-t plane bounded by $t = \theta + s$, $t = \theta - s$, and $s = \delta$. By Fubini's theorem,

$$\int_0^\delta \int_{(\theta - s, \theta + s)} P_r'(s) \, d\mu(t) \, ds = \int_{(\theta - \delta, \theta + \delta)} \int_{|\theta - t|}^\delta P_r'(s) \, ds \, d\mu(t).$$

If $(\theta - s, \theta + s)$ or $(\theta - \delta, \theta + \delta)$ extends outside of $[-\pi, \pi]$, we extend P_r and μ periodically. Thus we obtain

$$\int_0^\delta \mu(I(\theta; s)) P_r'(s) \, ds = \int_{I(\theta; \delta)} [P_r(\delta) - P_r(\theta - t)] \, d\mu(t).$$

(Note that $P_r(-x) = P_r(x)$.) Thus the second integral contributing to F is given by

$$\frac{1}{2\pi} \int_{I(\theta; \delta)} P_r(\theta - t) \, d\mu(t) = \frac{1}{2\pi} \left[P_r(\delta) \mu(I(\theta; \delta)) \right.$$

$$\left. - \int_0^\delta \mu(I(\theta; s)) P_r'(s) \, ds \right]$$

$$< \frac{1}{2\pi} \left[P_r(\delta) 2\delta A - \int_0^\delta 2sA P_r'(s) \, ds \right]$$

$$\text{since } P_r'(s) \le 0 \text{ for } 0 \le s \le \pi$$

$$= \frac{A}{\pi} \left(\delta P_r(\delta) - [s P_r(s)]_0^\delta + \int_0^\delta P_r(s) \, ds \right)$$

$$= \frac{A}{\pi} \int_0^\delta P_r(s) \, ds \le \frac{A}{\pi} \int_0^\pi P_r(s) \, ds$$

$$= \frac{A}{2\pi} \int_{-\pi}^\pi P_r(s) \, ds = A. \quad \|$$

A1.3   Fatou's Radial Limit Theorem

Let $f \in L^1[-\pi, \pi]$, and define

$$F(re^{i\theta}) = \frac{1}{2\pi} \int_{-\pi}^{\pi} P_r(\theta - t)f(t)\, dt, \qquad 0 \le r < 1, \quad -\pi \le \theta \le \pi.$$

Then as $r \to 1$, $F(re^{i\theta}) \to f(\theta)$ a.e. [Lebesgue measure].

PROOF   Without loss of generality, assume f real valued. Let $\mu(E) = \int_E f(t)\, dt$, $E \in \mathcal{B}[-\pi, \pi]$, and extend μ periodically to $\mathcal{B}(R)$. (This amounts to transferring μ to the Borel sets of $\{z : |z| = 1\}$.)

Fix θ, and suppose that A is greater than the upper derivative $(\bar{D}\mu)(\theta)$ (see RAP, 2.3.5, p. 74). Then for some $\delta \in (0, \pi)$ we have $(2s)^{-1}\mu(I(\theta; s)) < A$ for all $s \in (0, \delta]$. We apply A1.2, along with the observation that $P_r(\delta) \to 0$ as $r \to 1$, to obtain $\limsup_{r \to 1} F(re^{i\theta}) \le A$. Therefore

$$\limsup_{r \to 1} F(re^{i\theta}) \le (\bar{D}\mu)(\theta).$$

This argument applied to $-\mu$ yields

$$\liminf_{r \to 1} F(re^{i\theta}) \ge (\underline{D}\mu)(\theta)$$

where $\underline{D}\mu$ denotes the lower derivative of μ. But $\bar{D}\mu = \underline{D}\mu = D\mu = f$ a.e. (see RAP, 2.3.8, p. 75), and the result follows.   ‖

A1.4   The Space H^2

If f is analytic on the open unit disk D, we say that f belongs to the Hardy–Lebesgue space H^2 iff

$$\sup_{0 \le r < 1} \frac{1}{2\pi} \int_{-\pi}^{\pi} |f(re^{i\theta})|^2\, d\theta < \infty. \tag{1}$$

If $f(z) = \sum_{n=0}^{\infty} a_n z^n$, $|z| < 1$, then

$$\frac{1}{2\pi} \int_{-\pi}^{\pi} |f(re^{i\theta})|^2\, d\theta = \sum_{n=0}^{\infty} |a_n|^2 r^{2n},$$

so

$$f \in H^2 \qquad \text{iff} \qquad \sum_{n=0}^{\infty} |a_n|^2 < \infty; \tag{2}$$

furthermore, in the defining condition (1), "sup" may be replaced by "lim."
If $f \in H^2$, define the *boundary function* of f by

$$f^*(e^{i\theta}) = \sum_{n=0}^{\infty} a_n e^{in\theta}. \tag{3}$$

Since $\sum_{n=0}^{\infty} |a_n|^2 < \infty$, the series converges in $L^2(\lambda)$ where λ is Lebesgue measure on $[-\pi, \pi]$. By the Cauchy–Schwarz inequality, $f^*(e^{i\theta})$ is Lebesgue integrable. Now by the Parseval relation (RAP, 3.2.13, p. 122)

$$\frac{1}{2\pi} \int_{-\pi}^{\pi} |f^*(e^{i\theta}) - f(re^{i\theta})|^2 \, d\theta$$

$$= \frac{1}{2\pi} \int_{-\pi}^{\pi} \left| \sum_{n=0}^{\infty} a_n(1 - r^n)e^{in\theta} \right|^2 d\theta$$

$$= \sum_{n=0}^{\infty} |a_n|^2 (1 - r^n)^2$$

$$\to 0 \qquad \text{as} \quad r \to 1. \tag{4}$$

We claim that if $0 \le r < 1$, $-\pi \le \theta \le \pi$,

$$f(re^{i\theta}) = \frac{1}{2\pi} \int_{-\pi}^{\pi} P_r(\theta - t)f^*(e^{it}) \, dt, \tag{5}$$

and hence by A1.3.

$$f(re^{i\theta}) \to f^*(e^{i\theta}) \quad \text{a.e. } [\lambda] \qquad \text{as} \quad r \to 1. \tag{6}$$

To prove this, note that if $0 < s < 1$, $f(sz)$ is analytic on $\{z : |z| \le 1\}$; therefore by the Poisson integral formula (see Ash, 1971, p. 107),

$$f(sre^{i\theta}) = \frac{1}{2\pi} \int_{-\pi}^{\pi} P_r(\theta - t)f(se^{it}) \, dt.$$

Thus

$$\left| f(sre^{i\theta}) - \frac{1}{2\pi} \int_{-\pi}^{\pi} P_r(\theta, t)f^*(e^{it}) \, dt \right|^2$$

$$= \frac{1}{2\pi} \left| \int_{-\pi}^{\pi} P_r(\theta - t)[f(se^{it}) - f^*(e^{it})] \, dt \right|^2$$

$$\le \int_{-\pi}^{\pi} P_r^2(\theta - t) \, dt \times \frac{1}{2\pi} \int_{-\pi}^{\pi} |f(se^{it}) - f^*(e^{it})|^2 \, dt$$

$$\to 0 \qquad \text{as} \quad s \to 1, \text{ by (4)}.$$

But $f(sre^{i\theta}) \to f(re^{i\theta})$ as $s \to 1$, proving (6).

Because of Eq. (6), we may replace $f^*(e^{i\theta})$ by $f(e^{i\theta})$ without ambiguity; from now on we do this.

The following result is needed in prediction theory.

A1.5   Theorem

If $f \in H^2$, then

$$2\pi \ln |f(0)| \leq \int_{-\pi}^{\pi} \ln |f(e^{i\theta})| \, d\theta.$$

If, in addition, f is not identically 0, then

$$\int_{-\pi}^{\pi} \ln |f(e^{i\theta})| \, d\theta > -\infty;$$

in particular, $|f(e^{i\theta})| > 0$ a.e. [Lebesgue measure].

PROOF If $f(0) = 0$, the inequality to be proved is immediate. (Note that the integral always exists because $\ln |f(e^{i\theta})| \leq |f(e^{i\theta})|$, which is integrable.) Thus assume $f(0) \neq 0$. By Jensen's formula (see Ash, 1971, p. 112),

$$\ln |f(0)| \leq \frac{1}{2\pi} \int_{-\pi}^{\pi} \ln |f(re^{i\theta})| \, d\theta, \qquad 0 \leq r < 1.$$

By definition of H^2, the functions $\theta \to f(re^{i\theta})$, $0 \leq r < 1$, are L^2-bounded, and hence uniformly integrable (RAP, 7.6.9, p. 301); since $\ln^+ x \leq x/e$, the functions $\theta \to \ln^+ |f(re^{i\theta})|$ are also uniformly integrable. By A1.4, Eq. (6), and RAP, 7.5.2, p. 295,

$$\lim_{r \to 1} \frac{1}{2\pi} \int_{-\pi}^{\pi} \ln^+ |f(re^{i\theta})| \, d\theta = \frac{1}{2\pi} \int_{-\pi}^{\pi} \ln^+ |f(e^{i\theta})| \, d\theta.$$

By Fatou's lemma,

$$\liminf_{r \to 1} \frac{1}{2\pi} \int_{-\pi}^{\pi} \ln^- |f(re^{i\theta})| \, d\theta \geq \frac{1}{2\pi} \int_{-\pi}^{\pi} \ln^- |f(e^{i\theta})| \, d\theta.$$

Therefore

$$\frac{1}{2\pi} \int_{-\pi}^{\pi} \ln |f(e^{i\theta})| \, d\theta \geq \liminf_{r \to 1} \frac{1}{2\pi} \int_{-\pi}^{\pi} \ln |f(re^{i\theta})| \, d\theta$$

$$\geq \ln |f(0)|.$$

To prove the second statement, note that multiplication by a power of z does not change the absolute value of the boundary function, so we may assume without loss of generality that $f(0) \neq 0$, and hence $\ln |f(0)| > -\infty$. But if $f(e^{i\theta}) = 0$ on a set of positive Lebesgue measure, then $\int_{-\pi}^{\pi} \ln |f(e^{i\theta})| \, d\theta = -\infty$, a contradiction. ‖

We have constructed a boundary function f for each $f \in H^2$; under certain conditions, a partial reversal of the process is possible.

A1.6   Theorem

Let $f_0(e^{i\theta})$, $-\pi \le \theta \le \pi$, be nonnegative and Lebesgue integrable, with $\int_{-\pi}^{\pi} \ln f_0(e^{i\theta})\, d\theta > -\infty$. Then there exists an $f \in H^2$ with $(2\pi)^{-1} |f(e^{i\theta})|^2 = f_0(e^{i\theta})$ a.e.

PROOF   Define

$$g(z) = \frac{1}{2\pi} \int_{-\pi}^{\pi} Q_z(t) \ln f_0(e^{it})\, dt, \qquad h = \operatorname{Re} g.$$

By A1.1, g is analytic and h is harmonic on $\{z : |z| < 1\}$, and

$$h(re^{i\theta}) = \frac{1}{2\pi} \int_{-\pi}^{\pi} P_r(\theta - t) \ln f_0(e^{it})\, dt.$$

(Note that since $-\infty < \int_{-\pi}^{\pi} \ln f_0(e^{it})\, dt \le \int_{-\pi}^{\pi} f_0(e^{it})\, dt < \infty$, $\ln f_0(e^{it})$ is integrable.) Since $-\ln$ is a convex function and $(2\pi)^{-1} \int_{-\pi}^{\pi} P_r(\theta - t)\, dt = 1$, we may apply Jensen's inequality to obtain

$$h(re^{i\theta}) \le \ln \left[\frac{1}{2\pi} \int_{-\pi}^{\pi} P_r(\theta - t) f_0(e^{it})\, dt \right].$$

Set $f(z) = \sqrt{2\pi} \exp\left[\tfrac{1}{2} g(z)\right]$; then

$$\frac{1}{2\pi} |f(re^{i\theta})|^2 = \exp\left[\operatorname{Re} g(re^{i\theta})\right]$$

$$= \exp\left[h(re^{i\theta})\right]$$

$$\le \frac{1}{2\pi} \int_{-\pi}^{\pi} P_r(\theta - t) f_0(e^{it})\, dt.$$

By Fubini's theorem and the fact that $(2\pi)^{-1} \int_{-\pi}^{\pi} P_r(\theta - t)\, d\theta = 1$,

$$\frac{1}{2\pi} \int_{-\pi}^{\pi} |f(re^{i\theta})|^2\, d\theta \le \int_{-\pi}^{\pi} f_0(e^{it})\, dt < \infty.$$

Therefore $f \in H^2$; by A1.4, Eq. (6), we have, a.e.,

$$\frac{1}{2\pi} |f(e^{i\theta})|^2 = \frac{1}{2\pi} \lim_{r \to 1} |f(re^{i\theta})|^2$$

$$= \lim_{r \to 1} \exp\left[h(re^{i\theta})\right]$$

$$= f_0(e^{i\theta}) \quad \text{a.e., by A1.3.} \quad \|$$

The power series expansion of f may be found as follows.

A1.7   Theorem

In A1.6,

$$f(z) = (2\pi)^{1/2} \exp\left[\tfrac{1}{2}a_0 + \sum_{k=1}^{\infty} a_k z^k\right], \qquad |z| < 1,$$

where

$$a_k = \frac{1}{2\pi} \int_{-\pi}^{\pi} \ln f_0(e^{i\theta}) e^{-ik\theta} \, d\theta.$$

PROOF   By the proof of A1.6,

$$f(z) = (2\pi)^{1/2} \exp\left[\tfrac{1}{2}g(z)\right]$$

$$= (2\pi)^{1/2} \exp\left[\frac{1}{2\pi} \int_{-\pi}^{\pi} \frac{1}{2}\left(\frac{e^{it} + z}{e^{it} - z}\right) \ln f_0(e^{it}) \, dt\right].$$

But

$$\frac{e^{it} + z}{e^{it} - z} = 1 + \frac{2z}{e^{it} - z} = 1 + 2\sum_{k=1}^{\infty} e^{-ikt} z^k,$$

and the result follows.   ‖

Appendix 2

Fourier Transforms on the Real Line

We develop here some properties of Fourier transforms that are needed in the discussion of prediction theory in Chapter 2. Throughout this appendix m will denote Lebesgue measure on $(-\infty, \infty)$, and $L^P(m)$ will be denoted simply by L^P. If $g \in L^1$, the *Fourier transform* of g is defined by

$$G(\lambda) = \int_{-\infty}^{\infty} e^{it\lambda} g(t) \, dt, \qquad -\infty < \lambda < \infty.$$

If g has Fourier transform G, then the function $t \to e^{i\alpha t}g(t)$ has Fourier transform $\lambda \to G(\lambda + \alpha)$; some results of this type are now listed.

A2.1   Some Basic Properties

If $f, g \in L^1$ with Fourier transforms F, G, we have the following relations between functions h and Fourier transforms H (for example, entry (c) says that if h is the convolution of f and g, that is, $h(t) = \int_{-\infty}^{\infty} f(s)g(t - s) \, ds$, then $H(\lambda) = F(\lambda)G(\lambda)$).

$h(t)$	$H(\lambda)$
(a) $e^{i\alpha t}g(t)$	$G(\lambda + \alpha)$
(b) $g(t - \alpha)$	$e^{i\alpha\lambda}G(\lambda)$
(c) $(f * g)(t)$	$F(\lambda)G(\lambda)$
(d) $\overline{g(-t)}$	$\overline{G(\lambda)}$ (the bar denotes complex conjugate)
(e) $g(t/\alpha), \quad \alpha > 0$	$\alpha G(\alpha\lambda)$
(f) $itg(t)$	$G'(\lambda)$ (assuming $t \to tg(t) \in L^1(m)$)

242

PROOF   All results follow from the definition of the Fourier transform and basic measure-theoretic techniques, for example, Fubini's theorem in (c) and the dominated convergence theorem in (f).   ‖

From previous work on characteristic functions (see RAP, 8.1.4, p. 324) we might expect that if G as well as g belongs to L^1, then

$$g(t) = \frac{1}{2\pi} \int_{-\infty}^{\infty} G(\lambda)e^{-it\lambda} \, d\lambda \quad \text{a.e. } [m]. \tag{1}$$

Now in order to establish this formula we might try to write the integral as

$$\frac{1}{2\pi} \int_{-\infty}^{\infty} g(s) \left[\int_{-\infty}^{\infty} e^{i(s-t)\lambda} \, d\lambda \right] ds$$

but the integral in brackets does not exist. If we insert a "convergence factor," for example, $h_\alpha(\lambda) = e^{-\alpha|\lambda|}$, which for small positive α is close to 1 over a large interval, we then have

$$\frac{1}{2\pi} \int_{-\infty}^{\infty} G(\lambda)h_\alpha(\lambda)e^{-it\lambda} \, d\lambda$$

$$= \frac{1}{2\pi} \int_{-\infty}^{\infty} g(s) \left[\int_{-\infty}^{\infty} e^{i(s-t)\lambda}h_\alpha(\lambda) \, d\lambda \right] ds$$

$$= \frac{1}{2\pi} (g * H_\alpha)(t) \tag{2}$$

where H_α is the Fourier transform of h_α. (By direct computation, we may verify that

$$H_\alpha(t) = \frac{2\alpha}{\alpha^2 + t^2} = H_\alpha(-t).)$$

If $G \in L^1$ and we let $\alpha \to 0$ in (2), the dominated convergence theorem yields

$$\frac{1}{2\pi} \lim_{\alpha \to 0} (g * H_\alpha)(t) = \frac{1}{2\pi} \int_{-\infty}^{\infty} G(\lambda)e^{-it\lambda} \, d\lambda$$

so to establish (1) it suffices to show that $(2\pi)^{-1}g * H_\alpha \to g$ a.e. $[m]$. We start with bounded continuous functions.

A2.2   Lemma

If g is bounded and continuous, then as $\alpha \to 0$, $(2\pi)^{-1}(g * H_\alpha)(t) \to g(t)$ for all t.

PROOF   Since $(2\pi)^{-1} \int_{-\infty}^{\infty} H_\alpha(t) \, dt = 1$, we have

$$\frac{1}{2\pi} (g * H_\alpha)(t) - g(t)$$

$$= \frac{1}{2\pi} \int_{-\infty}^{\infty} [g(t - s) - g(t)] H_\alpha(s) \, ds$$

$$= \frac{1}{\pi} \int_{-\infty}^{\infty} [g(t - s) - g(t)] \frac{\alpha}{\alpha^2 + s^2} \, ds$$

$$= \frac{1}{\pi} \int_{-\infty}^{\infty} [g(t - \alpha u) - g(t)] \frac{1}{1 + u^2} \, du$$

$\to 0$ by the dominated convergence theorem. $\|$

If $g \in L^p$, $1 \le p < \infty$, we have L^p convergence in A2.2; we prove this after a preliminary result.

A2.3   Lemma

Let $g_s(t) = g(t - s)$. If $g \in L^p$, $1 \le p < \infty$, the map $s \to g_s$ is a bounded, uniformly continuous function from R to L^p.

PROOF   Since Lebesgue measure is translation-invariant, $\|g_s\|_p = \|g\|_p$ for all s, proving boundedness. To prove uniform continuity, let $\varepsilon > 0$ be given. Let h be a continuous function vanishing off some interval $[-A, A]$ such that $\|g - h\|_p < \varepsilon/3$ (see RAP, p. 188, Problem 3). Since h is uniformly continuous, we can find $\delta \in (0, A)$ such that

$$|h(s) - h(t)| < \frac{\varepsilon}{3(3A)^{1/p}} \qquad \text{wherever} \qquad |s - t| < \delta.$$

Then for $|s - t| < \delta$,

$$\|h_s - h_t\|_p^p = \int_{-\infty}^{\infty} |h(u - s) - h(u - t)|^p \, du$$

$$\le \left(\frac{\varepsilon}{3}\right)^p \frac{1}{3A} (2A + \delta) < \left(\frac{\varepsilon}{3}\right)^p$$

hence

$$\|g_s - g_t\|_p \le \|g_s - h_s\|_p + \|h_s - h_t\|_p + \|h_t - g_t\|_p$$

$$= 2\|g - h\|_p + \|h_s - h_t\|_p$$

by translation-invariance of Lebesgue measure

$$< \varepsilon. \quad \|$$

A2.4   Lemma

If $g \in L^p$, $1 \le p < \infty$, then as $\alpha \to 0$, $(2\pi)^{-1} g * H_\alpha \to g$ in L^p. (Thus by Minkowski's inequality, $g * H_\alpha \in L^p$.)

PROOF   First note that $(g * H_\alpha)(t)$ exists and is finite for each t, by Hölder's inequality. As in A2.2, we write

$$\frac{1}{2\pi}(g * H_\alpha)(t) - g(t) = \frac{1}{2\pi} \int_{-\infty}^{\infty} [g(t - s) - g(t)]H_\alpha(s)\, ds.$$

Since $(2\pi)^{-1} \int_{-\infty}^{\infty} H_\alpha(s)\, ds = 1$, we may apply Jensen's inequality (RAP, 7.3.5, p. 287) to obtain

$$\left| \frac{1}{2\pi}(g * H_\alpha)(t) - g(t) \right|^p \le \frac{1}{2\pi} \int_{-\infty}^{\infty} |g(t - s) - g(t)|^p H_\alpha(s)\, ds.$$

(Specifically, let P be the probability measure on $\mathscr{B}(R)$ with density $(2\pi)^{-1} H_\alpha$, and define a random variable X by $X(s) = g(t - s) - g(t)$; Jensen's inequality implies that $|E(X)|^p \le E(|X|^p)$.)

By Fubini's theorem,

$$\left\| \frac{1}{2\pi} g * H_\alpha - g \right\|_p^p \le \frac{1}{2\pi} \int_{-\infty}^{\infty} \left[\int_{-\infty}^{\infty} |g(t - s) - g(t)|^p\, dt \right] H_\alpha(s)\, ds$$

$$= \frac{1}{2\pi} \int_{-\infty}^{\infty} \|g_s - g\|_p^p H_\alpha(s)\, ds.$$

If $f(s) = \|g_s - g\|_p^p$, then by A2.3, f is bounded and continuous; also,

$$\frac{1}{2\pi} \int_{-\infty}^{\infty} f(s) H_\alpha(s)\, ds = (f * H_\alpha)(0)$$

$$\to f(0) = 0 \qquad \text{by A2.2.}$$

Thus $\|(2\pi)^{-1} g * H_\alpha - g\|_p^p \to 0$ as $\alpha \to 0$.   ‖

A2.5   Inversion Theorem

If $g \in L^1$ and the Fourier transform G also belongs to L^1, then

$$g(t) = \frac{1}{2\pi} \int_{-\infty}^{\infty} G(\lambda)e^{-it\lambda}\, d\lambda \quad \text{a.e. } [m].$$

PROOF   By the discussion before A2.2, it suffices to find a sequence of positive numbers $\alpha_n \to 0$ such that $(2\pi)^{-1} g * H_{\alpha_n} \to g$ a.e. $[m]$. But by A2.4, $(2\pi)^{-1} g * H_\alpha \to g$ in L^1 as $\alpha \to 0$, and hence $\{(2\pi)^{-1} g * H_\alpha\}$ has a subsequence converging to g a.e. $[m]$.   ‖

We now look at the properties of the map that associates with each g its Fourier transform G. The main results are as follows.

A2.6   Fourier–Plancherel Theorem

There exists a map Φ from L^2 onto L^2 such that:

(a) If $g \in L^1 \cap L^2$, then $\Phi(g) = G$;

(b) $\dfrac{1}{\sqrt{2\pi}}\, \Phi$ is an isometric isomorphism;

(c) If $g \in L^2$, $L^2 \displaystyle\lim_{A \to \infty} \int_{-A}^{A} g(t)e^{it\lambda}\, dt = \Phi(g)(\lambda)$;

(d) If $g \in L^2$, $L^2 \displaystyle\lim_{A \to \infty} \frac{1}{2\pi} \int_{-A}^{A} \Phi(g)(\lambda)e^{-it\lambda}\, d\lambda = g(t)$.

Thus if we wish to define the Fourier transform of a function $g \in L^2$, and we require that (i) the definition agree with the one given previously when $g \in L^1 \cap L^2$, and (ii) the map from a function to its Fourier transform be a bounded linear operator on L^2, then (because $L^1 \cap L^2$ is dense in L^2) there is only one possibility: the Fourier transform of g is $\Phi(g)$.

PROOF   We first show that if $g \in L^1 \cap L^2$, then $G \in L^2$, so we may define $\Phi(g) = G$ on $L^1 \cap L^2$, establishing (a). Fix $g \in L^1 \cap L^2$ and let $f = g * \tilde{g}$, where $\tilde{g}(t) = \overline{g(-t)}$. Then $f \in L^1$ and

$$f(t) = \int_{-\infty}^{\infty} g(t-s)\overline{g(-s)}\, ds = \int_{-\infty}^{\infty} g(t+u)\overline{g(u)}\, du = \langle g_{-t}, g \rangle_{L^2}. \quad (1)$$

By A2.3 and the continuity of the inner product, f is continuous; also, $|f(t)| \le \|g_{-t}\|_2 \|g\|_2 = \|g\|_2^2$, so f is bounded. Now by Eq. (2) in the discussion preceding A2.2,

$$\frac{1}{2\pi}(f * H_\alpha)(0) = \frac{1}{2\pi} \int_{-\infty}^{\infty} F(\lambda)h_\alpha(\lambda)\, d\lambda$$

$$= \frac{1}{2\pi} \int_{-\infty}^{\infty} |G(\lambda)|^2 h_\alpha(\lambda)\, d\lambda \qquad \text{by A2.1(c) and (d).}$$

By A2.2 and the monotone convergence theorem,

$$f(0) = (2\pi)^{-1} \int_{-\infty}^{\infty} |G(\lambda)|^2\, d\lambda,$$

that is (by (1) above), $\|g\|_2^2 = (2\pi)^{-1}\|G\|_2^2$. Thus $G \in L^2$.

If we define $\Phi(g) = G$, $g \in L^1 \cap L^2$, then $(2\pi)^{-1/2}\Phi$ is linear and norm-preserving, and thus extends uniquely to a norm-preserving linear map on L^2. If we can show that Φ maps onto L^2, (b) will be proved. This will be done after (d) is established.

To prove (c), let $g \in L^2$, and define $g_A = g I_{[-A, A]}$. Then $g_A \in L^1 \cap L^2$, and

$$\left\| \int_{-A}^{A} g(t) e^{it\lambda} \, dt - \Phi(g)(\lambda) \right\|_2$$

$$= \| \Phi(g_A) - \Phi(g) \|_2 \qquad \text{by (a)}$$

$$= \sqrt{2\pi} \, \| g_A - g \|_2 \qquad \text{since } (2\pi)^{-1/2} \Phi \text{ is norm-preserving}$$

$$\to 0 \qquad \text{as} \quad A \to \infty.$$

For (d), note that

$$\left\| \frac{1}{2\pi} \int_{-A}^{A} \Phi(g)(\lambda) e^{-it\lambda} \, d\lambda - g(t) \right\|_2$$

$$= \left\| \frac{1}{2\pi} \int_{-A}^{A} \overline{\Phi(g)(\lambda)} e^{it\lambda} \, d\lambda - \overline{g(t)} \right\|_2$$

$$= \left\| \frac{1}{2\pi} \Phi([\overline{\Phi(g)}]_A) - \bar{g} \right\|_2$$

which approaches 0 as $A \to \infty$ iff $\Phi(\overline{\Phi(g)}) = 2\pi \bar{g}$ (by continuity of Φ). But if g and $\Phi(g) \in L^1 \cap L^2$, this follows from (a) and A2.5.

Now let $g \in L^1 \cap L^2$; by A2.4, $g * H_\alpha \in L^1 \cap L^2$. Furthermore, $\Phi(g * H_\alpha) \in L^1 \cap L^2$ by A2.1(c) and the Cauchy–Schwarz inequality. Thus by the above argument,

$$\Phi(\overline{\Phi(g * H_\alpha)}) = 2\pi \, \overline{g} * H_\alpha .$$

Now let $\alpha \to 0$; by A2.4 and continuity of Φ,

$$\Phi(\overline{\Phi(g)}) = 2\pi \bar{g} \tag{2}$$

whenever $g \in L^1 \cap L^2$. Since $L^1 \cap L^2$ is dense in L^2, (d) is proved.

Finally, Eq. (2) with g replaced by $\bar{g}$ shows that Φ is onto. $\parallel$

References

ASH, R. B., "Information Theory." Wiley, New York, 1965.
ASH, R. B., "Basic Probability Theory." Wiley, New York, 1970.
ASH, R. B., "Complex Variables." Academic Press, New York, 1971.
ASH, R. B., "Real Analysis and Probability." Academic Press, New York, 1972.
BILLINGSLEY, P., "Ergodic Theory and Information." Wiley, New York, 1965.
BLUMENTHAL, R. M., and GETOOR, R. K., "Markov Processes and Potential Theory." Academic Press, New York, 1968.
BREIMAN, L., The Individual Ergodic Theorem of Information Theory, *Ann. Math. Statist.* **28**, No. 3, 809–811, 1957; correction, *ibid.* **31**, No. 3, 809–810.
BREIMAN, L., "Probability," Addison-Wesley, Reading, Massachusetts, 1968.
CHUNG, K. L., A Note on the Ergodic Theorem of Information Theory, *Ann. Math. Statist.* **32**, 612–614, 1961.
CRAMÉR, H., and LEADBETTER, M. R., "Stationary and Related Stochastic Processes." Wiley, New York, 1967.
DOLEANS-DADE, C., and MEYER, P. A., "Integrales Stochastiques" (lecture notes). Springer-Verlag, Berlin and New York, 1969.
DYNKIN, E. B., "Markov Processes" (two volumes). Springer, Berlin, 1965.
FREEDMAN, D., "Brownian Motion and Diffusion," Holden-Day, San Francisco, 1971.
GALLAGER, R. G., "Information Theory and Reliable Communication." Wiley, New York, 1968.
GARSIA, A. M., "Topics in Almost Everywhere Convergence." Markham, Chicago, 1970.
GIKHMAN, I. I., and SKOROKHOD, A. V., "Introduction to the Theory of Random Processes." Saunders, Philadelphia, 1969.
HALMOS, P. R., "Lectures on Ergodic Theory." Mathematical Society of Japan, Tokyo, 1956.
HANNAN, E. J., "Multiple Time Series." Wiley, New York, 1970.
HOFFMAN, K., "Banach Spaces of Analytic Functions." Prentice-Hall, Englewood Cliffs, New Jersey, 1962.
ITÔ, K., and MCKEAN, H. P. Jr., "Diffusion Processes and Their Sample Paths." Academic Press, New York, 1965.

JACOBS, K., "Lecture Notes on Ergodic Theory." Univ. of Aarhus, 1962.

KIEFFER, J., "A Generalization of the Shannon–McMillan Theorem and its Application to Information Theory." Thesis, Univ. of Illinois, Urbana, Illinois, 1970.

MCKEAN, H. P. Jr., "Stochastic Integrals." Academic Press, New York, 1969.

MCMILLAN, B., The Basic Theorems of Information Theory, *Ann. Math. Statist.* **24**, No. 2, 196–219, 1953.

ORNSTEIN, D., Bernoulli Shifts with the Same Entropy Are Isomorphic, *Advan. Math.* **4**, 337–352, 1970.

PAPOULIS, A., "Probability, Random Variables, and Stochastic Processes." McGraw-Hill, New York, 1967.

RIESZ, F., and SZ.-NAGY, B., "Functional Analysis." Ungar, New York, 1955.

ROZANOV, YU. A., "Stationary Random Processes." Holden-Day, San Francisco, 1967.

RUDIN, W., "Real and Complex Analysis." McGraw-Hill, New York, 1966.

SHANNON, C. E., A Mathematical Theory of Communication, *Bell System Tech. J.* **27**, 379–423, 623–656, 1948. Reprinted in Shannon, C. E., and Weaver, W., "The Mathematical Theory of Communication." Univ. of Illinois Press, Urbana, Illinois, 1949.

TITCHMARSH, E. C., "The Theory of Functions." Oxford, London, 1939.

WIDDER, D. V., "The Laplace Transform." Princeton Univ. Press, Princeton, New Jersey, 1941.

WONG, E., "Stochastic Processes in Information and Dynamical Systems." McGraw-Hill, New York, 1971.

YAGLOM, A. M., "Stationary Random Functions." Prentice-Hall, Englewood Cliffs, New Jersey, 1962.

Solutions to Problems

Chapter 1

Section 1.1

1. Assume each Y_t measurable, and let $A_i \in \mathscr{F}_{t_i}$, $i = 1, \ldots, n$. Define

$$A = \left\{ x \in \prod_t \Omega_t : x_{t_i} \in A_i, 1 \le i \le n \right\}.$$

Then

$$\{\omega : Y(\omega) \in A\} = \bigcap_{i=1}^{n} \{\omega : Y_{t_i}(\omega) \in A_i\} \in \mathscr{F}.$$

Thus $Y^{-1}(A) \in \mathscr{F}$ for each measurable rectangle A, hence for every $A \in \prod_t \mathscr{F}_t$. Conversely, if Y is measurable, then Y_t, the composition of Y with the projection of Ω on Ω_t, is measurable.

2. If $0 \le x < t$,

$$P\{U_t \le x\} = \sum_{n=1}^{\infty} P\{t - x \le Y_n \le t < Y_{n+1}\}$$

$$= P\{Z_{t-x} \le x\} \qquad \text{(see 1.1.5, Eq. (2))}$$

$$= F_1(x).$$

If $x \ge t$, $P\{U_t \le x\} = 1$ since $W_t \le t$.

3. By 1.1.6, Z_t and U_t are independent; Z_t has density $f(x) = \lambda e^{-\lambda x}$, $x \geq 0$, hence V_t has a density g given by the convolution formula

$$g(x) = \int_{-\infty}^{\infty} f(x - y) \, dF_{U_t}(y).$$

If $0 \leq x < t$, Problem 2 yields

$$g(x) = \int_0^x \lambda e^{-\lambda(x-y)} \lambda e^{-\lambda y} \, dy = \lambda^2 x e^{-\lambda x}.$$

If $x \geq t$,

$$g(x) = \int_0^t \lambda e^{-\lambda(x-y)} \lambda e^{-\lambda y} \, dy + f(x - t)P\{W_t = t\}$$

$$= \lambda^2 t e^{-\lambda x} + \lambda e^{-\lambda(x-t)} e^{-\lambda t}$$

$$= \lambda(1 + \lambda t)e^{-\lambda x}.$$

Now $E(Z_t) = 1/\lambda$ and $E(U_t) = \int_0^\infty P\{U_t > y\} \, dy$ (RAP, p. 280, Problem 2); thus

$$E(U_t) = \int_0^t e^{-\lambda y} \, dy = \lambda^{-1}(1 - e^{-\lambda t}).$$

Therefore

$$E(V_t) = \lambda^{-1}(2 - e^{-\lambda t}) \to 2/\lambda \text{ as } t \to \infty.$$

This is intuitively reasonable because for large t, Z_t, and U_t have approximately the same distribution.

4. (a) $\displaystyle\sum_{n=1}^{\infty} P\{Y_n \leq t\} = \sum_{n=1}^{\infty} \sum_{k=n}^{\infty} P\{N(t) = k\}$

$$= \sum_{k=1}^{\infty} \sum_{n=1}^{k} P\{N(t) = k\}$$

$$= \sum_{k=1}^{\infty} kP\{N(t) = k\} = E[N(t)].$$

Alternatively, $N(t) = \sum_{n=1}^{\infty} I_{\{Y_n \leq t\}}$; take the expectation term by term to obtain the desired result.

(b) $\displaystyle\sum_{n=1}^{\infty} P\{Y_n \leq t\} = \sum_{n=1}^{\infty} \int_0^t f_{Y_n}(x) \, dx$

$$= \int_0^t \sum_{n=1}^{\infty} f_{Y_n}(x) \, dx = \int_0^t \frac{1}{m} \, dx = \frac{t}{m}$$

(see 1.1.5, Eq. (10)).

5. Intuitively,

$$P\{y_1 \le Y_1 \le y_1 + dy_1, \ldots, y_n \le Y_n \le y_n + dy_n\}$$

$$= P\{y_1 \le T_1 \le y_1 + dy_1, y_2 - y_1 \le T_2 \le y_2 - y_1$$

$$+ dy_2, \ldots, y_n - y_{n-1} \le T_n \le y_n - y_{n-1} + dy_n\}$$

$$= \lambda e^{-\lambda y_1}\, dy_1 \lambda e^{-\lambda(y_2 - y_1)}\, dy_2 \cdots \lambda e^{-\lambda(y_n - y_{n-1})}\, dy_n$$

$$= \lambda^n e^{-\lambda y_n}\, dy_1 \cdots dy_n \quad (0 \le y_1 \le \cdots \le y_n).$$

Formally, the result is clear when $n = 1$ since $Y_1 = T_1$. Having established the result when $n = r - 1$, we have, for $0 \le y_1 \le \cdots \le y_r$,

$$P\{Y_1 \le y_1, \ldots, Y_r \le y_r\} = \int \cdots \int\limits_{\substack{x_1 \le y_1 \\ x_1 + x_2 \le y_2 \\ \vdots \\ x_1 + \cdots + x_r \le y_r \\ \text{all } x_i \ge 0}} \lambda^r e^{-\lambda(x_1 + \cdots + x_r)}\, dx_1 \cdots dx_r$$

$$= \int_0^{y_1} \lambda e^{-\lambda x_1} \left[\int \cdots \int\limits_{\substack{x_2 \le y_2 - x_1 \\ x_2 + x_3 \le y_3 - x_1 \\ \vdots \\ x_2 + \cdots + x_r \le y_r - x_1 \\ \text{all } x_i \ge 0}} \lambda^{r-1} e^{-\lambda(x_2 + \cdots + x_r)}\, dx_2 \cdots dx_n \right] dx_1$$

$$= \int_0^{y_1} \lambda e^{-\lambda x_1} P\{Y_1 \le y_2 - x_1, \ldots, Y_{r-1} \le y_r - x_1\}\, dx_1.$$

By induction hypothesis, this equals

$$\int_0^{y_1} \lambda e^{-\lambda x_1} \left[\int_{-\infty}^{y_2 - x_1} \cdots \int_{-\infty}^{y_r - x_1} g(y_2', \ldots, y_r')\, dy_2' \cdots dy_r' \right] dx_1$$

where $g(y_2', \ldots, y_r') = \lambda^{r-1} e^{-\lambda y_r'}$, $0 \le y_2' \le \cdots \le y_r'$. Let $y_j' = z_j' - x_1$ to obtain

$$\int_0^{y_1} \lambda e^{-\lambda x_1} \int_{-\infty}^{y_2} \cdots \int_{-\infty}^{y_r} g(z_2' - x_1, \ldots, z_r' - x_1)\, dz_2' \cdots dz_r'\, dx_1.$$

Hence

$$f(y_1, \ldots, y_r) = \lambda e^{-\lambda y_1} g(y_2 - y_1, \ldots, y_r - y_1)$$

$$= \lambda e^{-\lambda y_1} \lambda^{r-1} e^{-\lambda(y_r - y_1)} = \lambda^r e^{-\lambda y_r},$$

$$y_1 \ge 0, \quad 0 \le y_2 - y_1 \le \cdots \le y_r - y_1,$$

that is,

$$0 \le y_1 \le \cdots \le y_r.$$

6. $P\{Y_1 \le b_1, \ldots, Y_n \le b_n \mid N(t) = n\}$

$$= P\{Y_1 \le b_1, \ldots, Y_n \le b_n \mid Y_n \le t < Y_{n+1}\}$$

$$= \frac{P\{Y_1 \le b_1, \ldots, Y_n \le b_n, Y_{n+1} > t\}}{P\{N(t) = n\}},$$

$$0 \le b_1 \le \cdots \le b_n \le t$$

$$= \frac{1}{P\{N(t) = n\}} \int_{y_1 = 0}^{b_1} \cdots \int_{y_n = 0}^{b_n} \int_{y_{n+1} = t}^{\infty} f(y_1, \ldots, y_{n+1})\, dy_1 \cdots dy_{n+1}$$

$$= \frac{\int_0^{b_1} dy_1 \int_{y_1}^{b_2} dy_2 \cdots \int_{y_{n-1}}^{b_n} dy_n \int_t^{\infty} \lambda^{n+1} e^{-\lambda y_{n+1}}\, dy_{n+1}}{e^{-\lambda t}(\lambda t)^n / n!}$$

(see Problem 5)

$$= \int_0^{b_1} \cdots \int_0^{b_n} \frac{n!\, \lambda^n e^{-\lambda t}}{e^{-\lambda t}(\lambda t)^n}\, g(y_1, \ldots, y_n)\, dy_1 \cdots dy_n$$

where $g(y_1, \ldots, y_n) = 1$ if $0 \le y_1 \le \cdots \le y_n \le t$, and 0 elsewhere,

$$= \int_0^{b_1} \cdots \int_0^{b_n} h(y_1, \ldots, y_n)\, dy_1 \cdots dy_n$$

where $h(y_1, \ldots, y_n) = n!/t^n$, $0 \le y_1 \le \cdots \le y_n$, and 0 elsewhere. But h is the joint density of the order statistics of n independent random variables, each uniformly distributed on $[0, t]$ (see RAP, 5.9.3, p. 218); the result follows.

7.   Let $g(t) = e^{-t}f(t)$, integrable by hypothesis. Since $\int_0^{\infty} e^{-\beta t} g(t)\, dt = 0$ for all $\beta > -1$, and $g = 0$ a.e. iff $f = 0$ a.e., we may assume f integrable and $\int_0^{\infty} e^{-\alpha t} f(t)\, dt = 0$ for all $\alpha \ge 0$. If $\alpha > 0$, we may differentiate under the integral sign to obtain

$$\int_0^{\infty} (-t)^n e^{-\alpha t} f(t)\, dt = 0.$$

Thus $\int_0^{\infty} p(t) e^{-\alpha t} f(t)\, dt = 0$ for all polynomials p. If $\mathscr{H}$ is the class of all bounded Borel measurable functions h on $[0, \infty]$ (not $[0, \infty)$) such that $\int_0^{\infty} h(t) f(t)\, dt = 0$, then all functions $h(t) = p(t) e^{-\alpha t}$, p a polynomial, $\alpha > 0$, belong to $\mathscr{H}$, as do all constant functions. By the Stone–Weierstrass theorem, $C[0, \infty] \subset \mathscr{H}$, and since every lower semicontinuous function is the limit of a sequence of continuous functions, $\mathscr{H}$ contains all indicators of open sets. We may apply RAP, 4.1.4, p. 169, to conclude that $\mathscr{H}$ contains all indicators of Borel sets, hence all bounded Borel measurable functions. Therefore

$$\int_0^\infty [f(t)I_{\{t \le a, \, |f(t)| \le b\}}(t)] f(t) \, dt = 0.$$

Thus $f = 0$ a.e. on $\{t \le a, \, |f(t)| \le b\}$. Since f is integrable, it is finite a.e., so we may let $a, b \to \infty$ to conclude that $f = 0$ a.e.

8. If $\varepsilon > 0$, $X \in M$, then

$$P\{|X| \ge \varepsilon\} = P\{g(|X|) \ge g(\varepsilon)\} \qquad \text{since } g \text{ is increasing}$$

$$\le \frac{E[g(|X|)]}{g(\varepsilon)} \qquad \text{by Chebyshev's inequality.}$$

Since $g(\varepsilon) > 0$ for $\varepsilon > 0$, d-convergence implies convergence in probability. Now

$$E[g(|X|)] = \int_{\{|X| < \varepsilon\}} g(|X|) \, dP + \int_{\{|X| \ge \varepsilon\}} g(|X|) \, dP$$

$$\le g(\varepsilon) + (\sup |g|) P\{|X| \ge \varepsilon\}.$$

Since g is bounded and $g(\varepsilon) \to g(0) = 0$ as $\varepsilon \to 0$, convergence in probability implies d-convergence. (The property $g(x + y) \le g(x) + g(y)$ is used in showing that d is a metric.)

9. Let $X_n \to X$ in probability but not a.e. If such a function d exists, then for some $\varepsilon > 0$ we have $d(X_n, X) \ge \varepsilon$ for infinitely many n, say for $n \in T$. Since $X_n \xrightarrow{P} X$, there is a subsequence $\{X_n, n \in T'\}$ of $\{X_n, n \in T\}$ converging to X a.e. But then $d(X_n, X) \to 0$ as $n \to \infty$, $n \in T'$, a contradiction.

10. Suppose $(\Omega, \mathscr{F}, P)$ is such a space. Let X_t, $t \in T$, be independent random variables, with $P\{X_t = 0\} = P\{X_t = 1\} = \frac{1}{2}$ for all t, and card $T > $ card $2^\Omega \, (\ge $ card $\mathscr{F})$. If $t_1 \ne t_2$, the sets $\{X_{t_1} = 1\}$ and $\{X_{t_2} = 1\}$ cannot be equal a.e., for if so,

$$\tfrac{1}{2} = P\{X_{t_1} = 1\} = P\{X_{t_1} = 1, X_{t_2} = 1\} = \tfrac{1}{4},$$

a contradiction. Thus $\mathscr{F}$ must contain at least card T sets, contradicting the choice of T.

11. By RAP, p. 280, Problem 2, and integration by parts,

$$E(X^r) = \int_0^\infty [1 - F(x)] \, d(x^r) = [x^r(1 - F(x))]_0^\infty + \int_0^\infty x^r \, dF(x).$$

Since the last term is $E(X^r)$, the result follows.

Section 1.2

1. Since $\min (s, t)$ is real, we need only consider real numbers $a_1, \ldots, a_n$. Then if $t_1 \leq t_2 \leq \cdots \leq t_n$,

$$\sum_{j, k=1}^{n} a_j \min (t_j, t_k) a_k$$

$$= t_1(a_1^2 + 2a_1 a_2 + \cdots + 2a_1 a_n)$$
$$+ t_2(a_2^2 + 2a_2 a_3 + \cdots + 2a_2 a_n)$$
$$+ t_3(a_3^2 + 2a_3 a_4 + \cdots + 2a_3 a_n) + \cdots$$
$$+ t_{n-1}(a_{n-1}^2 + 2a_{n-1} a_n) + t_n a_n^2.$$

But

$$t_n a_n^2 + t_{n-1}(a_{n-1}^2 + 2a_{n-1} a_n) \geq t_{n-1}(a_n + a_{n-1})^2,$$
$$t_{n-1}(a_n + a_{n-1})^2 + t_{n-2}(a_{n-2}^2 + 2a_{n-2} a_{n-1} + 2a_{n-2} a_n)$$
$$\geq t_{n-2}(a_n + a_{n-1} + a_{n-2})^2.$$

Continue in this fashion to obtain $t_1(a_1 + \cdots + a_n)^2 \geq 0$.

Another approach is to observe that if f is any function from a set T to a Hilbert space H, then $g(s, t) = \langle f(s), f(t) \rangle$ is nonnegative definite, since

$$\sum_{i, j=1}^{n} a_i g(t_i, t_j) \bar{a}_j = \left\| \sum_{i=1}^{n} a_i f(t_i) \right\|^2 \geq 0.$$

In the present case, we have $T = [0, \infty)$, $H = L^2(R)$, $f(t) = I_{[0, t]}$, $g(s, t) = s \wedge t$.

2. We may write

$$X(s)X(s + t) = \sum_{i, j=1}^{\infty} V_i V_j I_{\{Y_{i-1} \leq s < Y_i\}} I_{\{Y_{j-1} \leq s+t < Y_j\}}.$$

If $i \neq j$, $E(|V_i V_j|) = E(|V_i|)E(|V_j|) \leq \sigma^2$ by the Cauchy–Schwarz inequality, and

$$\sum_{i, j=1}^{\infty} E[I_{\{Y_{i-1} \leq s < Y_i\}} I_{\{Y_{j-1} \leq s+t < Y_j\}}] = 1.$$

Thus term by term integration is allowed. We obtain

$$E[X(s)X(s + t)] = \sigma^2 \sum_{i=1}^{\infty} P\{Y_{i-1} \leq s < Y_i, Y_{i-1} \leq s + t < Y_i\}$$

$$= \sigma^2 P\{\text{no transition in } (s, s + t]\}$$

$$= \sigma^2 P\{Z_s > t\} = \sigma^2 (1 - F_1(t)).$$

Also, $X(t) = \sum_{i=1}^{\infty} V_i I_{\{Y_{i-1} \le t < Y_i\}}$, hence $E[X(t)] \equiv 0$. Since $K(-t) = K(t)$ we have $K(t) = \sigma^2(1 - F_1(|t|))$.

3. The random variables $X_1, \ldots, X_n$ are independent iff

$$h(u_1, \ldots, u_n) = \prod_{k=1}^{n} h_k(u_k),$$

where h_k is the characteristic function of X_k. The "only if" part is immediate, so assume $h(u_1, \ldots, u_n) = \prod_{k=1}^{n} h_k(u_k)$. If $F_1, \ldots, F_n$ are the distribution functions of the X_i, and

$$F(x_1, \ldots, x_n) = \prod_{k=1}^{n} F_k(x_k),$$

then F has characteristic function $\prod_{k=1}^{n} h_k(u_k) = h(u_1, \ldots, u_n)$. By RAP, p. 327, Problem 1, F must be the joint distribution function of $X_1, \ldots, X_n$, proving independence.

4. The eigenvalues of K are $\lambda_1 = 0$, $\lambda_2 = 2\sigma^2$, with orthonormalized eigenvectors

$$\begin{pmatrix} 1/\sqrt{2} \\ -1/\sqrt{2} \end{pmatrix} \quad \text{and} \quad \begin{pmatrix} 1/\sqrt{2} \\ 1/\sqrt{2} \end{pmatrix}.$$

Thus

$$W = \begin{bmatrix} 1/\sqrt{2} & 1/\sqrt{2} \\ -1/\sqrt{2} & 1/\sqrt{2} \end{bmatrix},$$

and $X = WY$, where $Y = \begin{pmatrix} Y_1 \\ Y_2 \end{pmatrix}$, $Y_1 \equiv 0$, Y_2 normal $(0, 2\sigma^2)$. Therefore $X_1 = X_2 = Y_2/\sqrt{2}$. The characteristic function of (X_1, X_2) is $\exp\left[-\frac{1}{2}(u_1 + u_2)^2\sigma^2\right]$.

5. (a) Let X be normal $(0, 1)$, and define Y as follows: Let Z be independent of X, with $P\{Z = 0\} = P\{Z = 1\} = \frac{1}{2}$. If $Z = 0$, let $Y = X$; if $Z = 1$, let $Y = -X$. Then $P\{Y \le y\} = \frac{1}{2}P\{X \le y\} + \frac{1}{2}P\{-X \le y\} = P\{X \le y\}$, so Y is normal $(0, 1)$. But if $Z = 0$, then $X + Y = 2X$, and if $Z = 1$, then $X + Y = 0$. Therefore $P\{X + Y = 0\} = \frac{1}{2}$, so $X + Y$ is not Gaussian. By 1.2.2(f), (X, Y) is not Gaussian.

(b) Let X be normal $(0, 1)$ and define Y as follows. For a fixed $c \ge 0$, let $Y = X$ if $|X| \le c$; $Y = -X$ if $|X| > c$. As in part (a), Y is normal $(0, 1)$. Now

$$E(XY) = E(X^2 I_{\{|X| \le c\}}) - E(X^2 I_{\{|X| > c\}}),$$

so that $E(XY) = -1$ for $c = 0$, and $E(XY) \to 1$ as $c \to \infty$. Since $E(XY)$ is continuous in c by the monotone convergence theorem, we will have

$E(XY) = 0$ for some c. But $P\{X > c, Y > c\} = 0 \neq P\{X > c\}P\{Y > c\}$, and hence X and Y are not independent.

6. Let $X = X_1 + iX_2$, $Y = X_3 + iX_4$, where (X_1, X_2, X_3, X_4) is Gaussian with 0 mean and covariance matrix

$$K = \begin{bmatrix} 1 & 0 & 0 & -1 \\ 0 & 1 & -1 & 0 \\ 0 & -1 & 1 & 0 \\ -1 & 0 & 0 & 1 \end{bmatrix}.$$

The quadratic form of K is $x_1^2 + x_2^2 + x_3^2 + x_4^2 - 2x_1 x_4 - 2x_2 x_3 = (x_1 - x_4)^2 + (x_2 - x_3)^2 \geq 0$, hence K is a covariance matrix. Now

$$E(X\bar{Y}) = E(X_1 X_3 + X_2 X_4) + iE(X_2 X_3 - X_1 X_4) = 0 + i(-1 + 1) = 0.$$

But X and Y are not independent, for if they were, X_1 and X_4 would be independent, hence uncorrelated. Since $E(X_1 X_4) = -1$, this is a contradiction.

7. (a) (i) is obvious.

(ii) Since $\bar{K}$ is symmetric and nonnegative definite, it is a covariance. By (i), Re $K = \frac{1}{2}(K + \bar{K})$ is a covariance.

(iii) If $X_j(t)$ is defined on $(\Omega_j, \mathscr{F}_j, P_j)$ and has covariance $K_j, j = 1, 2$, consider $(\Omega_1 \times \Omega_2, \mathscr{F}_1 \times \mathscr{F}_2, P_1 \times P_2)$, that is, observe the processes $X_1(t)$ and $X_2(t)$ independently. Formally, define $X(t, (\omega_1, \omega_2)) = X_1(t, \omega_1)X_2(t, \omega_2)$. Then, assuming $E[X_j(t)] \equiv 0$ without loss of generality,

$$E[X(s)\overline{X(t)}] = E[X_1(s)\overline{X_1(t)}]E[X_2(s)\overline{X_2(t)}]$$
$$= K_1(s, t)K_2(s, t).$$

Now $K(t, s) = \bar{K}(s, t)$, hence Im $K(t, s) = -$Im $K(s, t)$. Thus Im K is not symmetric, hence is not a covariance, unless it vanishes identically.

(b) Let $n \to \infty$ in $\sum_{j, k=1}^r a_j K_n(t_j, t_k)\bar{a}_k \geq 0$ to obtain the desired result.

8. The joint characteristic function of $X_1, \ldots, X_n$ is

$$h(u_1, \ldots, u_n) = E[\exp i(u_1 X_1 + \cdots + u_n X_n)];$$

since $Y = \sum_{j=1}^n u_j X_j$ is Gaussian by hypothesis, $E[e^{itY}] = e^{itm}e^{-t^2\sigma^2/2}$, $m = E(Y)$, $\sigma^2 = \text{Var } Y$. Set $t = 1$ to obtain, with $b_j = EX_j$, $K_{jk} = \text{Cov}(X_j, X_k)$,

$$h(u_1, \ldots, u_n) = \exp\left[i \sum_{j=1}^n u_j b_j - \frac{1}{2} \sum_{j, k=1}^n u_j K_{jk} u_k\right].$$

9. The analysis of 1.2.9 shows that $X_n(t_j) - X_n(t_{j-1}) \xrightarrow{d} B(t_j) - B(t_{j-1})$ for each j. Since the random walk and the Brownian motion have independent increments, the joint characteristic function of $X_n(t_1), \ldots, X_n(t_k)$ is

$$h_n(u_1, \ldots, u_k) = E\left[\exp i \sum_{j=1}^{k} u_j X_n(t_j)\right]$$

$$= E\left[\exp i\left(\sum_{j=1}^{k} u_j\right) X_n(t_1)\right]$$

$$\times E\left[\exp i \sum_{j=2}^{k} u_j(X_n(t_2) - X_n(t_1))\right] \cdots$$

$$E[\exp iu_k(X_n(t_k) - X_n(t_{k-1}))]$$

$$\to h(u_1, \ldots, u_k),$$

the joint characteristic function of $B(t_1), \ldots, B(t_k)$, by Lévy's theorem. The result follows.

10. (a) Since

$$E[X(t)] = \sum_{n=0}^{\infty} E[X(t)I_{\{M(t)=n\}}]$$

and

$$E[X(t)I_{\{M(t)=n\}}] = E\left[\sum_{j=1}^{n} Y_j I_{\{M(t)=n\}}\right] = nE(Y_1)P\{M(t) = n\},$$

we have $E[X(t)] = E[M(t)]E(Y_1)$. (To make sure that $E[X(t)]$ exists, apply this argument to $X^+(t)$ and $X^-(t)$ separately.)

(b) As in (a),

$$E[X^2(t)] = \sum_{n=0}^{\infty} E\left[\left(\sum_{j=1}^{n} Y_j\right)^2\right] P\{M(t) = n\}$$

$$= \sum_{n=0}^{\infty} [n \operatorname{Var} Y_1 + n^2(E(Y_1))^2] P\{M(t) = n\}$$

$$= E[M(t)] \operatorname{Var} Y_1 + E[M^2(t)](EY_1)^2.$$

The result follows from (a).

A similar calculation shows that the characteristic function of $M(t)$ is $\sum_{n=0}^{\infty} [h(u)]^n P\{M(t) = n\}$ where h is the characteristic function of Y_1. (This does not require that $M(t)$ or Y_1 have finite mean.)

(c)   Since $\{M(t)\}$ has independent increments, so does $\{X(t)\}$. As in 1.2.9, if $s \le t$,

$$\text{Cov}\,[X(s), X(t)] = \text{Var}\,X(s)$$

$$= \lambda s\,\text{Var}\,Y_1 + [E(Y_1)]^2\lambda s \qquad \text{by (b)}$$

$$= E(Y_1^2)\lambda s.$$

11. (a)   To have n customers arrive in $[0, t + h]$, there are three possibilities:

n customers arrive in $[0, t]$ and none in $(t, t + h]$;
$n - 1$ customers arrive in $[0, t]$ and 1 in $(t, t + h]$;
less than $n - 1$ customers arrive in $[0, t]$ and
more than 1 in $(t, t + h]$.

Thus by assumptions (1) and (2),

$$p_n(t + h) = p_n(t)[1 - \lambda(t)h + o(h)]$$

$$+ p_{n-1}(t)[\lambda(t)h + o(h)] + o(h), \qquad n \ge 1,$$

and

$$p_0(t + h) = p_0(t)[1 - \lambda(t)h + o(h)].$$

Divide by h and let $h \downarrow 0$ to obtain the desired equations. By assumption (3), $p_0(0) = 1$ and $p_n(0) = 0$, $n \ge 1$; thus we have initial conditions for each equation.

(b)   Since $N(t + h)$ is the sum of the independent random variables $N(t)$ and $N(t + h) - N(t)$, the characteristic function of $N(t + h) - N(t)$ is

$$\exp\,[L(t + h)(e^{iu} - 1)]/\exp\,[L(t)(e^{iu} - 1)]$$

$$= \exp\,[(L(t + h) - L(t))(e^{iu} - 1)],$$

as desired.

(c)   As in 1.2.9, $\text{Cov}\,[N(s), N(t)] = \text{Var}\,N(s \wedge t) = L(s \wedge t)$.

(d)   Since L is strictly increasing, it maps $[0, \infty)$ one-to-one onto an interval $[0, a)$, $0 < a \le \infty$. Let f be the inverse of L; since f is strictly increasing, the process $X(t) = N(f(t))$, $0 \le t < a$, has independent increments. Also, $X(t + h) - X(t)$ has the Poisson distribution with parameter $L(f(t + h)) - L(f(t)) = t + h - t = h$, as desired.

(e)   Define the process by specifying the finite dimensional distributions; if $0 \le t_1 < \cdots < t_n$, let $N(t_1)$, $N(t_2) - N(t_1)$, ..., $N(t_n) - N(t_{n-1})$ be independent, with $N(t_j) - N(t_{j-1})$ Poisson with parameter $L(t_j) - L(t_{j-1})$ (set $N(0) \equiv 0$). This determines the joint distribution of $N(t_1), \ldots, N(t_n)$, and it is not hard to check that the consistency conditions are satisfied. The

Kolmogorov extension theorem yields a process with the desired distributions. Assumptions (1) and (3) are satisfied by definition; also,

$$P\{N(t + h) - N(t) = 0\} = \exp\left[-\int_t^{t+h} \lambda(x)\, dx\right]$$

$$= e^{-h\lambda(y_h)}$$

$$\text{for some } y_h \in [t, t + h]$$

$$= 1 - h\lambda(y_h) + o(h)$$

$$= 1 - h\lambda(t) + o(h)$$

$$\text{by continuity of } \lambda.$$

Similarly,

$$P\{N(t + h) - N(t) = 1\} = \exp\left[-\int_t^{t+h} \lambda(x)\, dx\right]\int_t^{t+h} \lambda(x)\, dx$$

$$= [1 - \lambda(t)h + o(h)]h\lambda(y_h)$$

$$= h\lambda(t) + o(h),$$

and the result follows.

12. (a) Cover $[0, \infty)$ by disjoint intervals $I_1, I_2, \ldots$ of length $\delta > 0$, choose $y_j \in I_j, j = 1, 2, \ldots$, and define

$$X_\delta(t) = \sum_j h(t - y_j)\, \Delta N_j$$

where ΔN_j is the number of arrivals in I_j. Since h has at most countably many discontinuities, the probability is 0 that $t - A_n$ will be a discontinuity point of h for some n. Also, with probability 1 there will be at most one arrival in each I_j for sufficiently small $\delta = \delta(\omega)$; it follows that $X_\delta(t) \to X(t)$ a.e. as $\delta \to 0$, t fixed. Now the joint characteristic function of $X_\delta(t_1), \ldots, X_\delta(t_n)$ is

$$E(\exp[i(u_1 X_\delta(t_1) + \cdots + u_n X_\delta(t_n))])$$

$$= E\left(\exp\left[i\sum_j \sum_{k=1}^n u_k h(t_k - y_j)\, \Delta N_j\right]\right).$$

The ΔN_j are independent and Poisson with parameter $\int_{I_j} \lambda(x)\, dx$; thus we obtain

$$\prod_j \exp\left[\int_{I_j} \lambda(x)\, dx\left(\exp\left[i\sum_{k=1}^n u_k h(t_k - y_j)\right] - 1\right)\right]$$

$$= \exp\left[\sum_j \lambda(\theta_j)\left(\exp\left[i\sum_{k=1}^n u_k h(t_k - y_j)\right] - 1\right) \text{length } I_j\right]$$

$$\text{for some } \theta_j \in I_j.$$

The sum on j is actually finite, and λ is uniformly continuous on bounded intervals. Thus we have a Riemann sum that converges to the desired integral as $\delta \to 0$. But the sum also converges to the joint characteristic function of $X(t_1), \ldots, X(t_n)$, since

$$\sum_{k=1}^{n} u_k X_\delta(t_k) \to \sum_{k=1}^{n} u_k X(t_k) \quad \text{a.e.}$$

The result follows.

(b) If $|h| \le M$, then $|X(t)| \le MN(t)$, where $N(t)$ is the number of arrivals in $[0, t]$. Since $N(t)$ has a Poisson distribution, the result follows.

(c) If $\Phi(u_1, \ldots, u_n) = \int_{R^n} \exp\left[i(u_1 x_1 + \cdots + u_n x_n)\right] dF(x_1, \ldots, x_n)$, then

$$\left(\frac{\partial \Phi}{\partial u_j}\right)_{\text{all } u_i = 0} = iE(X_j) \qquad \text{and} \qquad \left(\frac{\partial^2 \Phi}{\partial u_i \, \partial u_j}\right)_{\text{all } u_i = 0} = -E(X_i X_j).$$

These formulas, along with part (a), yield the desired results.

(d) By (a) and (c), the joint characteristic function of $Y_k(t_1), \ldots, Y_k(t_n)$ is

$$\exp\left[\int_0^\infty k\lambda_0(x)\left(\exp\left[i\sum_{j=1}^{n} u_j \frac{h(t_j - x)}{\sqrt{k}}\right] - 1\right) dx\right.$$

$$\left. -i\sum_{j=1}^{n} u_j \int_0^\infty k\lambda_0(x)\frac{h(t_j - x)}{\sqrt{k}} dx\right]$$

$$= \exp\left[-\frac{1}{2}\int_0^\infty \lambda_0(x)\left(\sum_{j=1}^{n} u_j h(t_j - x)\right)^2 dx + o(1)\right]$$

$$\to \exp\left[-\frac{1}{2}\sum_{j,k=1}^{n} u_j u_k \int_0^\infty \lambda_0(x)h(t_j - x)h(t_k - x) \, dx\right],$$

the joint characteristic function of a Gaussian process with 0 mean and the desired covariance function.

13. (a) In the integral defining G_T, let $s = u + v$, $t = u - v$; in the s-t plane, the region of integration is the square with vertices $(0, 0)$, $(T, -T)$, $(2T, 0)$, (T, T). Since $|\partial(u, v)/\partial(s, t)| = \frac{1}{2}$,

$$G_T(x) = \frac{1}{2T}\int_{-T}^{T} e^{-itx} K(t) \int_{|t|}^{2T - |t|} ds \, dt$$

$$= \int_{-T}^{T} \left(1 - \frac{|t|}{T}\right) e^{-itx} K(t) \, dt.$$

(b) By (a), the left side of the desired equality is

$$\frac{1}{2\pi}\int_{-T}^{T} \left[\int_{-M}^{M} \left(1 - \frac{|x|}{M}\right) e^{i(u-t)x} \, dx\right] g_T(t) \, dt.$$

The integral in brackets equals

$$2 \int_0^M \left(1 - \frac{x}{M}\right) \cos (u - t)x \, dx = \frac{2(1 - \cos (u - t)M)}{M(u - t)^2} = f_M(u - t).$$

(c)   We have $f_M(u - t) \to 0$ as $M \to \infty$ if $u \neq t$, and

$$\frac{1}{2\pi} \int_R f_M(x) \, dx = \frac{1}{\pi} \int_R \frac{\sin^2 y}{y^2} \, dy = 1.$$

(To verify this, integrate $(1 - e^{i2z})/z^2$ along the real axis, with a small semi-circle around the pole at 0, and complete the contour by a large semicircle in the upper half-plane.) Since g_T is continuous, the result follows.

(d)   By (b) and (c), g_T is a pointwise limit of characteristic functions as $M \to \infty$, and since g_T is continuous, it is a characteristic function by Lévy's theorem. Since g_T converges pointwise to K as $T \to \infty$, and K is continuous, K is also a characteristic function.

Section 1.3

1.   (a)   Symmetry is clear. For nonnegative definiteness,

$$\sum_{i, j=1}^n \sum_{k=1}^\infty a_i g_k(t_i) \overline{g_k(t_j)} a_j = \sum_{k=1}^\infty \left| \sum_{i=1}^n a_i g_k(t_i) \right|^2 \geq 0.$$

(b)   Symmetry is clear. Now

$$\sum_{j, k=1}^n a_j K(t_j - t_k) \bar{a}_k = \sum_{j, k=1}^n a_j \left(\int_{-\infty}^\infty e^{it_j\lambda} e^{-it_k\lambda} f(\lambda) \, d\lambda \right) \bar{a}_k$$

$$= \int_{-\infty}^\infty \left| \sum_{j=1}^n a_j e^{it_j\lambda} \right|^2 f(\lambda) \, d\lambda \geq 0.$$

If f is even, then $\int_{-\infty}^\infty f(\lambda) \sin t\lambda \, d\lambda = 0$, hence K is real valued, so the process can be taken as real valued.

2.   Let the $X(t)$, $-\infty < t < \infty$, be independent, each with 0 mean and variance σ^2. Then the process is stationary with covariance $K(t) = 0, t \neq 0$; $K(0) = \sigma^2$.

3.   By 1.3.2, $(X(t + h) - X(t))/h$ converges in L^2 to a limit as $h \to 0$ iff

$$\frac{E[(X(t + h) - X(t))\overline{(X(t + h') - X(t))}]}{hh'}$$

approaches a unique finite limit as $h, h' \to 0$, that is, iff

$$[K(t + h, t + h') - K(t, t + h') - \overline{K(t, t + h)} + K(t, t)]/hh'$$

approaches a limit. For $K(s, t) = \min (s, t)$, set $h = h' > 0$ to obtain $h^{-2}[t + h - t - t + t] = h^{-1} \to \infty$. Thus $\{X(t)\}$ is not L^2-differentiable, but is L^2-continuous since $\min (s, t)$ is continuous.

4. Since $X(s) \overset{L^2}{\to} X(s)$ and

$$\frac{X(t_0 + h) - X(t_0)}{h} \overset{L^2}{\to} X'(t_0),$$

1.3.1 yields

$$\frac{K(s, t_0 + h) - K(s, t_0)}{h} \to E[X(s)\overline{X'(t_0)}],$$

hence $\partial K(s, t_0)/\partial t$ exists and equals $E[X(s)\overline{X'(t_0)}]$. But

$$\frac{X(t_0 + h) - X(t_0)}{h} \overset{L^2}{\to} X'(t_0), \qquad X'(t_0) \overset{L^2}{\to} X'(t_0),$$

so by 1.3.1, $h^{-1}(E[X(t_0 + h)\overline{X'(t_0)}] - E[X(t_0)\overline{X'(t_0)}]) \to E[\,|X'(t_0)|^2]$, that is, if $K_2 = \partial K/\partial t$, then

$$h^{-1}[K_2(t_0 + h, t_0) - K_2(t_0, t_0)] \to E[\,|X'(t_0)|^2].$$

Thus $\partial^2 K/\partial s\, \partial t$ exists at (t_0, t_0) and equals $E[\,|X'(t_0)|^2]$.

5. By the argument of Problem 3, if the process is L^2-differentiable at the point t, then

$$[K(h - h') - \overline{K(h')} - K(h) + K(0)]/hh'$$

approaches a limit as $h, h' \to 0$. But this is independent of t, hence, again by the criterion of Problem 3, the process is L^2-differentiable everywhere.

6. If the formula is valid for a given (n, m), then

$$X^{(n)}(s + t) \overset{L^2}{\to} X^{(n)}(s + t), \qquad \frac{X^{(m)}(s + h) - X^{(m)}(s)}{h} \overset{L^2}{\to} X^{(m+1)}(s).$$

By 1.3.1 and the induction hypothesis,

$$\frac{(-1)^m}{h}[K^{(n+m)}(t - h) - K^{(n+m)}(t)] \to E[X^{(n)}(s + t)\overline{X^{(m+1)}(s)}].$$

Also

$$\frac{X^{(n)}(s + t + h) - X^{(n)}(s + t)}{h} \overset{L^2}{\to} X^{(n+1)}(s + t), \qquad X^{(m)}(s) \overset{L^2}{\to} X^{(m)}(s).$$

Thus

$$\frac{(-1)^m}{h}\left[K^{(n+m)}(t+h)-K^{(n+m)}(t)\right] \to E[X^{(n+1)}(s+t)\overline{X^{(m)}(s)}].$$

Therefore the result holds for $(n, m+1)$ and $(n+1, m)$; since it holds for $n = m = 0$ by definition of the covariance function of a stationary process, it holds for all (n, m).

7. (a) By 1.3.10,

$$\left\|\int_a^b g(t)X(t)\,dt\right\|_2^2 = \int_a^b \int_a^b g(s)\overline{g(t)}K(s, t)\,ds\,dt$$

$$= \int_a^b \int_a^b \int_{-\infty}^{\infty} g(s)e^{is\lambda}\overline{g(t)}e^{-it\lambda}f(\lambda)\,d\lambda\,ds\,dt$$

$$= \int_{-\infty}^{\infty} |G(\lambda)|^2 f(\lambda)\,d\lambda$$

by Fubini's theorem,

where $G(\lambda) = \int_{-\infty}^{\infty} g(s)e^{is\lambda}\,ds$ is the Fourier transform of g,

$$\leq (\sup |f|)\|G\|_2^2$$

$$= (\sup |f|)2\pi\|g\|_2^2 \text{ by the Fourier–Plancherel}$$

theorem (A2.6).

Thus $g \to \int_a^b g(t)X(t)\,dt$ defines a bounded linear operator on the space of continuous complex-valued functions with compact support (with L^2 norm). Since such functions are dense in $L^2(-\infty, \infty)$ (see RAP, p. 188, Problem 3 or RAP, p. 34, Problem 12), the result follows.

(b) Let $\{g_n\}$, $\{h_n\}$ be sequences of continuous functions with compact support converging to g and h in $L^2(-\infty, \infty)$. Then by 1.3.10,

$$E\left[\int_{-\infty}^{\infty} g_n(s)X(s)\,ds \int_{-\infty}^{\infty} h_n(t)X(t)\,dt\right]$$

$$= \int_{-\infty}^{\infty} \int_{-\infty}^{\infty} g_n(s)\overline{h_n(t)}K(s, t)\,ds\,dt$$

$$= \int_{-\infty}^{\infty} G_n(\lambda)\overline{H_n(\lambda)}f(\lambda)\,d\lambda \qquad \text{as in (a)}.$$

As $n \to \infty$, the left side approaches

$$E\left[\int_{-\infty}^{\infty} g(s)X(s)\,ds \int_{-\infty}^{\infty} h(t)X(t)\,dt\right]$$

by continuity of the linear operator defined in (a), and continuity of the inner product in a Hilbert space. The right side approaches

$$\int_{-\infty}^{\infty} G(\lambda)\overline{H(\lambda)}f(\lambda)\, d\lambda$$

by the Cauchy–Schwarz inequality. But this integral is

$$\int_{-\infty}^{\infty} \int_{-\infty}^{\infty} g(s)\overline{h(t)}K(s,\, t)\, ds\, dt,$$

proving the first equality of 1.3.10; the second equality is similar.

8. (a) By 1.3.10 and 1.3.11,

$$E\left[\left| \frac{1}{h}\int_{t}^{t+h} X(s)\, ds - X(t) \right|^2\right]$$

$$= \frac{1}{h^2}\int_{t}^{t+h} \int_{t}^{t+h} K(s,\, t)\, ds\, dt - \frac{1}{h}\int_{t}^{t+h} K(s,\, t)\, ds$$

$$- \frac{1}{h}\int_{t}^{t+h} \bar{K}(s,\, t)\, ds + K(t,\, t)$$

$$\to 0 \qquad \text{as } h \to 0$$

by the mean value theorem for integrals.

(b) By (a), both sides of the equation have the same L^2-derivative, so the result follows if we can prove that if X is a function from $[a, b]$ to a Banach space B, and $X'(t) = 0$ on $[a, b]$, then X is constant. But if f is any continuous linear functional on B, then

$$\frac{1}{h}[f(X(t + h)) - f(X(t))] = f\left[\frac{X(t + h) - X(t)}{h}\right] \to 0,$$

and therefore $f \circ X$ is a constant. But if $X(t_1) \neq X(t_2)$, the Hahn–Banach theorem yields an f for which $f(X(t_1)) \neq f(X(t_2))$; therefore X is constant.

Section 1.4

1. Let $C = \{(t, \omega) : \sum_{k=1}^{\infty} Z_k(\omega)e_k(t) \neq X(t, \omega)\}$. Then for each t, the section $C(t)$ has probability 0. It follows (see RAP, p. 107, Problem 4) that for almost every ω, $C(\omega)$ has Lebesgue measure 0.

2. The series $\sqrt{2} \sum_{n=1}^{\infty} Z_n^* \sin(n - \frac{1}{2})\pi t/(n - \frac{1}{2})\pi$, where the Z_n^* are independent and normal $(0, 1)$, yields an L^2 process $\{X(t), 0 \le t \le 1\}$ with covariance min (s, t) (see 1.4.4). Let

$$Y(t) = c\sqrt{2} \sum_{n=1}^{\infty} Z_n^* \frac{\sin(n - \frac{1}{2})\pi t/a}{(n - \frac{1}{2})\pi}.$$

Then $\{Y(t), 0 \le t \le a\}$ is an L^2 process with covariance c^2 min $(s/a, t/a) = (c^2/a)$ min (s, t), $s, t \in [0, a]$. Choose c so that $c^2/a = \sigma^2$.

3. By Mercer's theorem, $K(t, t) = \sum_{n=1}^{\infty} \lambda_n e_n^2(t)$. Since the series converges uniformly, we may integrate term by term to obtain

$$\int_a^b K(t, t)\, dt = \sum_{n=1}^{\infty} \lambda_n \int_a^b e_n^2(t)\, dt = \sum_{n=1}^{\infty} \lambda_n.$$

4. (a) $\int_0^1 s t e(t)\, dt = \lambda e(s)$, $0 \le s \le 1$, has orthonormalized solution $e(s) = \sqrt{3}\, s$, $\lambda = \frac{1}{3}$. The Karhunen–Loève expansion is $X(t) = Z\sqrt{3}\, t$, $0 \le t \le 1$, where $E(Z) = 0$, $E(Z^2) = \frac{1}{3}$.

(b) If $e_n'(t) = t^n - 2/(n + 2)$, $n = 1, 2, \ldots$, then $\int_0^1 s t e_n'(t)\, dt = 0$. If we consider the functions $s, 1, s^2, s^3, \ldots$ and orthonormalize by the Gram–Schmidt process to get $e_1(s), e_2(s), \ldots$ ($e_1(s) = \sqrt{3}\, s$), the eigenspace of 0 is spanned by $e_2, e_3, \ldots$ (since $e_1, e_2, \ldots$ span $L^2[0, 1]$).

5. (a) Since

$$S_n(t) = \operatorname{Im} \sum_{k=2^n+1}^{2^{n+1}} \frac{\exp\left[i(k - \frac{1}{2})\pi t\right]}{(k - \frac{1}{2})\pi} Z_k^*$$

and $|\operatorname{Im} z| \le |z|$, we may write

$$|S_n(t)|^2 \le \sum_{k=2^n+1}^{2^{n+1}} \frac{1}{\pi^2(k - \frac{1}{2})^2} |Z_k^*|^2 + \frac{2}{\pi^2} \operatorname{Re} \sum_{p=1}^{2^{n+1}-2^n-1} \sum_{h=2^n+1}^{2^{n+1}-p} U_{hp}(t)$$

where

$$U_{hp}(t) = \frac{\exp\left[i(h - \frac{1}{2})\pi t - i(h + p - \frac{1}{2})\pi t\right]}{(h - \frac{1}{2})(h + p - \frac{1}{2})} Z_h^* Z_{h+p}^*.$$

The real part of the complex exponential is $\cos \pi t p$, which does not depend on the index of summation h. Therefore

$$|S_n(t)|^2 \le \sum_{k=2^n+1}^{2^{n+1}-p} \frac{1}{\pi^2(k - \frac{1}{2})^2} |Z_k^*|^2$$

$$+ \frac{2}{\pi^2} \sum_{p=1}^{2^{n+1}-2^n-1} \left| \sum_{h=2^n+1}^{2^{n+1}-p} \frac{Z_h^* Z_{h+p}^*}{(h - \frac{1}{2})(h + p - \frac{1}{2})} \right|.$$

Call the expression inside the absolute value signs W_p. If $h < k$, then $E(Z_h^* Z_{h+p}^* Z_k^* Z_{k+p}^*) = 0$ by independence. Since

$$E[\,|\,Z_h^* Z_{h+p}^*\,|^2) = E(\,|\,Z_h^*\,|^2)E(\,|\,Z_{h+p}^*\,|^2) = 1,$$

we have

$$E(W_p^2) \le \sum_{h=2^n+1}^{2^{n+1}-p} \frac{1}{(h-\frac{1}{2})^2 (h+p-\frac{1}{2})^2}.$$

Now

$$|\,E(T_n)\,|^2 \le E(T_n^2)$$

$$\le \frac{1}{\pi^2} \sum_{k=2^n+1}^{2^{n+1}} \frac{1}{(k-\frac{1}{2})^2} + \frac{2}{\pi^2} \sum_{p=1}^{2^{n+1}-2^n-1} E(\,|\,W_p\,|),$$

and since $E(\,|\,W_p\,|) \le [E(W_p^2)]^{1/2}$, the result follows.

(b) By part (a) and the fact that $2^{n+1} - 2^n = 2^n$,

$$\pi^2 E(T_n^2) \le 2^n \frac{1}{(2^n)^2} + 2(2^n)\left[2^n \frac{1}{(2^n)^2(2^n)^2} \right]^{1/2}$$

$$= \frac{1}{2^n} + \frac{2}{2^{n/2}} \le \frac{3}{2^{n/2}}.$$

Thus

$$E(T_n) \le [E(T_n^2)]^{1/2} \le \frac{\sqrt{3}}{\pi} \left[\frac{1}{2^{1/4}} \right]^n,$$

so that $\sum_{n=0}^{\infty} E(T_n) < \infty$.

Section 1.5

1. (a) We may write $\hat{Y}(r) = h_0 X(r) + h_1 X(r-1)$, where

$$\begin{bmatrix} K_X(0) & K_X(1) \\ K_X(1) & K_X(0) \end{bmatrix} \begin{bmatrix} h_0 \\ h_1 \end{bmatrix} = \begin{bmatrix} K_Z(0) \\ K_Z(1) \end{bmatrix}.$$

But $K_X(0) = K_Z(0) + K_W(0) = 4 + 1 = 5$, $K_X(n) = K_Z(n) = 0$, $n \ne 0$. Thus $h_0 = \frac{4}{5}$, $h_1 = 0$, so $\hat{Y}(r) = \frac{4}{5}X(r)$. Since $Z(n) + W(n)$ determines $Z(n)$ uniquely, $Y^*(r)$ is given by $g(X(r))$ where $g(3) = g(1) = 2$, $g(-1) = g(-3) = -2$.

(b) $\sigma^2 = \|Y\|^2 - \|\hat{Y}(r)\|^2 = E[Z^2(r)] - \frac{16}{25}E[X^2(r)] = 4 - \frac{16}{5} = \frac{4}{5}$.

(c) $Y^*(r) = Y(r) = Z(r)$, so $\sigma^{*2} = 0$.

2. (a) For each i we have $\hat{X}_i(k) \in S_{k-1}$ and $\hat{X}_i(k) - X_i(k) \perp S_{k-1}$; therefore $\hat{X}(k) \in S_{k-1}^n$ and $X(k) - \hat{X}(k) \perp S_{k-1}^n$.

    (b) (i) Observe that $\hat{X}(k)$ is $\sigma(S_k)$-measurable.

       (ii) For $j = 0, 1, \ldots, k-1$, we have

$$E[(X(k) - \hat{X}(k))V^*(j)]$$
$$= E\{E[(X(k) - \hat{X}(k))V^*(j) \mid V(0), \ldots, V(k-1)]\}$$
$$= E\{V^*(j)E[(X(k) - \hat{X}(k)) \mid V(0), \ldots, V(k-1)]\}$$
$$= 0.$$

Thus $X(k) - \hat{X}(k)$ and $V(0), \ldots, V(k-1)$ are orthogonal, and hence independent; similarly, $V(k) - E[V(k) \mid V(0), \ldots, V(k-1)]$ and $V(0), \ldots, V(k-1)$ are independent. Now by RAP, p. 262, Problem 2, $E(Y \mid X, Z) = E(Y \mid X)$ if (X, Y) and Z are independent. Set $X = V(k) - E\{V(k) \mid V(0), \ldots, V(k-1)\}$, $Y = X(k) - \hat{X}(k)$, $Z = (V(0), \ldots, V(k-1))$ to obtain the desired result.

       (iii) We have

$$V(k) - E[V(k) \mid V(0), \ldots, V(k-1)]$$
$$= M(k)X(k) + W(k)$$
$$\quad - E[M(k)X(k) + W(k) \mid V(0), \ldots, V(k-1)]$$
$$= M(k)\{X(k) - E[X(k) \mid V(0), \ldots, V(k-1)]\}$$
$$\quad + W(k) - E[W(k) \mid V(0), \ldots, V(k-1)]$$
$$= M(k)(X(k) - \hat{X}(k)) + W(k).$$

       (iv) Since $E(X \mid Y)$ belongs to the subspace spanned by the components of Y, we may write $E(X \mid Y) = CY$ for some matrix C. Also, $X - CY$ is orthogonal to Y, so that $E[(X - CY)Y^*] = E(XY^*) - CE(YY^*) = 0$. The result follows.

       (v) By (ii) and (iii), the summand is

$$\Phi(k)E[X(k) - \hat{X}(k) \mid V(k) - M(k)\hat{X}(k)].$$

Apply (iv) with $X = X(k) - \hat{X}(k)$, $Y = V(k) - M(k)\hat{X}(k) = M(k) \times (X(k) - \hat{X}(k)) + W(k)$, and use the fact that $X(k) = \hat{X}(k)$ and $W(k)$ are orthogonal.

       (vi) We have, for $j = 0, \ldots, k$,

$$E[U(k)V^*(j)] = E\{U(k)[M(j)X(j) + W(j)]\}$$
$$= M(j)E[U(k)X(j)]$$
$$= 0$$

since $X(j)$ can be expressed in terms of $X(0)$ and $U(i)$, $i < j$. Thus $U(k)$ and $(V(0), \ldots, V(k))$ are independent, so that

$$E[U(k) \mid S_k] = E[U(k)] = 0.$$

(c) By (b) we have

$$
\begin{aligned}
X(k+1) - \hat{X}(k+1) &= \Phi(k)(X(k) - \hat{X}(k)) + U(k) \\
&\quad - A(k)[V(k) - M(k)\hat{X}(k)] \\
&= \Phi(k)(X(k) - \hat{X}(k)) + U(k) \\
&\quad - A(k)M(k)(X(k) - \hat{X}(k)) \\
&\quad - A(k)W(k)
\end{aligned}
$$

(see biii).

Set $\Phi_0(k) = \Phi(k) - A(k)M(k)$; by orthogonality (note that $\hat{X}(k)$ is in the space spanned by the components of $V(j)$, $j \le k - 1$, and $X(k)$ is a linear combination of $X(0)$ and the $U(j)$, $j \le k - 1$),

$$P(k+1) = \Phi_0(k)P(k)\Phi_0^*(k) + Q(k) + A(k)P(k)A^*(k).$$

(d) Since each component of $\hat{X}(j+1 \mid j)$ is in the space spanned by the components of $V(0), \ldots, V(j)$, the same is true of the random vector $\Phi(j+k-1) \cdots \Phi(j+1)\hat{X}(j+1 \mid j)$. If $k = 2$, we have, for $i \le j$,

$$
\begin{aligned}
E[\{X(j+2) - \Phi(j+1)\hat{X}(j+1 \mid j)\}V^*(i)] \\
= E[\{\Phi(j+1)[X(j+1) - \hat{X}(j+1 \mid j)] + U(j+1)\}V^*(i)] \\
= 0.
\end{aligned}
$$

If $k > 2$, the result follows by induction.

Chapter 2

Section 2.1

1. Properties (a) and (b) of 2.1.3 are immediate; to prove (c), note that $Z(\bigcup_{n=1}^{\infty} E_n) - \sum_{k=1}^{n} Z(E_k)$ is the indicator of $A_n = \bigcup_{k=n+1}^{\infty} E_k$, and $\|I_{A_n}\|^2 = \mu(A_n) \to 0$ as $n \to \infty$ by continuity of μ. The identity $\int_S f\, dZ = f$ is clear for simple functions and extends to all functions in $L^2(S, \mathscr{S}, \mu)$ by 2.1.2(e).

2. If $E_1, E_2 \in \mathscr{S}$,

$$\langle Z(E_1), Z(E_2) \rangle = \left\langle \sum_j e_j I_{E_1}(x_j), \sum_k e_k I_{E_2}(x_k) \right\rangle$$

$$= \sum_k I_{E_1}(x_k) I_{E_2}(x_k) \|e_k\|^2$$

by orthogonality

$$= m(E_1 \cap E_2),$$

proving 2.1.3(a); 2.1.3(b) is immediate. To prove 2.1.3(c), write

$$Z\left(\bigcup_{n=1}^{\infty} E_n\right) - \sum_{i=1}^{n} Z(E_i)$$

$$= \sum_{k=1}^{\infty} e_k \sum_{n=1}^{\infty} I_{E_n}(x_k) - \sum_{i=1}^{n} \sum_{k=1}^{\infty} e_k I_{E_i}(x_k)$$

$$= \sum_{k=1}^{\infty} e_k \left[\sum_{n=1}^{\infty} I_{E_n}(x_k) - \sum_{i=1}^{n} I_{E_i}(x_k) \right]$$

since the limit of a finite sum

is the sum of the limits

$$= \sum_{k=1}^{\infty} e_k I_{A_n}(x_k) \qquad \text{where} \quad A_n = \bigcup_{i=n+1}^{\infty} E_i .$$

For a fixed N, we have $x_1, \ldots, x_N \notin A_n$ for large n, since each x_i can be in at most one E_j. Thus

$$\left\| Z\left(\bigcup_{n=1}^{\infty} E_n\right) - \sum_{i=1}^{n} Z(E_i) \right\|^2 \leq \sum_{k > N} \|e_k\|^2 \qquad \text{for large } n$$

$$\to 0 \qquad \text{as} \quad N \to \infty,$$

as desired.

Now $\int_S f \, dZ = \sum_{k=1}^{\infty} f(x_k) e_k$ is clear for simple functions; in general, if $f_1, f_2, \ldots$ are simple functions converging in $L^2(S, \mathscr{S}, m)$ to f, then

$$\left\| \int_S f \, dZ - \sum_{k=1}^{\infty} f(x_k) e_k \right\| \leq \left\| \int_S f \, dZ - \int_S f_n \, dZ \right\|$$

$$+ \left\| \int_S f_n \, dZ - \sum_{k=1}^{\infty} f(x_k) e_k \right\|.$$

The first term on the right-hand side approaches 0 by 2.1.2(e), and, by the result for simple functions, the square of the second term is

$$\sum_{k=1}^{\infty} \|e_k\|^2 \, |f_n(x_k) - f(x_k)|^2 = \int_S |f_n - f|^2 \, dm \to 0.$$

3. (a) Since

$$Z(E_1 \cup E_2) = Z(E_1) + Z(E_2 - E_1) = Z(E_2) + Z(E_1 - E_2),$$

we have

$$\|Z(E_1) - Z(E_2)\|^2 = m(E_1 - E_2) + m(E_2 - E_1) = m(E_1 \triangle E_2)$$

$$= m(E_1) + m(E_2) - 2m(E_1 \cap E_2).$$

(b) This follows from (a).

(c) Note that $\|Z(E_2)\|^2 = m(E_2) \le m(E_1) = \|Z(E_1)\|^2$.

(d) This follows from (c).

(e) This is a consequence of finite additivity (2.1.3(b)).

(f) This follows from (e).

(g) We have $Z(E_1 \cup E_2) = Z(E_2) + Z(E_1 - E_2)$, and the result follows from (e).

(h) Write $Z(E_1 \triangle E_2) = Z(E_1 - (E_1 \cap E_2)) + Z(E_2 - (E_1 \cap E_2))$, and apply (f).

4. If Z is a mov and $E_1 \cap E_2 = \varnothing$, then $Z(E_1) \perp Z(E_2)$ by 2.1.3(a). Conversely, if the condition (call it 2.1.3(a')) is satisfied, let $m(E) = \|Z(E)\|^2$; by 2.1.3(a') and 2.1.3(c), m is a measure on $\mathscr{S}$. If $E_1, E_2 \in \mathscr{S}$, then by 2.1.3(b),

$$\langle Z(E_1)Z(E_2)\rangle$$

$$= \langle Z(E_1 - E_2) + Z(E_1 \cap E_2), Z(E_2 - E_1) + Z(E_1 \cap E_2)\rangle$$

$$= \langle Z(E_1 \cap E_2), Z(E_1 \cap E_2)\rangle \qquad \text{by 2.1.3(a')}$$

$$= m(E_1 \cap E_2).$$

5. If $a < a' < b' < b$, then

$$\left\| \int_{[a,\,b]} f \, dZ_{ab} - \int_{[a',\,b']} f \, dZ_{a'b'} \right\|^2$$

$$= \left\| \int_{[a,\,b]} (f - f I_{[a',\,b']}) \, dZ_{ab} \right\|^2$$

$$= \int_{[a,\,b]} |f - f I_{[a',\,b']}|^2 \, dm$$

$$\le \int_{R - [a',\,b']} |f|^2 \, dm \to 0 \qquad \text{as} \quad a' \to -\infty, \, b' \to \infty.$$

6. Condition (a) of 2.1.3 follows from the fact that ψ is an isometric isomorphism and Z is a mov. To prove (c), write

$$\left\| Z'\left(\bigcup_{n=1}^{\infty} E_n \right) - \sum_{k=1}^{n} Z'(E_k) \right\|$$

$$= \left\| \psi \left[Z\left(\bigcup_{n=1}^{\infty} E_n \right) - \sum_{k=1}^{n} Z(E_k) \right] \right\|$$

$$= \left\| Z\left(\bigcup_{n=1}^{\infty} E_n \right) - \sum_{k=1}^{n} Z(E_k) \right\|$$

$$\to 0 \qquad \text{as} \quad n \to \infty.$$

The proof of (b) is similar.

Section 2.2

1. By RAP, p. 127, Problem 9, we may expand $e^{it\lambda}$ in a Fourier series on $[-2\pi W, 2\pi W]$; for each t,

$$e^{it\lambda} = \sum_{n=-\infty}^{\infty} c_n(t) e^{in\lambda/2W} \qquad \text{(convergence in } L^2[-2\pi W, 2\pi W])$$

where

$$c_n(t) = \frac{1}{4\pi W} \int_{-2\pi W}^{2\pi W} e^{it\lambda} e^{-in\lambda/2W} \, d\lambda$$

$$= \frac{\sin (2\pi Wt - n\pi)}{2\pi Wt - n\pi}.$$

Since $e^{it\lambda}$, $-2\pi W \leq \lambda \leq 2\pi W$, is of bounded variation, the Fourier series converges boundedly on $(-2\pi W, 2\pi W)$ (Titchmarsh, 1939, p. 408), and since μ assigns measure 0 to the endpoints of the interval, the series converges in $L^2(\mu)$. Since

$$\int_R e^{it\lambda} \, dZ(\lambda) = X(t) \quad \text{and} \quad \int_R e^{in\lambda/2W} \, dZ(\lambda) = X(n/2W),$$

the result follows from 2.1.2(e).

2. Since $X(n/2W) = \cos (n\pi + \theta) = (-1)^n X(0)$, the infinite series, if it converges, must yield a random variable proportional to $X(0)$. But $X(-1/4W) = \cos (-\frac{1}{2}\pi + \theta) = \sin \theta$, which is orthogonal to $X(0)$. Thus when $t = -1/4W$, the series cannot converge to $X(t)$.

3. Since the X_n are mutually orthonormal, it follows from 2.2.8 that $\{X_n\}$ is purely nondeterministic. Now if $m < n$,

$$X_n = e^{in\theta} = (e^{i\theta})^n = (X_m/X_{m-1})^n.$$

4. Write $H_t = H_{-\infty} \oplus H'_t$, where $H_t = H_t(X)$; then $P_t = P_{-\infty} + P'_t$, where P'_t is the projection on H'_t. If $t_n \downarrow -\infty$, and $Y \in L^2\{X(t), t \in T\}$, the random variables $P'_{t_k} Y - P'_{t_{k+1}} Y$ are orthogonal (note that $P'_{t_{k+1}} P'_{t_k} = P'_{t_{k+1}}$, so $(P'_{t_j} - P'_{t_{j+1}})(P'_{t_k} - P'_{t_{k+1}}) = 0$ for $j > k$). Thus for all n,

$$\left\| \sum_{k=1}^{n} (P'_{t_k} Y - P'_{t_{k+1}} Y) \right\|^2 = \| P'_{t_1} Y - P'_{t_{n+1}} Y \|^2$$

$$\leq 4 \| Y \|^2 < \infty.$$

Therefore $\sum_{n=1}^{\infty} [P'_{t_n} Y - P'_{t_{n+1}} Y]$ converges in L^2, and hence $P'_{t_n} Y$ approaches a limit Y_0, necessarily in $\bigcap_t H_t = H_{-\infty}$. But $P'_{t_n} Y \in H'_{t_n} \perp H_{-\infty}$, so $Y_0 = 0$. Thus $P_t Y = P_{-\infty} Y + P'_t Y \to P_{-\infty} Y$ as $t \to -\infty$.

5. (a) Let $T_t^{(i)}$ be the shift operator on $H_i(X)$, $i = 1, 2$. Define T_t^* on $H_1(X) \oplus H_2(X)$ $(= H(X)$ by 2.2.6) by

$$T_t^*(Y_1 + Y_2) = T_t^{(1)}(Y_1) + T_t^{(2)}(Y_2).$$

Since

$$T_t^*(X(r)) = T_t^{(1)}(X_1(r)) + T_t^{(2)}(X_2(r)) = X_1(r + t) + X_2(r + t)$$

$$= X(r + t), \qquad T_t^* \equiv T_t.$$

The result now follows if we set $Y_1 = X_1(r)$, $Y_2 = 0$, and then $Y_1 = 0$, $Y_2 = X_2(r)$.

(b) This is immediate when φ is a finite linear combination of complex exponentials, and the result follows from 2.1.2(e).

(c) Since $X_i(0) \in H_i(X) \subset H(X)$, by 2.2.2(c) we can find $\varphi_i \in L^2(\mu_X)$ such that

$$X_i(0) = \int_S \varphi_i(\lambda) \, dZ_X(\lambda).$$

Since $T_t X_i(0) = X_i(t)$ by (a), the result follows from (b).

6. Since $X_1(t), X_2(t) \in H(X)$, we have $E[X_i(t)] = 0$, $i = 1, 2$, and by Problem 5c and the technique of 2.2.2(a), each $\{X_i(t), t \in T\}$ has a continuous covariance function and hence a spectral measure. Let μ_i be the spectral measure of $\{X_i(t)\}$, and μ the spectral measure of $\{X(t)\}$. If Z_i is the spectral mov of $\{X_i(t)\}$, then by orthogonality of $\{X_1(t)\}$ and $\{X_2(t)\}$, $Z_1 + Z_2$ is a mov with associated measure $\mu_1 + \mu_2$. But

$$\int_S e^{it\lambda} \, d(Z_1 + Z_2)(\lambda) = X_1(t) + X_2(t) = X(t),$$

so by the uniqueness part of the spectral decomposition theorem, the spectral mov of $\{X(t)\}$ is $Z = Z_1 + Z_2$. Since $\mu(E) = \|Z(E)\|^2$, the result follows by orthogonality.

Section 2.3

1. (a) We compute

$$E[X(m)\overline{W(n)}] = \sum_{j=0}^{\infty} \overline{b}_j E[X(m)\overline{X(n-j)}]$$

$$= \sum_{j=0}^{\infty} \overline{b}_j \int_{-\pi}^{\pi} e^{i(m-n+j)\lambda} \frac{1}{2\pi \,|B(e^{-i\lambda})|^2} \, d\lambda$$

$$= \frac{1}{2\pi} \int_{-\pi}^{\pi} e^{i(m-n)\lambda} \frac{\overline{B(e^{-i\lambda})}}{|B(e^{-i\lambda})|^2} \, d\lambda$$

$$= \frac{1}{2\pi} \int_{-\pi}^{\pi} \frac{e^{i(m-n)\lambda}}{B(e^{-i\lambda})} \, d\lambda.$$

Let $z = e^{-i\lambda}$, $dz = -iz \, d\lambda$, to obtain the desired result; note that changing the path from clockwise to counterclockwise absorbs the minus sign.

(b) If $B(z)$ has no zeros for $|z| \leq 1$, and $m < n$, the integral of part (a) is 0 by Cauchy's theorem.

2. By the discussion of 2.3.5 and 2.3.6, the coefficients a_j are determined by the Fourier expansion $\varphi(\lambda) = \sum_{j=-\infty}^{\infty} a_j e^{-ij\lambda}$, where $|\varphi(\lambda)|^2 = 2\pi f_Y(\lambda) = |B(e^{-i\lambda})|^2$. Thus we may choose $\varphi(\lambda) = [B(e^{-i\lambda})]^{-1}$, and since $[B(z)]^{-1}$ is analytic on an open set $U \supset \{z : |z| \leq 1\}$, we must have $a_j = 0$ for $j < 0$.

3. We have $B(z) = 1 - \alpha z$, so $[B(z)]^{-1} = \sum_{j=0}^{\infty} \alpha^j z^j$, $|z| < |\alpha|^{-1}$. By Problem 2, the process may be represented as $Y(n) = \sum_{j=0}^{\infty} \alpha^j W(n-j)$, where $\{W(n)\}$ is white noise. By 2.3.5, the spectral density is $f_Y(\lambda) = (2\pi)^{-1} |1 - \alpha e^{-i\lambda}|^{-2}$; the covariance function is, for $n \geq 0$,

$$K_Y(n) = E[Y(n+m)\overline{Y(m)}]$$

$$= \sum_{j,k=0}^{\infty} \alpha^j \overline{\alpha}^k E[W(n+m-j)\overline{W(m-k)}]$$

$$= \sum_{k=0}^{\infty} \alpha^{n+k} \overline{\alpha}^k$$

$$= \frac{\alpha^n}{1 - |\alpha|^2}.$$

4. By 2.3.4, the spectral mov of $\{X_0(n)\}$ is given by

$$Z_0(E) = \int_E B(e^{-i\lambda})\, dZ_X(\lambda),$$

hence by 2.1.5,

$$Z_X(E) = \int_E [B(e^{-i\lambda})]^{-1}\, dZ_0(\lambda), \qquad E \in \mathscr{S}$$

$$= \int_E \frac{A(e^{-i\lambda})}{B(e^{-i\lambda})}\, dZ_W(\lambda).$$

Therefore

$$X(n) = \int_S e^{in\lambda}\, dZ_X(\lambda)$$

$$= \int_S e^{in\lambda} \frac{A(e^{-i\lambda})}{B(e^{-i\lambda})}\, dZ_W(\lambda).$$

Since $W(n) = \int_S e^{in\lambda}\, dZ_W(\lambda)$, 2.1.2(b) yields

$$E[X(m)\overline{W(n)}] = \int_S e^{i(m-n)\lambda} \frac{A(e^{-i\lambda})}{B(e^{-i\lambda})} \frac{d\lambda}{2\pi}$$

$$= \frac{1}{2\pi i} \int_{|z|=1} z^{n-m-1} \frac{A(z)}{B(z)}\, dz$$

$$= 0$$

if $m < n$, as in Problem 1.

To show that $\{X(n)\}$ is a one-sided moving average, proceed as in Problem 2, with $\varphi(\lambda) = A(e^{-i\lambda})/B(e^{-i\lambda})$. Since $A(z)/B(z)$ is analytic on an open set $U \supset \{z : |z| \le 1\}$, the result follows.

5. By 2.3.4, we have

$$Z_{X_1}(E) = \int_E \varphi_1(\lambda)\, dZ_Y(\lambda), \; Z_{X_2}(E) = \int_E \varphi_2(\lambda)\, dZ_{X_1}(\lambda),$$

and the result follows by 2.1.5.

6. By 2.3.4, $\varphi(\lambda) = \frac{1}{3}(1 + e^{-i\lambda} + e^{-i2\lambda}) = \frac{1}{3}e^{-i\lambda}(1 + 2\cos\lambda)$, which is near 0 for λ near $\pm 2\pi/3$.

Section 2.4

1. The stochastic difference equation shows that $X(n + \tau)$ can be written as $a^{\tau} X(n) + Z(n)$, where $Z(n) \perp X(m)$ for $m \leq n$, and hence $Z(n) \perp H_n(X)$. It follows that $P_n X(n + \tau) = a^{\tau} X(n)$.

2. (a) By 2.3.6, $f(\lambda) = (2\pi)^{-1} |B(e^{-i\lambda})|^{-2}$, where

$$B(z) = 1 - (\alpha_1 + \alpha_2)z + \alpha_1 \alpha_2 z^2 = (1 - \alpha_1 z)(1 - \alpha_2 z).$$

(b) Let

$$C(z) = 1/B(z) = \sum_{j=0}^{\infty} \alpha_1^j z^j \sum_{k=0}^{\infty} \alpha_2^k z^k = C_1(z)C_2(z)$$

$$= \sum_{r=0}^{\infty} \sum_{j=0}^{r} \alpha_1^j \alpha_2^{r-j} z^r.$$

By 2.4.1, the transfer function of the optimal predictor is

$$\varphi(\lambda) = e^{i\tau\lambda} \frac{C_\tau(e^{-i\lambda})}{C(e^{-i\lambda})}$$

$$= e^{i\tau\lambda} C_1^{-1}(e^{-i\lambda}) C_2^{-1}(e^{-i\lambda}) \sum_{r=\tau}^{\infty} \alpha_2^r e^{-ir\lambda} \left[\frac{1 - (\alpha_1/\alpha_2)^{r+1}}{1 - (\alpha_1/\alpha_2)} \right].$$

The sum can be written as

$$\sum_{r=\tau}^{\infty} \frac{\alpha_2^{r+1} - \alpha_1^{r+1}}{\alpha_2 - \alpha_1} e^{-ir\lambda},$$

hence

$$\varphi(\lambda) = e^{i\tau\lambda} \frac{\alpha_2}{\alpha_2 - \alpha_1} \frac{\alpha_2^{\tau} e^{-i\tau\lambda}}{C_1(e^{-i\lambda})} - e^{i\tau\lambda} \frac{\alpha_1}{\alpha_2 - \alpha_1} \frac{\alpha_1^{\tau} e^{-i\tau\lambda}}{C_2(e^{-i\lambda})}$$

$$= \frac{1}{\alpha_2 - \alpha_1} [\alpha_2^{\tau+1}(1 - \alpha_1 e^{-i\lambda}) - \alpha_1^{\tau+1}(1 - \alpha_2 e^{-i\lambda})].$$

Thus

$$P_n X(n + \tau) = \int_{[-\pi, \pi]} e^{in\lambda} \varphi(\lambda) \, dZ_X(\lambda)$$

$$= \frac{1}{\alpha_2 - \alpha_1} [(\alpha_2^{\tau+1} - \alpha_1^{\tau+1}) X(n)$$

$$- \alpha_1 \alpha_2 (\alpha_2^{\tau} - \alpha_1^{\tau}) X(n - 1)].$$

Note that when $\alpha_2 = 0$, then $P_n X(n + \tau) = \alpha_1^{\tau} X(n)$, in agreement with the result for the first order autoregressive scheme in Problem 1.

(c)   By (b) we have

$$\delta(1) = E[\,|X(n+1) - P_n X(n+1)|^2]$$
$$= E[\,|X(n+1) - (\alpha_1 + \alpha_2)X(n) + \alpha_1\alpha_2 X(n-1)|^2]$$
$$= E\,|W(n+1)|^2$$
$$= 1.$$

Thus $\delta(1)$ is the variance of the innovation.

3.   Let $\hat{H}_n(X)$ be the subspace of $L^2(\mu_X)$ spanned by $e^{im\lambda}$, $m \leq n$. By definition of φ_τ, we have $\varphi_\tau(\lambda) \in \hat{H}_0(X)$ and $e^{i\tau\lambda} - \varphi_\tau(\lambda) \perp \hat{H}_0(X)$. Therefore $e^{in\lambda}\varphi_\tau(\lambda) \in \hat{H}_n(X)$ and $e^{i(n+\tau)\lambda} - e^{in\lambda}\varphi_\tau(\lambda) \perp \hat{H}_n(X)$, as desired.

4.   (a)   If R has degree r and S has degree s, then since $f(\lambda)$ is real,

$$f(\lambda) = \overline{f(\lambda)} = \bar{A}e^{in\lambda}\frac{\prod(e^{i\lambda} - \bar{\alpha}_j)}{\prod(e^{i\lambda} - \bar{\beta}_k)}$$
$$= \bar{A}e^{i(n+r-s)\lambda}\frac{\prod(1 - \bar{\alpha}_j e^{-i\lambda})}{\prod(1 - \bar{\beta}_k e^{-i\lambda})}.$$

Thus $f(\lambda)$ can be expressed in the form

$$\frac{R(e^{-i\lambda})}{S(e^{-i\lambda})} = Be^{ip\lambda}\frac{\prod(e^{-i\lambda} - 1/\bar{\alpha}_j)}{\prod(e^{-i\lambda} - 1/\bar{\beta}_k)}.$$

By the identity theorem (see Ash 1971, p. 40), this holds with $e^{-i\lambda}$ replaced by any complex z such that $S(z) \neq 0$; the result follows.

(b)   The desired factorization follows from (a) and the observation that

$$(e^{-i\lambda} - \alpha)(e^{-i\lambda} - 1/\bar{\alpha}) = \frac{|e^{-i\lambda} - \alpha|^2}{-\bar{\alpha}e^{i\lambda}}.$$

(c)   Write $\alpha_j = e^{-i\theta_j}$; the power series expansion of the exponential function shows that near $\lambda = \theta_j$,

$$e^{-i\lambda} - e^{-i\theta_j} = c_j(\lambda - \theta_j) + o(\lambda - \theta_j)$$

where $|c_j| = 1$. If α_j had odd multiplicity, $f(\lambda)$ would change sign at $\lambda = \theta_j$, contradicting the fact that $f(\lambda) \geq 0$. Since $|\alpha_j| = 1$ implies that $\alpha_j = 1/\bar{\alpha}_j$, the argument of part (b) yields the desired result.

Section 2.5

1.   (a)   If $x \in \bigcap_n (A_n \oplus B)$, then for each n we have $x = P_{A_n}x + P_B x$, where P indicates projection. Let $n \to \infty$ to obtain, as in Problem 4, Section 2.2, $P_{A_n}x \to P_A x$, where $A = \bigcap_n A_n$. Thus $x \in A \oplus B$; the reverse inclusion is clear.

(b)   If $m < n$, we have

$$H_{-\infty}(X) \subset H_m(Y_1) \oplus H_m(Y_2) \subset H_m(Y_1) \oplus H_n(Y_2).$$

Fix n; by part (a), $H_{-\infty}(X) \subset H_{-\infty}(Y_1) \oplus H_n(Y_2)$, and the result follows by another application of part (a).

Section 2.6

1.   Since h is Riemann integrable on any bounded interval, we have, for any sequence of partitions whose maximal subinterval length approaches 0,

$$\sum_k h(s_k)e^{i(t-s_k)\lambda} \, \Delta s_k \to \int_{-T}^{T} h(s)e^{i(t-s)\lambda} \, ds \qquad \text{for each real } \lambda.$$

Since $h \in L^1(m)$, the left side is bounded by a constant independent of λ, so by the dominated convergence theorem, we have convergence in $L^2(\mu_Y)$. By 2.1.2(e),

$$\sum_k h(s_k) \int_{-\infty}^{\infty} e^{i(t-s_k)\lambda} \, dZ_Y(\lambda) \, \Delta s_k \to \int_{-\infty}^{\infty} \int_{-T}^{T} h(s)e^{i(t-s)\lambda} \, ds \, dZ_Y(\lambda).$$

But $\int_{-\infty}^{\infty} e^{i(t-s_k)\lambda} \, dZ_Y(\lambda) = Y(t - s_k)$, and therefore by 1.3.9,

$$\int_{-T}^{T} h(s)Y(t - s) \, ds = \int_{-\infty}^{\infty} \int_{-T}^{T} h(s)e^{i(t-s)\lambda} \, ds \, dZ_Y(\lambda).$$

Now as $T \to \infty$, $\int_{-T}^{T} h(s)e^{i(t-s)\lambda} \, ds$ converges in $L^2(m)$ to $e^{it\lambda}H(-\lambda)$, and since the spectral density of $\{Y(t)\}$ is bounded, convergence also occurs in $L^2(\mu_Y)$. By 2.1.2(e),

$$L^2 \lim_{T \to \infty} \int_{-T}^{T} h(s)Y(t - s) \, ds = \int_{-\infty}^{\infty} e^{it\lambda}H(-\lambda) \, dZ_Y(\lambda),$$

as desired.

2.   (a)   Linearity is clear; now

$$E[|X(g)|^2] = \int_{-\infty}^{\infty} |G(\lambda)|^2 f(\lambda) \, d\lambda \qquad \text{by 2.1.2(c)}$$

$$\leq M \int_{-\infty}^{\infty} |G(\lambda)|^2 \, d\lambda \qquad \text{where } M \text{ is a bound on } f$$

$$= 2\pi M \|g\|_2^2 \qquad \text{by the Fourier–Plancherel}$$

theorem (Appendix 2, Theorem A2.6).

Thus X is continuous.

(b) The solution to Problem 1, with $Y(t - s)$ replaced by $X(s)$, $e^{i(t-s)\lambda}$ by $e^{is\lambda}$, and h by g shows that

$$\int_{-\infty}^{\infty} g(s)X(s)\, ds = \int_{-\infty}^{\infty} G(\lambda)\, dZ_X(\lambda),$$

as desired.

3. Let $\{Z(\lambda), -\infty < \lambda < \infty\}$ be an L^2 process with 0 mean and orthogonal increments such that $E[\,|Z(\lambda_2) - Z(\lambda_1)|^2] = |\lambda_2 - \lambda_1|$, and define

$$Z_n(E) = \int_E \sqrt{f_n(\lambda)}\, dZ(\lambda)$$

(see 2.1.6 for the definition of the integral). By 2.1.6, Z_n is a stochastic mov whose associated measure μ_n has density f_n with respect to Lebesgue measure m, and

$$\int_{-\infty}^{\infty} h\, dZ_n = \int_{-\infty}^{\infty} h\sqrt{f_n}\, dZ$$

for $h \in L^2(\mu_n) \supset L^\infty(m) \cup L^2(m)$. (Approximate by simple functions in the usual way.) Now define

$$W_n(t) = \int_{-\infty}^{\infty} e^{it\lambda}\, dZ_n(\lambda)$$

$$= \int_{-\infty}^{\infty} e^{it\lambda}\sqrt{f_n(\lambda)}\, dZ(\lambda).$$

Then $\{W_n(t)\}$ has spectral density f_n, and if $g \in L^2(m)$,

$$W_n(g) = \int_{-\infty}^{\infty} g(t)W_n(t)\, dt$$

$$= \int_{-\infty}^{\infty} G(\lambda)\, dZ_n(\lambda) \qquad \text{by Problem 2}$$

$$= \int_{-\infty}^{\infty} G(\lambda)\sqrt{f_n(\lambda)}\, dZ(\lambda).$$

To show that $\{W_m(g)\}$ converges, write (using 2.1.2(c))

$$E[\,|W_m(g) - W_n(g)|^2] = \int_{-\infty}^{\infty} |G(\lambda)|^2\, |\sqrt{f_n(\lambda)} - \sqrt{f_m(\lambda)}|^2\, d\lambda$$

$$\to 0 \qquad \text{as } n, m \to \infty$$

by the dominated convergence theorem.

Finally, if $g, h \in L^2(m)$,

$$E[W_n(g)\overline{W_n(h)}] = \int_{-\infty}^{\infty} G(\lambda)\overline{H(\lambda)}f_n(\lambda)\, d\lambda \qquad \text{by 2.1.2(b)}$$

$$\to f_0 \int_{-\infty}^{\infty} G(\lambda)\overline{H(\lambda)}\, d\lambda$$

by the dominated convergence theorem

$$= 2\pi f_0 \int_{-\infty}^{\infty} g(t)\overline{h(t)}\, dt$$

by the Fourier–Plancherel theorem.

4. (a) If $s, t \geq 0$, $K(s, t) = \sigma^2 \min(s, t)$ by 1.2.9; if $s, t \leq 0$, $K(s, t) = \sigma^2 \min(-s, -t)$; if $s \leq 0 \leq t$, $K(s, t) = 0$. In each case, $K(s, t)$ agrees with the desired expression.

(b) If $0 \leq t - s \leq 1/n$, independence of increments yields

$$E[W_n(s)\overline{W_n(t)}] = n^2 E\left[\left|B\left(s + \frac{1}{n}\right) - B(t)\right|^2\right]$$

$$= n^2\sigma^2\left(s + \frac{1}{n} - t\right)$$

$$= n^2\sigma^2\left(\frac{1}{n} - |s - t|\right).$$

Handling the other cases similarly, we see that $\{W_n(t)\}$ is stationary, with covariance

$$K_n(t) = \begin{cases} n\sigma^2(1 - n|t|), & |t| \leq 1/n \\ 0, & |t| > 1/n. \end{cases}$$

The spectral density is

$$f_n(\lambda) = \frac{1}{2\pi} \int_{-\infty}^{\infty} e^{-it\lambda} K_n(t)\, dt$$

$$= \frac{n\sigma^2}{\pi} \int_0^{1/n} (1 - nt)\cos \lambda t\, dt$$

$$= \frac{\sigma^2}{2\pi}\left[\frac{\sin \lambda/2n}{\lambda/2n}\right]^2.$$

(c) If g is continuous with compact support and $W_n(g) = \int_{-\infty}^{\infty} g(t)W_n(t)\, dt$, then (as in 1.3.10) if $m \leq n$,

$$E[W_m(g)\overline{W_n(g)}] = \int_{-\infty}^{\infty}\int_{-\infty}^{\infty} g(s)g(t)K_{mn}(t - s)\, ds\, dt$$

where $K_{mn}(t)$ is 0 for $t \leq -1/n$ or $t \geq 1/m$, $K_{mn}(t) = m\sigma^2$, $0 \leq t \leq 1/m -$ $1/n$, and the remaining values of K_{mn} are determined by linear interpolation. Since $\int_{-\infty}^{\infty} K_{mn}(t)\, dt = \sigma^2$ and $K_{mn}(t) \to 0$ as $m, n \to \infty$ except at $t = 0$, it follows that $E[W_m(g)\overline{W_n(g)}] \to \sigma^2 \|g\|_2^2$ as $m, n \to \infty$. Since the norms of the operators $g \to W_n(g)$ are uniformly bounded (see the solution to Problem 2), this result also holds for $g \in L^2(m)$. By 1.3.2, the sequence $\{W_n(g)\}$ converges. A similar argument shows that if $g, h \in L^2(m)$,

$$E[W_m(g)\overline{W_n(h)}] \to \sigma^2 \int_{-\infty}^{\infty} g(t)\overline{h(t)}\, dt.$$

Therefore $\{W_n(t)\}$ converges to white noise, with $f_0 = \sigma^2/2\pi = \lim_{n\to\infty} f_n(\lambda)$.

(d) First assume W Gaussian.

The process $\{Z(\lambda)\}$ is Gaussian because an arbitrary linear combination

$$\sum_{i=1}^{n} a_i Z(\lambda) = W\left[\sum_{i=1}^{n} a_i(I_{(0,\,\lambda_i^+]} + I_{(-\lambda_i^-,\,0]}) \right]$$

is Gaussian. Set $B_1(\lambda) = Z(\lambda)$, $\lambda \geq 0$; $B_2(\lambda) = Z(-\lambda)$, $\lambda \geq 0$. Then

(i) $\{B_1(\lambda), \lambda \geq 0\}$ is Gaussian.
(ii) $E[B_1(\lambda)] = E[W(I_{(0,\,\lambda]})] = 0$, $\lambda \geq 0$.
(iii) If $0 \leq \lambda \leq \mu$,

$$E[B_1(\lambda)B_2(\mu)] = E[Z(\lambda)(Z(\lambda) + Z(\mu) - Z(\lambda))]$$

$$= E[Z^2(\lambda)] = 2\pi f_0 \lambda \qquad \text{as in 2.6.4.}$$

(Note that since the approximating processes $\{W_n(t)\}$ of 2.6.4 are real in this case, and $I_{(0,\,\lambda]}$ is a real function, the process $\{Z(\lambda)\}$ is real.) Therefore $\{B_1(\lambda)\}$, and similarly $\{B_2(\lambda)\}$ are Brownian motions; they are independent by the orthogonality of $Z(\lambda)$ $(= Z(\lambda) - Z(0))$ and $Z(\mu)$ for $\mu \leq 0 \leq \lambda$.

If $\{Z(\lambda)\}$ is extended Brownian motion, it is Gaussian, so if g is a step function then $W(g) = \int_{-\infty}^{\infty} g\, dZ = \sum_i a_i[Z(t_{i+1}) - Z(t_i)]$ is Gaussian. The general case follows by the usual approximation technique and 1.4.2.

5. (a) If $g \in L^2(m)$, with Fourier transform G, set $\tilde{g}(\lambda) = (2\pi)^{-1}G(-\lambda)$; then $\tilde{g}$ has Fourier transform g. Now define

$$Z_n(E) = W(\tilde{I}_{E \cap [-n,\,n]}), \qquad E \in \mathscr{S},$$

By Section 2.1, Problems 1 and 6, and the fact that $g \to \tilde{g}$ is a multiple of an isometric isomorphism, Z_n is a mov with associated measure

$$\mu_n(E) = f_0 m(E \cap [-n,\,n]);$$

thus μ_n has density $f_n = f_0 I_{[-n,\,n]}$. Define

$$W_n(t) = \int_{-\infty}^{\infty} e^{it\lambda}\, dZ_n(\lambda);$$

then $\{W_n(t)\}$ is a stationary L^2 process, and for $g \in L^2(m)$,

$$W_n(g) = \int_{-\infty}^{\infty} G(\lambda)\, dZ_n(\lambda) \qquad \text{by Problem 2}$$

$$= W(GI_{[-n,\,n]}) \qquad \text{(approximate G by simple functions)}$$

$$\rightarrow W(\tilde{G}) \qquad \text{by continuity of } W$$

$$= W(g) \qquad \text{since both g and $\tilde{G}$ have Fourier transform } G.$$

If $g, h \in L^2(m)$, then by 2.1.2(b),

$$E[W_n(g)\overline{W_n(h)}] = \int_{-\infty}^{\infty} G(\lambda)\overline{H(\lambda)} f_n(\lambda)\, d\lambda$$

$$\rightarrow f_0 \int_{-\infty}^{\infty} G(\lambda)\overline{H(\lambda)}\, d\lambda$$

by the dominated convergence theorem

$$= 2\pi f_0 \int_{-\infty}^{\infty} g(t)\overline{h(t)}\, dt$$

by the Fourier–Plancherel theorem.

(b)   For some $f_0 > 0$, the map $W'(h) = (2\pi f_0)^{-1/2}\hat{W}(H)$ is a norm-preserving linear map of $L^2(m)$ into $L^2(\Omega, \mathscr{F}, P)$, and by (a), $W(h) = (2\pi f_0)^{1/2}W'(h) = \hat{W}(H)$ is the desired white noise.

6.   Write

$$\varphi(\lambda) = \frac{A(i\lambda)}{B(i\lambda)} = \sum_{j=0}^{n-m} c_j(i\lambda)^j + \sum_{j=1}^{n'}\sum_{k=1}^{n_{j'}} \frac{d'_{jk}}{(i\lambda - r'_j)^k}$$

$$+ \sum_{j=1}^{n''}\sum_{k=1}^{n_{j''}} \frac{d''_{jk}}{(i\lambda - r''_j)^k}$$

where r'_j, r''_j are the roots of B with negative and positive real parts, respectively. Using the given formulas and Problem 1, we get

$$X(t) = \int_{-\infty}^{\infty} e^{it\lambda}\varphi(\lambda)\, dZ_Y(\lambda)$$

$$= \sum_{j=0}^{n-m} c_j Y^{(j)}(t) + \int_0^{\infty} G_-(\tau)Y(t - \tau)\, d\tau$$

$$+ \int_0^{\infty} G_+(\tau)Y(t + \tau)\, d\tau$$

where

$$G_-(\tau) = \sum_{j=1}^{n'} \sum_{k=1}^{nj'} d'_{jk} \frac{\tau^{k-1}}{(k-1)!} \exp [r'_j \tau],$$

$$G_+(\tau) = \sum_{j=1}^{n''} \sum_{k=1}^{nj''} d''_{jk} (-1)^k \frac{\tau^{k-1}}{(k-1)!} \exp [-r''_j \tau].$$

The three terms in the expression for $X(t)$ indicate dependence on present, past, and future values of $Y(t)$.

Section 2.7

1. (a) Since $f(\lambda)$ is real, we have

$$f(\lambda) = \overline{f(\lambda)} = \bar{A} \frac{\prod_j (\lambda - \bar{\alpha}_j)}{\prod_k (\lambda - \bar{\beta}_k)}.$$

By the identity theorem, this holds with λ replaced by any complex z such that $S(z) \neq 0$; the result follows.

(b) The desired factorization follows from (a) and the observation that $(\lambda - \alpha)(\lambda - \bar{\alpha}) = |\lambda - \alpha|^2$.

(c) If α_j were of odd multiplicity, f would change sign at α_j, contradicting $f(\lambda) \geq 0$. Since α_j is real iff $\alpha_j = \bar{\alpha}_j$, the argument of part (b) yields the desired result.

2. (a) We may take $\beta_1 = \alpha(1+i)/\sqrt{2}$, $\beta_2 = \alpha(-1+i)/\sqrt{2}$. The equations 2.7.3, with $C_\tau(\lambda) = c_0 + c_1 \lambda$, become

$$c_0 + c_1 \frac{(1+i)}{\sqrt{2}} \alpha = \exp [i\tau\alpha(1+i)/\sqrt{2}] = e^{-\alpha\tau/\sqrt{2}} e^{i\tau\alpha/\sqrt{2}}$$

$$c_0 + c_1 \frac{(-1+i)}{\sqrt{2}} \alpha = \exp [i\tau\alpha(-1+i)/\sqrt{2}] = e^{-\alpha\tau/\sqrt{2}} e^{-i\tau\alpha/\sqrt{2}}.$$

We obtain

$$c_0 = e^{-\alpha\tau/\sqrt{2}} [\sin \alpha\tau/\sqrt{2} + \cos \alpha\tau/\sqrt{2}],$$

$$c_1 = (i\sqrt{2}/\alpha) e^{-\alpha\tau/\sqrt{2}} \sin \alpha\tau/\sqrt{2}.$$

Thus the predictor is

$$e^{-\alpha\tau/\sqrt{2}} [(\sin \alpha\tau/\sqrt{2} + \cos \alpha\tau/\sqrt{2}) X(t) + ((\sqrt{2}/\alpha) \sin \alpha\tau/\sqrt{2}) X'(t)]$$

and the prediction error is

$$\sigma_\tau^2 = \int_{-\infty}^{\infty} [1 - |\varphi_\tau(\lambda)|^2] f(\lambda) \, d\lambda$$

$$= \int_{-\infty}^{\infty} [1 - c_0^2 - |c_1|^2 \lambda^2] \frac{d\lambda}{\lambda^4 + \alpha^4}$$

since c_0 is real and c_1 is imaginary.

A residue calculation yields

$$\int_{-\infty}^{\infty} \frac{d\lambda}{\lambda^4 + \alpha^4} = \frac{\pi}{\sqrt{2}\,\alpha^3}, \qquad \int_{-\infty}^{\infty} \frac{\lambda^2 \, d\lambda}{\lambda^4 + \alpha^4} = \frac{\pi}{\sqrt{2}\,\alpha}.$$

Thus

$$\sigma_\tau^2 = \frac{\pi}{\sqrt{2}\,\alpha^3}\left[1 - e^{-\alpha\tau\sqrt{2}}\left(1 + 2\sin\frac{\alpha\tau}{\sqrt{2}}\cos\frac{\alpha\tau}{\sqrt{2}}\right)\right]$$

$$-\frac{\pi}{\sqrt{2}\,\alpha}\frac{2}{\alpha^2}e^{-\alpha\tau\sqrt{2}}\sin^2\frac{\alpha\tau}{\sqrt{2}}$$

$$= \frac{\pi}{\sqrt{2}\,\alpha^3}\left[1 - e^{-\alpha\tau\sqrt{2}}(2 + \sin\sqrt{2}\,\alpha\tau - \cos\sqrt{2}\,\alpha\tau)\right].$$

(b) We may take $\beta_1 = i\sqrt{2}\,\alpha$, $\beta_2 = i\alpha/\sqrt{2}$; as in (a),

$$c_0 + c_1 i\sqrt{2}\,\alpha = e^{-\alpha\tau\sqrt{2}}$$

$$c_0 + c_1 i\alpha/\sqrt{2} = e^{-\alpha\tau/\sqrt{2}}.$$

Thus

$$c_0 = e^{-\alpha\tau/\sqrt{2}}(2 - e^{-\alpha\tau/\sqrt{2}})$$

$$c_1 = \frac{i\sqrt{2}}{\alpha}e^{-\alpha\tau/\sqrt{2}}(1 - e^{-\alpha\tau/\sqrt{2}}).$$

The predictor is

$$e^{-\alpha\tau/\sqrt{2}}\left[(2 - e^{-\alpha\tau/\sqrt{2}})X(t) + \frac{\sqrt{2}}{\alpha}(1 - e^{-\alpha\tau/\sqrt{2}})X'(t)\right]$$

and the prediction error is, as in (a),

$$\sigma_\tau^2 = \int_{-\infty}^{\infty}[1 - c_0^2 - |c_1|^2\lambda^2]\frac{d\lambda}{(\lambda^2 + 2\alpha^2)(\lambda^2 + \alpha^2/2)}$$

$$= (1 - c_0^2)\frac{\sqrt{2}\,\pi}{3\alpha^3} - |c_1|^2\frac{\sqrt{2}\,\pi}{3\alpha}$$

$$= \frac{\sqrt{2}\,\pi}{3\alpha^3}[1 - e^{-\alpha\tau\sqrt{2}}(4 - 4e^{-\alpha\tau/\sqrt{2}} + e^{-\alpha\tau\sqrt{2}})]$$

$$-\frac{\sqrt{2}\,\pi}{3\alpha}\frac{2}{\alpha^2}e^{-\alpha\tau\sqrt{2}}(1 - 2e^{-\alpha\tau/\sqrt{2}} + e^{-\alpha\tau\sqrt{2}})$$

$$= \frac{\sqrt{2}\,\pi}{3\alpha^3}[1 - e^{-\alpha\tau\sqrt{2}}(6 - 8e^{-\alpha\tau/\sqrt{2}} + 3e^{-\alpha\tau\sqrt{2}})].$$

(c)   We may take β_1 and β_2 as in (a), and $\alpha_1 = i\alpha$. The equations 2.7.3 become

$$c_0 + c_1 \frac{(1 + i)}{\sqrt{2}} \alpha = e^{-\alpha\tau/\sqrt{2}} e^{i\tau\alpha/\sqrt{2}} \left[\frac{(1 + i)}{\sqrt{2}} \alpha - i\alpha \right]$$

$$c_0 + c_1 \frac{(-1 + i)}{\sqrt{2}} \alpha = e^{-\alpha\tau/\sqrt{2}} e^{-i\tau\alpha/\sqrt{2}} \left[\frac{(-1 + i)}{\sqrt{2}} \alpha - i\alpha \right].$$

We obtain

$$c_0 = i\alpha e^{-\alpha\tau/\sqrt{2}} \left[-\cos \frac{\alpha\tau}{\sqrt{2}} + (\sqrt{2} - 1) \sin \frac{\alpha\tau}{\sqrt{2}} \right]$$

$$c_1 = e^{-\alpha\tau/\sqrt{2}} \left[\cos \frac{\alpha\tau}{\sqrt{2}} + (\sqrt{2} - 1) \sin \frac{\alpha\tau}{\sqrt{2}} \right].$$

Now

$$\varphi_\tau(\lambda) = \frac{C_\tau(\lambda)}{\lambda - \alpha_1} = \frac{c_0 + c_1 \lambda}{\lambda - i\alpha} = c_1 + \frac{ic_0 - \alpha c_1}{\alpha + i\lambda}.$$

Since $1/(\alpha + i\lambda)$ is the Fourier transform of $e^{-\alpha t}$, $t \geq 0$, the predictor is (see 2.6.1 and Problem 1, Section 2.6)

$$c_1 X(t) + (ic_0 - \alpha c_1) \int_0^\infty e^{-\alpha s} X(t - s) \, ds$$

$$= e^{-\alpha\tau/\sqrt{2}} \left[\left(\cos \frac{\alpha\tau}{\sqrt{2}} + (\sqrt{2} - 1) \sin \frac{\alpha\tau}{\sqrt{2}} \right) X(t) \right.$$

$$\left. - 2\alpha(\sqrt{2} - 1) \sin \frac{\alpha\tau}{\sqrt{2}} \int_0^\infty e^{-\alpha s} X(t - s) \, ds \right].$$

The prediction error is

$$\sigma_\tau^2 = \int_{-\infty}^\infty \left[1 - \frac{|c_0|^2 + c_1^2 \lambda^2}{\lambda^2 + \alpha^2} \right] \frac{\lambda^2 + \alpha^2}{\lambda^4 + \alpha^2} \, d\lambda$$

$$= \frac{\sqrt{2}\,\pi}{\alpha} - |c_0|^2 \frac{\pi}{\sqrt{2}\,\alpha^3} - c_1^2 \frac{\pi}{\sqrt{2}\,\alpha}$$

$$= \frac{\sqrt{2}\,\pi}{\alpha} [1 - e^{-\alpha\tau\sqrt{2}} (\sqrt{2} - 1)(\sqrt{2} + \cos \sqrt{2}\,\alpha\tau)].$$

3.   (a)   The best linear estimate is

$$\hat{S}(t) = \int_{-\infty}^\infty e^{it\lambda} \varphi(\lambda) \, dZ_X(\lambda)$$

where $\hat{S}(t) - S(t) \perp X(s)$ for all s, that is, $E[S(t)\bar{X}(s)] = E[\hat{S}(t)\bar{X}(s)]$ for all s. By orthogonality of signal and noise, this becomes

$$E[S(t)\bar{S}(s)] = \int_{-\infty}^{\infty} e^{i(t-s)\lambda} \varphi(\lambda) f_X(\lambda) \, d\lambda.$$

But

$$E[S(t)\bar{S}(s)] = \int_{-\infty}^{\infty} e^{i(t-s)\lambda} f_S(\lambda) \, d\lambda.$$

Since $K_X = K_S + K_N$ by orthogonality, we have $f_X = f_S + f_N$, and hence this yields the desired expression for $\varphi(\lambda)$. The estimation error is

$$\sigma^2 = E[\,|\,\hat{S}(t) - S(t)|^2] = E[(\hat{S}(t) - S(t))(-\bar{S}(t))]$$

since $\hat{S}(t) - S(t)$ is orthogonal to $H(X)$, in particular to $\hat{S}(t)$. Now since $X(t) = S(t) + N(t)$ and the complex exponentials span $L^2(\mu_X)$ (see the proof of 2.2.1),

$$\int f \, dZ_X = \int f \, dZ_S + \int f \, dZ_N \qquad \text{for all } f \in L^2(\mu_X).$$

Thus

$$\hat{S}(t) - S(t) = \int_{-\infty}^{\infty} e^{it\lambda}[\varphi(\lambda) - 1] \, dZ_S(\lambda) + \int_{-\infty}^{\infty} e^{it\lambda} \varphi(\lambda) \, dZ_N(\lambda)$$

hence by orthogonality of signal and noise,

$$\sigma^2 = -\int_{-\infty}^{\infty} e^{it\lambda}[\varphi(\lambda) - 1]e^{-it\lambda} f_S(\lambda) \, d\lambda$$

$$= \int_{-\infty}^{\infty} \frac{f_S(\lambda) f_N(\lambda)}{f_S(\lambda) + f_N(\lambda)} \, d\lambda.$$

(b)   Write $Y = \int_{-\infty}^{\infty} e^{it\lambda} \varphi_\tau(\lambda) \, dZ_X(\lambda)$; by the projection theorem and the basic isomorphism between $L^2\{X(t), \ -\infty < t < \infty\}$ and $L^2(\mu_X)$, φ_τ is required to satisfy (1), (2), and in addition, $Y - S(t + \tau) \perp X(s)$ for all $s \leq t$. Thus

$$\int_{-\infty}^{\infty} e^{it\lambda}[\varphi_\tau(\lambda) - e^{i\tau\lambda}] \, dZ_S(\lambda) + \int_{-\infty}^{\infty} e^{it\lambda} \varphi_\tau(\lambda) \, dZ_N(\lambda)$$

$$\perp \int_{-\infty}^{\infty} e^{is\lambda} \, dZ_S(\lambda) + \int_{-\infty}^{\infty} e^{is\lambda} \, dZ_N(\lambda), \qquad s \leq t.$$

Thus

$$\int_{-\infty}^{\infty} e^{i(t-s)\lambda}[\varphi_\tau(\lambda) - e^{i\tau\lambda}]f_S(\lambda)\, d\lambda$$

$$+ \int_{-\infty}^{\infty} e^{i(t-s)\lambda}\varphi_\tau(\lambda)f_N(\lambda)\, d\lambda = 0, \qquad s \le t.$$

Since $f_S + f_N = f_X$, this is equivalent to condition (3).

(c) This is proved exactly as in 2.7.2, with $e^{i\tau\lambda}f_S(\lambda) - \varphi_\tau(\lambda)f_X(\lambda)$ in place of $[e^{i\tau\lambda} - \varphi_\tau(\lambda)]f(\lambda)$.

(d) By the Pythagorean relation,

$$\sigma^2 = \|S(t+\tau)\|^2 - \|\hat{S}(t)\|^2 = \int_{-\infty}^{\infty} f_S(\lambda)\, d\lambda - \int_{-\infty}^{\infty} |\varphi_\tau(\lambda)|^2 f_X(\lambda)\, d\lambda,$$

as desired.

4. We want

$$\psi_\tau(\lambda) = e^{i\tau\lambda} \frac{A}{\lambda^2 + \alpha^2} - \varphi_\tau(\lambda)\left[\frac{A}{\lambda^2 + \alpha^2} + \frac{B}{\lambda^2 + \beta^2}\right]$$

$$= \frac{e^{i\tau\lambda}A(\lambda^2 + \beta^2) - \varphi_\tau(\lambda)(A+B)(\lambda^2 + \gamma^2)}{(\lambda^2 + \alpha^2)(\lambda^2 + \beta^2)}$$

to be analytic for Im $\lambda \ge 0$, so the numerator must vanish at $\lambda = i\alpha$ and $\lambda = i\beta$. Thus

$$\varphi_\tau(i\alpha) = \frac{Ae^{-\alpha\tau}}{A+B} \frac{\beta^2 - \alpha^2}{\gamma^2 - \alpha^2}, \qquad \varphi_\tau(i\beta) = 0. \tag{1}$$

Now $\varphi_\tau(\lambda)(\lambda^2 + \gamma^2)$ can have singularities only at $\lambda = i\alpha$ and $\lambda = i\beta$, because φ_τ must be analytic for Im $\lambda \le 0$ and therefore any other singularities would not be canceled by zeros of the denominator. Thus $\varphi_\tau(\lambda) = h(\lambda)/(\lambda^2 + \gamma^2)$ where h is entire; in fact since φ_τ must be analytic in the lower half-plane we may write

$$\varphi_\tau(\lambda) = \frac{C_\tau(\lambda)}{\lambda - i\gamma} \qquad \text{where} \quad C_\tau \text{ is entire.}$$

If we try a polynomial for $C_\tau(\lambda)$, condition (i) of (c) requires that $C_\tau(\lambda) = c_0 + c_1\lambda$, and such a choice also satisfies conditions (ii) and (iii) if $\tau \ge 0$. (Note that in this case, $e^{i\tau\lambda}$ remains bounded in the upper half-plane.) The constants c_0 and c_1 are determined by (1); we obtain

$$\varphi_\tau(\lambda) = \frac{\lambda - i\beta}{\lambda - i\gamma} \frac{Ae^{-\alpha\tau}}{A+B} \frac{\alpha + \beta}{\alpha + \gamma},$$

as desired. Now if $\tau \leq 0$, $e^{i\tau\lambda}$ grows exponentially in the upper half-plane, so in order to satisfy the condition $\psi_\tau(\lambda) = O(|\lambda|^{-1-\varepsilon})$ in the upper half-plane, the exponential term must be canceled. Thus it is reasonable to try to find a φ_τ of the form

$$\varphi_\tau(\lambda) = \frac{A(\lambda^2 + \beta^2)e^{i\tau\lambda} - \theta_\tau(\lambda)}{(A + B)(\lambda^2 + \gamma^2)}.$$

Then

$$\psi_\tau(\lambda) = \frac{\theta_\tau(\lambda)}{(\lambda^2 + \alpha^2)(\lambda^2 + \beta^2)}.$$

For φ_τ to be analytic in the lower half-plane, we require

$$A(\lambda^2 + \beta^2)e^{i\tau\lambda} - \theta_\tau(\lambda) = 0 \qquad \text{at} \quad \lambda = -i\gamma. \tag{2}$$

For ψ_τ to be analytic in the upper half-plane, we require

$$\theta_\tau(\lambda) = 0 \quad \text{at} \quad \lambda = i\alpha \qquad \text{and} \qquad \lambda = i\beta. \tag{3}$$

Since φ_τ is analytic in the lower half-plane, so is θ_τ (a singularity at $-i\gamma$ is ruled out by (2)), and since ψ_τ is analytic in the upper half-plane, so is θ_τ (singularities at $i\alpha$ and $i\beta$ are ruled out by (3)). Therefore θ_τ is entire. To get $\psi_\tau(\lambda) = O(|\lambda|^{-1-\varepsilon})$ in the upper half-plane we try a quadratic in λ for θ_τ; by (3), $\theta_\tau(\lambda) = c(\lambda - i\alpha)(\lambda - i\beta)$, where c is determined by (2). We get $c = (A(\gamma - \beta)/(\alpha + \gamma))e^{\gamma\tau}$, so that

$$\varphi_\tau(\lambda) = \frac{A}{A + B} \frac{(\alpha + \gamma)(\lambda^2 + \beta^2)e^{i\tau\lambda} - (\beta - \gamma)(\lambda - i\alpha)(\lambda - i\beta)e^{\gamma\tau}}{(\alpha + \gamma)(\lambda^2 + \gamma^2)},$$

as desired.

5. Since the $(m_j - 1)$th derivative of p is 0 at x_j, $(x - x_j)^{m_j}$ is a factor of p. Thus we may write

$$p(x) = A(x) = (x - x_1)^{m_1} \cdots (x - x_r)^{m_r}$$

where $A(x)$ is a polynomial. But $m_1 + \cdots + m_r = n > \deg p$, so that $A(x)$ must be 0.

Section 2.8

1. Set $f_n = \sum_{i=1}^n f(t_i)I_{[t_{i-1}, t_i]}$, where $a = t_0 < \cdots < t_n = b$. If μ is the measure on $\mathscr{B}[a, b]$ associated with the mov Z_Y determined by Y, then $f_n \to f$ in $L^2(\mu)$. Thus

$$\int_a^b f \, dY = \int_a^b f \, dZ_Y = L^2 \lim_{n \to \infty} \int_a^b f_n \, dZ_Y.$$

But

$$\int_a^b f_n \, dZ_Y = \sum_{i=1}^n f(t_i)[Y(t_i) - Y(t_{i-1})].$$

Using summation by parts, we obtain

$$\int_a^b f_n \, dZ_Y = f(t_n)Y(t_n) - f(t_1)Y(t_0) - \sum_{i=1}^{n-1} [f(t_{i+1}) - f(t_i)]Y(t_i).$$

By the mean value theorem, we obtain numbers ξ_i between t_i and t_{i+1} such that

$$\int_a^b f_n \, dZ_Y = f(t_n)Y(t_n) - f(t_1)Y(t_0) - \sum_{i=1}^{n-1} f'(\xi_i)Y(t_i) \, \Delta t_i$$

$$\xrightarrow{L^2} [f(t)Y(t)]_a^b - \int_a^b f'(t)Y(t) \, dt.$$

2. We have

$$\frac{d}{dt} \int_{t_0}^t f(s, t)Y(s) \, ds$$

$$= L^2 \lim_{h \to 0} \frac{1}{h} \left[\int_{t_0}^{t_0+h} f(s, t + h)Y(s) \, ds - \int_{t_0}^t f(s, t)Y(s) \, ds \right]$$

$$= L^2 \lim_{h \to 0} \left[\int_{t_0}^t \left(\frac{f(s, t + h) - f(s, t)}{h} \right) Y(s) \, ds \right.$$

$$\left. + \frac{1}{h} \int_t^{t+h} f(s, t + h)Y(s) \, ds \right].$$

We claim that this expression equals

$$\int_{t_0}^t \frac{\partial f}{\partial t}(s, t)Y(s) \, ds + f(t, t)Y(t).$$

To see this, we compute

$$E\left[\left| \int_{t_0}^t \left[\frac{f(s, t + h) - f(s, t)}{h} - \frac{\partial f}{\partial t}(s, t) \right] Y(s) \, ds \right|^2 \right]$$

$$\leq \left(\int_{t_0}^t \left| \frac{f(s, t + h) - f(s, t)}{h} - \frac{\partial f}{\partial t}(s, t) \right| [\mathrm{Var}\ Y(s)]^{1/2} \, ds \right)^2$$

by 1.3.10 and the Cauchy–Schwarz inequality

$\to 0$ \quad as $h \to 0$ by the mean value theorem and the dominated convergence theorem.

Similarly,

$$E\left[\left|\frac{1}{h}\int_t^{t+h} f(s, t+h)Y(s)\, ds - f(t, t)Y(t)\right|^2\right]$$

$$= E\left[\left|\frac{1}{h}\int_t^{t+h} [f(s, t+h)Y(s) - f(t, t)Y(t)]\, ds\right|^2\right]$$

$$\to 0 \qquad \text{as } h \to 0$$

by 1.3.10 and the mean value theorem for integrals.

3. We have $a_0 V'(t) + a_1 V(t) = B'(t)$, hence (see 2.8.1) $\varphi_0(t) = e^{-\alpha t}$, $h(t) = a_0^{-1} e^{-\alpha t}$, $t \geq 0$, where $\alpha = a_1/a_0$. Therefore,

$$V(t) = v_0 \varphi_0(t) + \int_0^t h(t-s)\, dB(s)$$

$$= v_0 e^{-\alpha t} + \int_0^t \frac{1}{a_0} e^{-\alpha(t-s)}\, dB(s).$$

It follows that $EV(t) = v_0 e^{-\alpha t}$ and

$$\text{Cov}\,(V(s), V(t)) = \frac{1}{a_0^2} E\left[\int_0^s e^{-\alpha(s-u)}\, dB(u) \int_0^t e^{-\alpha(t-u)}\, dB(u)\right]$$

$$= \frac{1}{a_0^2} \int_0^{s \wedge t} e^{-\alpha(s+t)-2u}\sigma^2\, du$$

$$= \frac{\sigma^2}{2a_0 a_1}(e^{-\alpha|t-s|} - e^{-\alpha(s+t)})$$

which approximates $(\sigma^2/2a_0 a_1)e^{-\alpha|t-s|}$ as $s, t \to \infty$.
   Now

$$X(t) = x_0 + \int_0^t V(s)\, ds$$

$$= x_0 + \int_0^t \left[v_0 e^{-\alpha s} + \int_0^s \frac{1}{a_0} e^{-\alpha(s-u)}\, dB(u)\right] ds$$

$$= x_0 + \frac{v_0 a_0}{a_1}(1 - e^{-\alpha t}) + \int_0^t \frac{1}{a_0} \int_u^t e^{-\alpha(s-u)}\, ds\, dB(u)$$

$$= x_0 + \frac{v_0 a_0}{a_1}(1 - e^{-\alpha t}) + \frac{1}{a_1} \int_0^t [1 - e^{-\alpha(t-u)}]\, dB(u).$$

Therefore

$$EX(t) = x_0 + \frac{v_0 a_0}{a_1}(1 - e^{-\alpha t})$$

and

Cov $(X(s), X(t))$

$$= \frac{1}{a_1^2} E\left[\int_0^s (1 - e^{-\alpha(s-u)})\, dB(u) \int_0^t (1 - e^{-\alpha(t-u)})\, dB(u)\right]$$

$$= \frac{1}{a_1^2}\int_0^{s\wedge t}[1 - e^{-\alpha(s-u)} - e^{-\alpha(t-u)} + e^{-\alpha(s+t-2u)}]\sigma^2\, du$$

$$= \frac{\sigma^2}{a_1^2}\left[s\wedge t - \frac{1}{\alpha}(1 + e^{-\alpha|t-s|} - e^{-\alpha s} - e^{-\alpha t})\right.$$

$$\left. + \frac{1}{2\alpha}(e^{-\alpha|t-s|} - e^{-\alpha(s+t)})\right].$$

4. (a) By 2.8.1, we have

$$X(t) = X(0)\varphi_0(t) + \cdots + X^{(n-1)}(0)\varphi_{n-1}(t)$$

$$+ \int_0^t h(t-s)\, dB(s), \qquad t \geq 0$$

$$= X(t_1)\varphi_0(t-t_1) + \cdots + X^{(n-1)}(t_1)\varphi_{n-1}(t-t_1)$$

$$+ \int_{t_1}^t h(t-s)\, dB(s), \qquad t \geq t_1.$$

Thus

$$\hat{X}(t_2) - X(t_2) = -\int_{t_1}^{t_2} h(t_2 - s)\, dB(s)$$

so

$$E[X(t)(\hat{X}(t_2) - X(t_2))]$$

$$= E\left[(c_0\varphi_0(t) + \cdots + c_{n-1}\varphi_{n-1}(t)\right.$$

$$+ \int_0^t h(t-s)\, dB(s))\left(-\int_{t_1}^{t_2} h(t_2 - s)\, dB(s)\right)\right]$$

$$= -\int_{[0,t]\cap[t_1,t_2]} h(t-s)h(t_2-s)\sigma^2\, ds$$

$$= 0 \qquad \text{if } 0 \leq t \leq t_1.$$

Therefore $\hat{X}(t_2) - X(t_2) \perp X(t)$, $0 \le t \le t_1$, as desired.

(b)   This is immediate from the uniqueness part of 2.8.1.

(c)   We have

$$E[\,|\,\hat{X}(t_2) - X(t_2)|^2\,] = E\left[\left|\int_{t_1}^{t_2} h(t_2 - s)\,dB(s)\right|^2\right]$$

$$= \int_{t_1}^{t_2} h^2(t_2 - s)\sigma^2\,ds$$

$$= \sigma^2 \int_0^{t_2 - t_1} h^2(u)\,du.$$

Chapter 3

Section 3.2

1.   Let N be a set of measure 0 such that if $\omega \notin N$, then $\omega \in A$ iff $T\omega \in A$. Now if $\omega \notin T^{-n}N$, then $T^n\omega \notin N$, hence $T^n\omega \in A$ iff $T^{n+1}\omega \in A$. Thus if $\omega \notin \bigcup_{n=0}^{\infty} T^{-n}N$, we have $\omega \in A$ iff $T^n\omega \in A$ for $n = 1, 2, \ldots$.

2.   In the first case, if A is an invariant set, then $\omega \in A$ iff $\omega + 1 \in A$, so that A is either the empty set or the entire space. In the second case, the even integers form a nontrivial invariant set.

3.   (a)   Let $\Omega = \{x_1, x_2, x_3, x_4\}$, $A = \{x_1, x_2\}$, $\mathscr{F} = \{\varnothing, \Omega, A, A^c\}$, $T(x_1) = T(x_2) = x_1$, $T(x_3) = T(x_4) = x_3$. Then T preserves any measure μ on $\mathscr{F}$, but $T(A) = \{x_1\} \notin \mathscr{F}$.

(b)   Consider the one-sided shift transformation (3.1.2d). Let $A = \{\omega : \omega_0 \in B\}$ where B is a nonempty Borel subset of R. Then $T(A) = \Omega$, so that $P(A) \ne P(TA)$ in general.

(c)   Consider the two-sided shift transformation (3.1.2e), but replace $\mathscr{F}$ by the σ-field generated by the measurable cylinders $\{\omega : (\omega_k, \ldots, \omega_{k+n-1}) \in B_n\}$, where k is required to be *nonnegative*. If $A = \{\omega : \omega_0 \in B\}$ where B is a nonempty proper Borel subset of R, then $T(A) = \{\omega : \omega_{-1} \in B\} \notin \mathscr{F}$. For let $\mathscr{C}$ be the class of sets $C \in \mathscr{F}$ such that if $\omega \in C$ and ω' differs from ω only in negative coordinates, then ω' is also in C; $\mathscr{C}$ is a σ-field that includes the generating class of measurable cylinders, hence $\mathscr{C} = \mathscr{F}$. But if $\omega \in T(A)$ and we change the -1th coordinate so that ω_{-1} is no longer in B, we get a new point ω' that cannot be in $T(A)$; therefore $T(A) \notin \mathscr{F}$.

4. If $T^{-1}A \subset A$, then $A^c \subset T^{-1}A^c$, and conversely; also, $A - T^{-1}A = T^{-1}A^c - A^c$. This shows that the two definitions of incompressibility are equivalent.

(a) 1 *implies* 2: Let A be wandering, and let $B = \bigcup_{n=0}^{\infty} T^{-n}A$. Then $T^{-1}B = \bigcup_{n=1}^{\infty} T^{-n}A \subset B$, so by (1), $\mu(B - T^{-1}B) = 0$. But $B - T^{-1}B = A$ since the $T^{-n}A$ are disjoint; hence $\mu(A) = 0$.

2 *implies* 3: Let $A^{(r)} = A \cap \bigcup_{n=1}^{\infty} T^{-n}A = \{\omega \in A : T^n\omega \in A \text{ for some } n \geq 1\}$. If $C = A - A^{(r)}$, then $T^{-n}C = \{\omega : T^n\omega \in A, \text{ but } T^k\omega \notin A, k > n\}$. Thus the $T^{-n}C$ are disjoint, so that C is wandering. By (2), $\mu(C) = 0$, and therefore T is recurrent.

3 *implies* 1: Let $T^{-1}A \subset A$. Then (by induction) $T^{-n}A \subset T^{-1}A$, $n \geq 1$. Thus $T^{-1}A = \bigcup_{n=1}^{\infty} T^{-n}A$. Now

$$A - T^{-1}A = A - \bigcup_{n=1}^{\infty} T^{-n}A = A - A^{(r)},$$

which has measure 0 by (3). Thus $\mu(A - T^{-1}A) = 0$, proving (1).

4 *implies* 3: Obvious.

1 *implies* 4: Let

$$A^{(i)} = A \cap \limsup_{n \geq 1} T^{-n}A = \{\omega \in A : T^n\omega \in A \text{ for infinitely many } n \geq 1\}.$$

If $A \in \mathscr{F}$, let $B = \bigcup_{n=0}^{\infty} T^{-n}A$. Then $T^{-1}B \subset B$, hence by (1), $\mu(B - T^{-1}B) = 0$. Similarly, $T^{-(k+1)}B \subset T^{-k}B$, hence

$$\mu(T^{-k}B - T^{-(k+1)}B) = 0.$$

But

$$T^{-k}B - T^{-(k+1)}B$$

$$= \bigcup_{n=k}^{\infty} T^{-n}A - \bigcup_{n=k+1}^{\infty} T^{-n}A$$

$$= \{\omega : T^n\omega \text{ enters } A \text{ for the last time at } n = k\}.$$

Since

$$A - A^{(i)} = A \cap \bigcup_{k=0}^{\infty} \{\omega : T^n\omega \in A \text{ for the last time at } n = k\},$$

we have

$$\mu(A - A^{(i)}) \leq \sum_{k=0}^{\infty} \mu(T^{-k}B - T^{-(k+1)}B) = 0.$$

(b) Any interval of length less than 1 is a nontrivial wandering set.

(c) Let A be a wandering set; since

$$\sum_{n=0}^{\infty} \mu(A) = \sum_{n=0}^{\infty} \mu(T^{-n}A) = \mu\left(\bigcup_{n=0}^{\infty} T^{-n}A\right) \le \mu(\Omega) < \infty,$$

$\mu(A)$ must be 0. Thus T is conservative, hence infinitely recurrent.

Section 3.3

1. For any ω, $f(T^n\omega) = 0$ for large n, hence $\hat{f} \equiv 0$. But if $-(n-1) < \omega \le 1$, then $T^k\omega \in (0, 1]$ for exactly one integer $k \in [0, n-1]$. Hence if $-(n-1) < \omega \le 1$,

$$\frac{1}{n}\sum_{k=0}^{n-1} f(T^k\omega) = \frac{1}{n}.$$

For other values of ω we have $n^{-1}\sum_{k=0}^{n-1} f(T^k\omega) = 0$. Thus

$$\int_{\Omega} \left| \frac{1}{n}\sum_{k=0}^{n-1} f(T^k\omega) - \hat{f}(\omega) \right|^p d\mu(\omega) = \int_{-(n-1)}^{1} \left(\frac{1}{n}\right)^p d\mu = n^{1-p},$$

hence there is no convergence in L^1.

2. If $\omega_0 = \omega_1 = \cdots = \omega_{n-1} = 1$, then $n^{-1}\sum_{k=0}^{n-1} f(T^k\omega) = 1$. Therefore

$$\left| \frac{1}{n}\sum_{k=0}^{n-1} f(T^k\omega) - \frac{1}{2} \right| = \frac{1}{2}$$

on a set of probability 2^{-n}. Thus

$$\left\| \frac{1}{n}\sum_{k=0}^{n-1} f(T^k) - \frac{1}{2} \right\|_{\infty} = \frac{1}{2} \nrightarrow 0.$$

3. Let A be an invariant set, and take $f = I_A$. Since μ is finite, f belongs to L^1, hence $f^{(n)} \xrightarrow{\text{a.e.}} c$ by hypothesis. But since A is invariant, $f^{(n)} = I_A$ for all n, so that I_A is constant a.e. If $I_A = 0$ a.e., then $\mu(A) = 0$; if $I_A = 1$ a.e., then $\mu(\Omega - A) = 0$.

For the counterexample, let $\Omega = \{(x, y) : x \text{ and } y \text{ are integers}\}$, and let μ be counting measure on all subsets of Ω. If $T(x, y) = (x + 1, y)$, then T is not ergodic; if A is a horizontal line, then A is invariant, but $\mu(A) = \mu(\Omega - A) = \infty$. But if $f \in L^1$, then $\sum_{x, y} |f(x, y)| < \infty$, hence

$$f^{(n)}(x, y) = \frac{1}{n}\sum_{k=0}^{n-1} f(x + k, y) \to 0.$$

4. First assume $f \ge 0$, and let $f^{(n)} \to \hat{f}$ a.e. Since f is finite valued, the argument of 3.3.8 shows that $\hat{f}$ is almost invariant. Since T is ergodic, $\hat{f} = c$ a.e. Let $f_r(\omega) = f(\omega)$ if $f(\omega) \le r$; $f_r(\omega) = r$ if $f(\omega) > r$. Then $0 \le f_r \uparrow f$, hence

$\int_\Omega f_r \, d\mu \to \int_\Omega f \, d\mu$. Now $f_r \in L^1$ since $\mu(\Omega) < \infty$, so that $f_r^{(n)}$ converges a.e. to a constant d; since $f_r \leq f$, we have $d \leq c$. But by 3.3.9, $\int_\Omega f_r \, d\mu = \int_\Omega \hat{f}_r \, d\mu = d\mu(\Omega) \leq c\mu(\Omega)$. Let $r \to \infty$ to obtain $\int_\Omega f \, d\mu \leq c\mu(\Omega) < \infty$.

In general, the existence of $\int_\Omega f \, d\mu$ implies that either f^+ or f^- belongs to L^1. If, say, $f^- \in L^1$, then $f^{-(n)}$ converges a.e. to a constant. But $f^{(n)}$ converges a.e. to a constant by the above argument (this part of the proof does not require $f \geq 0$). Therefore $f^{+(n)}$ converges a.e. to a constant, so that $f^+ \in L^1$. Therefore, $f = f^+ - f^- \in L^1$.

5. (a) (1) *implies* (2): By hypothesis, U has a left and a right inverse, hence U is one-to-one onto; also, $\langle f, g \rangle = \langle f, U^*Ug \rangle = \langle Uf, Ug \rangle$.

(2) *implies* (3): Take $g = f$.

(3) *implies* (2): Use the polarization identity (RAP, 3.2.17, p. 124).

(2) *implies* (1): Write $\langle U^*Uf, g \rangle = \langle Uf, U^{**}g \rangle = \langle Uf, Ug \rangle = \langle f, g \rangle$. Since f and g are arbitrary, $U^*U = I$. But then $(UU^*)U = U$; so if U is onto, then $UU^* = I$.

(b) This is done exactly as in (a).

(c) If $Uf = f$, then $U^*Uf = U^*f$, hence by (b), $f = U^*f$. Conversely, if $U^*f = f$, then

$$\|Uf - f\|^2 = \|Uf\|^2 - \langle f, Uf \rangle - \langle Uf, f \rangle + \|f\|^2$$
$$= 2\|f\|^2 - \langle U^*f, f \rangle - \langle f, U^*f \rangle$$
$$= 2\|f\|^2 - 2\langle f, f \rangle \qquad \text{by hypothesis}$$
$$= 0.$$

(d) If $f_k \in E$, $f_k \to f$, then

$$\|A_m f - A_n f\| \leq \|A_m f - A_m f_k\| + \|A_m f_k - A_n f_k\| + \|A_n f_k - A_n f\|$$
$$\leq 2\|f - f_k\| + \|A_m f_k - A_n f_k\|,$$

and it follows that $\{A_n f\}$ is a Cauchy sequence, so that $f \in E$, proving E closed. It is immediate that E is a subspace.

(e) If $f \in M$, then $A_n f \equiv f$, so $\hat{f} = f$. If $f = g - Ug \in N_0$, then $A_n f = n^{-1}(g - U^n g)$, hence $\|A_n f\| \leq 2n^{-1}\|g\| \to 0$.

(f) If $h \in H$, then

$$h \perp N \qquad \text{iff} \qquad h \perp N_0$$
$$\text{iff} \qquad \langle h, g - Ug \rangle = 0 \qquad \text{for all} \quad g \in H$$
$$\text{iff} \qquad \langle h - U^*h, g \rangle = 0 \qquad \text{for all} \quad g \in H$$
$$\text{iff} \qquad U^*h = h$$
$$\text{iff} \qquad Uh = h \qquad \text{by (c)}$$
$$\text{iff} \qquad h \in M.$$

Thus $M = N$, and the result follows.

(g) Since $E = H$, $A_n f$ converges to a limit $\hat{f}$. Write $f = f_1 + f_2$, where $f_1 \in M, f_2 \in N$. Now $A_n f_1 \to f_1$ by (e), and also $A_n f_2 \to 0$. For choose $g \in N_0$ such that $\| f_2 - g \| < \varepsilon$; then

$$\| A_n f_2 \| \le \| A_n (f_2 - g) \| + \| A_n g \| \le \| f_2 - g \| + \| A_n g \| < \varepsilon + \| A_n g \|.$$

By (e), $A_n g \to 0$, so $A_n f_2 \to 0$. Therefore $A_n f \to f_1 = Pf$.

(h) By definition of $\hat{S}$ and $\hat{T}$, we have $\hat{S}\hat{T} = \hat{T}\hat{S} = I$. By 3.3.1, $\hat{S}$ and $\hat{T}$ are isometries, and since they are invertible, they must be unitary operators. By (a), if $U = \hat{T}$, then $U^* = \hat{S}$.

6. If P is not an extreme point, so that a representation of the given form is possible, then P_1 is preserved by T, $P_1 \ll P$, and $P_1 \not\equiv P$ (if $P_1 \equiv P$, then $(1 - \lambda_1) P_1 = \lambda_2 P_2$, so that P_1 and P_2, being probability measures, must be identical). By 3.3.12, P is not ergodic.

Conversely, assume that P is not ergodic. If A is an invariant set with $0 < P(A) < 1$, then for each $B \in \mathscr{F}$,

$$P(B) = P(A)P(B \,|\, A) + P(A^c)P(B \,|\, A^c)$$
$$= \lambda_1 P_1(B) + \lambda_2 P_2(B).$$

By the end of the proof of 3.3.12, P_1 and P_2 are preserved by T, hence P_1, $P_2 \in K$. If $P_1 \equiv P_2$, then $P \equiv P_1$; but $P_1(A) = P(A \,|\, A) = 1 \ne P(A)$. Therefore $P_1 \ne P_2$, so that P is not an extreme point.

7. (a) *implies* (b): Apply the pointwise ergodic theorem.

(b) *implies* (c), (c) *implies* (d): Obvious.

(d) *implies* (e): By the pointwise ergodic theorem, $I_A^{(n)}$ converges to $\hat{I}_A$ almost everywhere, hence in probability. By (d), $\hat{I}_A = P(A)$ a.e.

(e) *implies* (f): Since $I_B^{(n)} \to \hat{I}_B$ a.e., we may multiply by I_A and integrate to obtain, by the dominated convergence theorem,

$$\frac{1}{n} \sum_{k=0}^{n-1} P(A \cap T^{-k}B) \to E(I_A \hat{I}_B).$$

But if $\mathscr{G}$ is the σ-field of almost invariant sets,

$$E(I_A \hat{I}_B) = E[E(I_A \hat{I}_B) \,|\, \mathscr{G}]$$
$$= E[\hat{I}_B E(I_A \,|\, \mathscr{G})] \qquad \text{by 3.3.8}$$
$$= E(\hat{I}_A \hat{I}_B).$$

Under the hypothesis (e), $\hat{I}_A = P(A)$ a.e., $\hat{I}_B = P(B)$ a.e., proving (f).

(f) *implies* (g): If $A, B \in \mathscr{F}$ and $\varepsilon > 0$, choose $A_0, B_0 \in \mathscr{F}_0$ such that

$P(A \triangle A_0)$, $P(B \triangle B_0)$, and $P[(A \cap T^{-k}B) \triangle (A_0 \cap T^{-k}B_0)]$ are less than ε for all k (see the proof of 3.2.7). Then

$$\left| \frac{1}{n} \sum_{k=0}^{n-1} P(A \cap T^{-k}B) - P(A)P(B) \right|$$

$$\leq \left| \frac{1}{n} \sum_{k=0}^{n-1} [P(A \cap T^{-k}B) - P(A_0 \cap T^{-k}B_0)] \right|$$

$$+ \left| \frac{1}{n} \sum_{k=0}^{n-1} [P(A_0 \cap T^{-k}B_0) - P(A_0)P(B_0)] \right|$$

$$+ |P(A_0)P(B_0) - P(A)P(B)|.$$

The first term is less than ε for all n, and the second is less than ε for large n by (f). Since the third term is less than 2ε, the result follows.

(g) *implies* (a): Let A be an invariant set, and set $B = A$; then $n^{-1} \sum_{k=0}^{n-1} P(A \cap T^{-k}B) = P(A)$. By (g), $P(A) = [P(A)]^2$, hence $P(A) = 0$ or 1; thus T is ergodic.

Section 3.4

1. It suffices to consider an arbitrary but fixed base r. Let $b = (i_1, \ldots, i_m)$, and let A_b be the set of all y in $[0, 1)$ such that the relative frequency of b in the first $n + m - 1$ digits of y converges to r^{-m} as $n \to \infty$. If λ is Lebesgue measure, we have, taking all $p_i = 1/r$ in 3.4.1,

$$\lambda(A_b) = P\{Y \in A_b\} = P\left\{ \frac{1}{n} \sum_{k=0}^{n-1} f(T^k\omega) \to r^{-m} \right\}$$

where T is the one-sided shift and $f(\omega) = I_{\{\omega: X_1(\omega) = i_1, \ldots, X_m(\omega) = i_m\}}$. By the pointwise ergodic theorem,

$$\frac{1}{n} \sum_{k=0}^{n-1} f(T^k\omega) \overset{\text{a.e.}}{\to} \int_\Omega f \, dP = P\{X_1 = i_1, \ldots, X_m = i_m\} = r^{-m}.$$

The set of normal numbers is $A = \bigcap_b A_b$, and therefore $\lambda(A) = 1$.

2. In Section 3.4, take $r = 2$, $p_0 = p_1 = \frac{1}{2}$. The corresponding space $(\Omega, \mathscr{F}, P)$ is isomorphic to $(\Omega_0, \mathscr{F}_0, \mu_0)$, where $\Omega_0 = [0, 1)$, $\mathscr{F}_0 = \mathscr{B}[0, 1)$, and μ_0 is Lebesgue measure; the isomorphism is given explicitly by $Y(\omega_1, \omega_2, \ldots) = \sum_{n=1}^\infty \omega_n 2^{-n}$. Now if $0 \leq Y(\omega) < \frac{1}{2}$, then $T(Y(\omega)) = 2Y(\omega) = .\omega_2 \omega_3 \ldots$, and if $\frac{1}{2} \leq Y(\omega) < 1$, then $T(Y(\omega)) = 2Y(\omega) - 1 = .\omega_2 \omega_3 \ldots$. If we define $T^*\omega = Y^{-1}T(Y(\omega))$, then $T^*(\omega_1, \omega_2, \ldots) = (\omega_2, \omega_3, \ldots)$, in other words,

T^* is a one-sided shift. By construction, T^* has the same action on Ω as T does on Ω_0, that is, the diagram of Figure P3.1 is commutative.

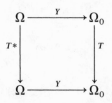

FIGURE P3.1

3. Take $p_0 = p_1 = \cdots = p_{r-1} = 1/r$, and proceed exactly as in Problem 2.

Section 3.5

1. Since $\mathcal{F}_n \downarrow \mathcal{F}_\infty$, we have, for each $A \in \mathcal{F}$, $P(A \mid \mathcal{F}_n) \to P(A \mid \mathcal{F}_\infty)$, almost everywhere and in L^1 (see RAP, 7.6.3, p. 299). Also, since $\mathcal{F}_\infty$ is trivial, $P(A \mid \mathcal{F}_\infty) = P(A)$ a.e. Now if B is a measurable cylinder,

$$\left| P(A \cap T^{-n}B) - P(A)P(B) \right| = \left| \int_{T^{-n}B} [P(A \mid \mathcal{F}_n) - P(A)] \, dP \right|$$

$$\leq \int_\Omega |P(A \mid \mathcal{F}_n) - P(A)| \, dP$$

$$\to 0 \qquad \text{since} \quad P(A \mid \mathcal{F}_n) \overset{L^1}{\to} P(A).$$

2. If T is mixing, let $A = \{X_0 = i\}$, $B = \{X_0 = j\}$. Then $T^{-n}B = \{X_n = j\}$, hence $P(A \cap T^{-n}B) = v_i p_{ij}^{(n)} \to P(A)P(B) = v_i v_j$ by hypothesis. Since $v_i > 0$, $p_{ij}^{(n)} \to v_j$, as desired.

Conversely, let $p_{ij}^{(n)} \to v_j$. If $A = \{X_0 = i_0, \ldots, X_k = i_k\}$, $B = \{X_0 = j_0, \ldots, X_l = j_l\}$, then

$$P(A \cap T^{-n}B) = v_{i_0} p_{i_0 i_1} \cdots p_{i_{k-1} i_k} p_{i_k j_0}^{(n-k)} p_{j_0 j_1} \cdots p_{j_{l-1} j_l}.$$

Since $p_{i_k j_0}^{(n-k)} \to v_{j_0}$, $P(A \cap T^{-n}B) \to P(A)P(B)$. But any measurable cylinder is a finite disjoint union of such sets B, so the mixing condition holds for $A = \{X_0 = i_0, \ldots, X_k = i_k\}$ and measurable cylinders B. Apply the same argument to A to show that the mixing condition holds for all measurable cylinders A and B; thus T is mixing.

3. (a) We have $E(R_n) = 1 + \sum_{k=2}^{n} P(B_k)$, and

$$P(B_k) = P\{X_k \neq 0, X_k + X_{k-1} \neq 0, \ldots, X_k + \cdots + X_2 \neq 0\}$$
$$= P\{S_1 \neq 0, S_2 \neq 0, \ldots, S_{k-1} \neq 0\}$$

since the X_i are iid.

Thus $n^{-1}E(R_n)$ is the arithmetic average of a sequence converging to $P(A)$, hence $n^{-1}E(R_n) \to P(A)$.

(b) The Z_k are iid, with $|Z_k| \leq N$. Since $R_{nN} \leq Z_1 + \cdots + Z_n$, the strong law of large numbers yields the desired result.

(c) For any positive integer n, $(k-1)N + 1 \leq n \leq kN$ for some k, hence $|R_n - R_{kN}| \leq N$; therefore

$$\frac{R_n}{n} \leq \frac{R_{kN}}{kN} \frac{kN}{n} + \frac{N}{n}.$$

Since $kN/n \to 1$ as $n \to \infty$, $\limsup_{n \to \infty} n^{-1}R_n \leq N^{-1}E(Z_1)$ by (b). Now $Z_1 = R_N$ and N is arbitrary; thus the result follows from (a).

(d) Since $V_k = 1$ if $X_{k+1} \neq 0$, $X_{k+1} + X_{k+2} \neq 0$, $\ldots$, and $V_k = 0$ otherwise, V_k can be expressed as $g(X_k, X_{k+1}, \ldots)$ where $g\colon R^\infty \to R^\infty$, measurable relative to $[\mathscr{B}(R)]^\infty$. The pointwise ergodic theorem therefore applies.

(e) The sum $\sum_{i=1}^{n} V_i$ is the number of states visited in the first n steps that are never revisited. If $i < j$, and S_i and S_j are never revisited, then $S_i \neq S_j$; thus $\sum_{i=1}^{n} V_i \leq R_n$. By (d), $\liminf_{n \to \infty} n^{-1}R_n \geq E(V_1)$ a.e. But since the X_i are iid,

$$E(V_1) = P\{X_1 \neq 0, X_1 + X_2 \neq 0, \ldots\} = P(A).$$

Section 3.6

1. Let $\{X_n\}$ and $\{X'_n\}$ be discrete ergodic sequences with entropies H and H'. Flip a coin; if the result is heads, let $X''_n = X_n$ for all n, and if tails, let $X''_n = X'_n$ for all n. If p is the probability of heads, the limit random variable is H with probability p, and H' with probability $1 - p$, so that the entropy of $\{X''_n\}$ is $pH + (1 - p)H'$. In part (a), choose $H = H'$, and in part (b), choose $H \neq H'$. There is no problem in realizing these choices; for example, if the X_n are independent and take on r values with equal probability, then $H = \log r$.

2. If k is a positive integer,

$$\int_{\{Z_n^p \geq k\}} Z_n^p \, dP = \sum_{i=k}^{\infty} \int_{\{i \leq Z_n^p < i+1\}} Z_n^p \, dP$$

$$\leq \sum_{i=k}^{\infty} (i+1) P\{Z_n^p \geq i\}.$$

If $Z_n^p \geq i$, then $-n^{-1} \log p(X_0, \ldots, X_{n-1}) \geq i^{1/p}$, hence $p(X_0, \ldots, X_{n-1}) \leq \exp_2 [-ni^{1/p}]$; thus

$$P\{Z_n^p \geq i\} = \sum \{p(x_0, \ldots, x_{n-1}):$$
$$-n^{-1} \log p(x_0, \ldots, x_{n-1}) \geq i^{1/p}\}$$
$$\leq \exp_2 [-ni^{1/p}] r^n$$

where r is the number of elements in the coordinate space. Therefore

$$\int_{\{Z_n^p \geq k\}} Z_n^p \, dP \leq r^n \sum_{i=k}^{\infty} (i+1) \exp_2 [-ni^{1/p}]$$

$$\to 0 \qquad \text{as } k \to \infty, \text{ uniformly in } n.$$

(Note that $r^n \exp_2 [-ni^{1/p}] = \exp_2 [n(\log r - i^{1/p})]$, and if i is large enough, this will be bounded above by $\exp_2 (-i^c)$ for some $c > 0$.)

3. (a)   If $\mathbf{x} = (x_0, \ldots, x_{n-1})$, then

$$E\left(\frac{q_m}{p}\right) = \sum_{p(\mathbf{x})>0} p(\mathbf{x}) \frac{q_m(\mathbf{x})}{p(\mathbf{x})} = \sum_{p(\mathbf{x})>0} q_m(\mathbf{x}) \leq \sum_{\text{all } \mathbf{x}} q_m(\mathbf{x}) = 1$$

and

$$E\left[-\ln \frac{q_m}{p}\right] = -\sum_{p(\mathbf{x})>0} p(\mathbf{x}) \ln \frac{q_m(\mathbf{x})}{p(\mathbf{x})}$$

(note that $p(\mathbf{x}) > 0$ implies $q_m(\mathbf{x}) > 0$, but not conversely). Now

$$-\sum_{p(\mathbf{x})>0} p(\mathbf{x}) \ln p(x_j \mid x_{j-m}, \ldots, x_{j-1})$$

$$= -\sum_{\text{all } \mathbf{x}} p(\mathbf{x}) \ln p(x_j \mid x_{j-m}, \ldots, x_{j-1})$$

$$= H(X_j \mid X_{j-m}, \ldots, X_{j-1})$$

$$= H(X_m \mid X_0, \ldots, X_{m-1}) \qquad \text{by stationarity.}$$

Similarly

$$- \sum_{p(\mathbf{x}) > 0} p(\mathbf{x}) \ln p(x_0, \ldots, x_{m-1}) = H(X_0, \ldots, X_{m-1}),$$

$$\sum_{p(\mathbf{x}) > 0} p(\mathbf{x}) \ln p(\mathbf{x}) = -H(X_0, \ldots, X_{n-1})$$

and the result follows.

(b) If $p(X_0, \ldots, X_{n-1}) > 0$ for all n,

$$-\frac{1}{n} \ln q_m = -\frac{1}{n} \ln p(X_0, \ldots, X_{m-1}) + \frac{1}{n} \sum_{k=0}^{n-m-1} f(X_k, X_{k+1}, \ldots)$$

where $f(X_0, X_1, \ldots) = -\ln p(X_m \mid X_0, \ldots, X_{m-1})$; note that

$$E[f(X_0, X_1, \ldots)] = H(X_m \mid X_0, \ldots, X_{m-1}).$$

Since the set on which $p(X_0, \ldots, X_{n-1}) = 0$ for some n has probability 0, we may regard the above equality as holding on the entire space. The pointwise ergodic theorem now gives the desired result.

(c) By (a), $E[n^{-1} \mid \ln (q_m/p) \mid] < \varepsilon/2$ for large n, and by (b),

$$E\left[\left| -\frac{1}{n} \ln q_m - H \right| \right] \leq E\left[\left| -\frac{1}{n} \ln q_m - H(X_m \mid X_0, \ldots, X_{m-1}) \right| \right]$$

$$+ \mid H(X_m \mid X_0, \ldots, X_{m-1}) - H \mid < \frac{\varepsilon}{2}$$

for large n.

Now if $p(x_0, \ldots, x_{n-1}) > 0$ for all n,

$$-\frac{1}{n} \ln p - H = \frac{1}{n} \ln \frac{q_m}{p} - \frac{1}{n} \ln q_m - H;$$

as in (b), this may be assumed to hold on the entire space, and the result follows.

Chapter 4

Section 4.1

1. Let A be the negligible set, T_0 the separating set. By separability,

$$\{\omega \notin A : X(\cdot, \omega) \text{ is continuous at } t_0\}$$

$$= \bigcap_{n=1}^{\infty} \bigcup_{m=1}^{\infty} \bigcap_{\substack{t \in T_0 \\ |t-t_0| < 1/m}} \left\{ \omega \notin A : d(X(t, \omega), X(t_0, \omega)) < \frac{1}{n} \right\}$$

and

$$\{\omega \notin A : X(\cdot, \omega) \text{ is uniformly continuous on } T\}$$

$$= \bigcap_{n=1}^{\infty} \bigcup_{m=1}^{\infty} \bigcap_{\substack{t_1, t_2 \in T_0 \\ |t_1 - t_2| < 1/m}} \left\{ \omega \notin A : d(X(t_1, \omega), X(t_2, \omega)) < \frac{1}{n} \right\}.$$

If $B \in \mathscr{B}(R)$, then $\{\omega : d(X(t_1, \omega), \quad X(t_2, \omega)) \in B\} = \{\omega : (X(t_1, \omega), X(t_2, \omega)) \in d^{-1}(B)\} \in \mathscr{F}$ by continuity of d. Since $\{\omega \in A : X(\cdot, \omega) \text{ is continuous at } t_0\} \subset A$, with $P(A) = 0$, the result follows.

2. Let A be the negligible set, T_0 the separating set. By definition of separability, if $\omega \notin A$, we have

$$\sup_{t \in T} X(t, \omega) = \sup_{t \in T_0} X(t, \omega), \qquad \limsup_{\substack{t \to t_0 \\ t \in T}} X(t, \omega) = \limsup_{\substack{t \to t_0 \\ t \in T_0}} X(t, \omega)$$

and the result follows.

3. Consider the Poisson process constructed from waiting times $T_1, T_2, \ldots$ as in Chapter 1 ($X(t) = n$ if $T_1 + \cdots + T_n \leq t < T_1 + \cdots + T_{n+1}$, where the T_i are independent, each with density $\lambda e^{-\lambda x}$, $x \geq 0$). The sample functions are right continuous step functions, and hence the process is separable by 4.1.5. Also, $P\{\omega : X(\cdot, \omega) \text{ is continuous on } [0, \infty)\} = 0$ since the sample functions are step functions. But for t fixed, $P\{X(s) \nrightarrow X(t) \text{ as } s \to t\} = P\{\text{there is a jump at } t\} = \sum_{k=1}^{\infty} P\{T_1 + \cdots + T_k = t\} = 0$ since $T_1 + \cdots + T_k$ has a density for each k.

Section 4.2

1. Let Ω consist of a single point $\{\omega\}$, and take $X(t, \omega) = f(t)$, where $f : R \to R$ is not Borel measurable, for example, $f = I_A$, $A \notin \mathscr{B}(R)$. If the process were measurable, then $X(t, \omega)$ would be Borel measurable in t for ω fixed, a contradiction.

2. Fix t, and let $h_n(s) = jt/n$ for $(j-1)t/n \leq s < jt/n, j = 1, 2, \ldots, n, n = 1, 2, \ldots, h_n(t) = t$. Then $h_n(s) \downarrow s, 0 \leq s \leq t$. Now

$$X(h_n(s), \omega) = \sum_{j=1}^{n} X\left(\frac{jt}{n}, \omega\right) I_{[(j-1)t/n, jt/n) \times \Omega}(s, \omega)$$

$$+ X(t, \omega) I_{\{t\} \times \Omega}(s, \omega),$$

so the map $(s, \omega) \to X(h_n(s), \omega)$ from $[0, t] \times \Omega$ to the state space S is $\mathscr{B}[0, t] \times \mathscr{F}(t)$-measurable. Let $n \to \infty$; $X(h_n(s), \omega) \to X(s, \omega)$ by right continuity, and the result follows.

3. (a) If $t \geq 0$, then

$$\{S + T > t\} = \{S > t\} \cup \{T > t\} \cup \{S + T > t, S \leq t, T \leq t\}.$$

Thus we must show that $\{S + T > t, S \leq t, T \leq t\} \in \mathcal{F}(t)$. But

$$\{S + T > t, S \leq t, T \leq t\}$$
$$= \bigcup_{\substack{r \text{ rational} \\ 0 < r < t}} \{r < T \leq t, S \leq t, S + r > t\}.$$

(Note that if $S(\omega) + T(\omega) > t$ and $S(\omega) \leq t$, then $S(\omega) + r > t$ for some rational r with $0 < r < T(\omega)$.) Now

$$\{r < T \leq t\} = \{T \leq t\} - \{T \leq r\} \in \mathcal{F}(t),$$
$$\{S > t - r\} = \Omega - \{S \leq t - r\} \in \mathcal{F}(t - r) \subset \mathcal{F}(t),$$

proving the result.

(In the discrete parameter case, the proof is much easier:

$$\{S + T \leq n\} = \bigcup_{k=0}^{n} \{S = k, T \leq n - k\} \in \mathcal{F}_n.)$$

(b) Since $S \wedge T \leq S$ and $S \wedge T \leq T$, $\mathcal{F}(S \wedge T) \subset \mathcal{F}(S) \cap \mathcal{F}(T)$ by 4.2.4(e). If $A \in \mathcal{F}(S) \cap \mathcal{F}(T)$, then

$$A \cap \{S \wedge T \leq t\} = [A \cap \{S \leq t\}] \cup [A \cap \{T \leq t\}] \in \mathcal{F}(t),$$

so $A \in \mathcal{F}(S \wedge T)$.

(c) By 4.2.4(d) with $A = \Omega$, $\{S \leq T\} \in \mathcal{F}(T)$, hence $\{S > T\} \in \mathcal{F}(T)$. If $R = S \wedge T$, then R is $\mathcal{F}(R)$-measurable by 4.2.4(b), and $\{S < T\} = \{R < T\}$ by definition of R. But R and T are both $\mathcal{F}(T)$-measurable (note $\mathcal{F}(R) \subset \mathcal{F}(T)$ by 4.2.4(e)), so $\{R < T\} \in \mathcal{F}(T)$; thus $\{S < T\} \in \mathcal{F}(T)$. Finally, $\{S = T\} = \Omega - (\{S > T\} \cup \{S < T\}) \in \mathcal{F}(T)$. The sets also belong to $\mathcal{F}(S)$ by symmetry.

4. First assume B open. Then by right-continuity,

$$\{\omega : T(\omega) < t\} = \bigcup_{\substack{r < t \\ r \text{ rational}}} \{\omega : X(r, \omega) \in B\} \in \mathcal{F}(t).$$

If $\delta > 0$ and N is a positive integer such that $N^{-1} < \delta$,

$$\{\omega : T(\omega) \leq t\} = \bigcap_{n=N}^{\infty} \{\omega : T(\omega) < t + n^{-1}\} \in \mathcal{F}(t + \delta),$$

and therefore $\{T \leq t\} \in \mathcal{F}(t^+)$, as desired.

Now assume B closed, and let $V_n = \{x \in S : d(x, B) < 1/n\}$; the V_n are open and $V_n \downarrow B$. If T_n is the hitting time of V_n and T is the hitting time of B, we claim that $T_n \uparrow T$.

To prove this note that since $V_n \supset V_{n+1}$, we have $T_n \leq T_{n+1}$. Therefore $T_n \uparrow T' \leq T$. If $T'(\omega) = \infty$, there is nothing further to prove, so assume $T'(\omega) = t < \infty$. For any $\varepsilon > 0$ we have $t - \varepsilon < T_n(\omega) < t + \varepsilon$ for sufficiently large n, so that $X(t_n, \omega) \in V_n$ for some $t_n \in (t - \varepsilon, t + \varepsilon)$. Pick a subsequence $t_{n_k} \to t_0$; by continuity, $X(t_{n_k}, \omega) \to X(t_0, \omega)$. But $X(t_{n_k}, \omega) \in V_{n_k}$, so that $X(t_{n_k}, \omega)$ is within distance $1/n_k$ of B; consequently $X(t_0, \omega) \in B$, so that $T(\omega) \leq t + \varepsilon$. Since ε is arbitrary, $T \leq T'$.

Finally, $\{T \leq t\} = \bigcap_{n=1}^{\infty} \{T_n \leq t\} \in \mathscr{F}(t^+)$.

Section 4.3

1. If $X_i > a$ and $X_n \geq X_i$, then $X_n > a$; therefore

$$P(A_i \cap B) \geq P(A_i \cap \{X_n \geq X_i\})$$
$$= P(A_i)P\{X_n \geq X_i\} \qquad \text{by independence}$$
$$= P(A_i)P\{Y_{i+1} + \cdots + Y_n \geq 0\}.$$

By RAP, 8.1.5(d), p. 325, $Y_{i+1} + \cdots + Y_n$ is symmetric, and hence $P\{Y_{i+1} + \cdots + Y_n > 0\} = P\{Y_{i+1} + \cdots + Y_n < 0\}$; therefore

$$P\{Y_{i+1} + \cdots + Y_n \geq 0\} \geq \tfrac{1}{2}.$$

Thus

$$P(B) = \sum_{i=1}^{n} P(A_i \cap B) \geq \frac{1}{2} \sum_{i=1}^{n} P(A_i) = \frac{1}{2} P\left\{ \max_{1 \leq i \leq n} X_i > a \right\}.$$

2. (a) If $T \leq n - 1$, $X_n - X_T \leq -\varepsilon$, and $X_n > a$, then $X_T > a + \varepsilon$. Since $X_{T-1} \leq a$ by definition of T, $Y_T > \varepsilon$, hence $Y_j > \varepsilon$ for some $j \leq n - 1$. (If $T = 1$, then $Y_T = X_T > a + \varepsilon > \varepsilon$ since $a > 0$.) Therefore,

$$\{T \leq n - 1, X_n - X_T \leq -\varepsilon\} \subset \{T \leq n - 1, X_n \leq a\} \cup \bigcup_{j=1}^{n-1} \{Y_j > \varepsilon\}$$

and the result follows.

(b) If $T \leq n - 1$, $X_n > a + 2\varepsilon$, $X_n - X_T \leq \varepsilon$, then $X_T > a + \varepsilon$, so $Y_T > \varepsilon$ as in (a), and hence $Y_j > \varepsilon$ for some $j \leq n - 1$.

(c) If $X_n > a + 2\varepsilon$, then $T \leq n$; if $T > n - 1$, then $T = n$; therefore

$X_{n-1} \le a$, so that $Y_n > 2\varepsilon$. (Note (a), (b), and (c) do not use symmetry or independence.)

(d) We have

$$P\left\{ \max_{1 \le i \le n} X_i > a, X_n \le a \right\} = P\{T \le n - 1, X_n \le a\}$$

$$\ge P\{T \le n - 1, X_n - X_T \le -\varepsilon\}$$

$$- \sum_{j=1}^{n-1} P\{Y_j > \varepsilon\} \qquad \text{by (a)}$$

$$= P\{T \le n - 1, X_n - X_T \ge \varepsilon\}$$

$$- \sum_{j=1}^{n-1} P\{Y_j > \varepsilon\}$$

by symmetry and independence. (Note that

$$P\{T = k, X_n - X_k \le -\varepsilon\} = P\{T = k\}P\{X_n - X_k \le -\varepsilon\}$$

$$\text{by independence}$$

$$= P\{T = k\}P\{X_n - X_k \ge \varepsilon\}$$

$$\text{by symmetry}$$

$$= P\{T = k, X_n - X_k \ge \varepsilon\}.)$$

Thus by (b),

$$P\left\{ \max_{1 \le i \le n} X_i > a, X_n \le a \right\} \ge P\{T \le n - 1, X_n > a + 2\varepsilon\}$$

$$- 2\sum_{j=1}^{n-1} P\{Y_j > \varepsilon\},$$

and the result follows from (c).

(e) By (d),

$$P\left\{ \max_{1 \le i \le n} X_i > a, X_n \le a \right\} \ge P\{X_n > a + 2\varepsilon\} - P\{Y_n > 2\varepsilon\}$$

$$- 2\sum_{j=1}^{n-1} P\{Y_j > \varepsilon\},$$

$$P\left\{ \max_{1 \le i \le n} X_i > a, X_n > a \right\} = P\{X_n > a\}.$$

Add these equations to obtain

$$P\left\{\max_{1\le i\le n} X_i > a\right\} \ge P\{X_n > a + 2\varepsilon\} + P\{X_n > a\}$$

$$- P\{Y_n > 2\varepsilon\} - 2\sum_{j=1}^{n-1} P\{Y_j > \varepsilon\}.$$

Since $P\{X_n > a + 2\varepsilon\} \le P\{X_n > a\}$, and

$$P\{Y_n > 2\varepsilon\} \le P\{Y_n > \varepsilon\} \le 2P\{Y_n > \varepsilon\},$$

we have

$$P\left\{\max_{1\le i\le n} X_i > a\right\} \ge 2P\{X_n > a + 2\varepsilon\} - 2\sum_{j=1}^{n} P\{Y_j > \varepsilon\}.$$

3. By Problem 1,

$$p = P\left\{\max_{1\le j\le n} B(jt/n) > a\right\} \le 2P\{B(t) > a\},$$

and by Problem 2(e),

$$p \ge 2P\{B(t) > a + 2\varepsilon\} - 2\sum_{j=1}^{n} P\left\{B\left(\frac{jt}{n}\right) - B\left(\frac{(j-1)t}{n}\right) > \varepsilon\right\}$$

$$= 2P\{B(t) > a + 2\varepsilon\} - 2nP\{B(t/n) > \varepsilon\}.$$

(The random variables $B(jt/n) - B((j-1)t/n)$ are independent and normal $(0, \sigma^2 t/n)$, and hence symmetric.) Now

$$nP\{B(t/n) > \varepsilon\} \sim \frac{n\sigma\sqrt{t/n}}{\sqrt{2\pi}\,\varepsilon} e^{-\varepsilon^2 n/2\sigma^2 t} \to 0 \qquad \text{as} \quad n \to \infty,$$

and by continuity of the sample functions,

$$P\left\{\max_{1\le j\le n} B(jt/n) > a\right\} \to P\left\{\max_{0\le s\le t} B(s) > a\right\}.$$

Therefore

$$2P\{B(t) > a + 2\varepsilon\} \le P\left\{\max_{0\le s\le t} B(s) > a\right\} \le 2P\{B(t) > a\}.$$

Since ε is arbitrary, we have $P\{\max_{0\le s\le t} B(s) > a\} = 2P\{B(t) > a\}$.

4. (a) If $Y(t) = c^{-1/2}B(ct)$, then $\{Y(t), t \geq 0\}$ is a Gaussian process with 0 mean and covariance function $K(s, t) = c^{-1}\sigma^2 \min (cs, ct) = \sigma^2 \min (s, t)$. Separability of $\{Y(t)\}$ follows from 4.1.5.

(b) The process $\{Z(t), t \geq 0\}$ is Gaussian with 0 mean and covariance function $K(s, t) = st\sigma^2 \min (1/s, 1/t) = \sigma^2 \min (s, t)$. By 4.1.5, $\{Z(t), t > 0\}$ is separable, and if T_0 is a separating set for $\{Z(t), t > 0\}$, then $T_0 \cup \{0\}$ is a separating set for $\{Z(t), t \geq 0\}$. Thus $\{Z(t), t \geq 0\}$ is separable.

5. The process $\{Z(t)\}$ of Problem 4(b) has continuous sample functions by 4.3.1, so that $Z(t') \to 0$ a.e. as $t' \to 0$. Let $t' = 1/t$ to obtain the desired result.

6. Let I be the indicator of $\{(t, \omega) : B(t, \omega) = c\}$. By Fubini's theorem,

$$\int_0^\infty \int_\Omega I \, dP \, dt = \int_\Omega \int_0^\infty I \, dt \, dP.$$

Since $P\{\omega : B(t, \omega) = c\} = 0$ for all t, we obtain, with λ denoting Lebesgue measure,

$$0 = \int_\Omega \lambda\{t : B(t, \omega) = c\} \, dP(\omega) = \int_\Omega \lambda(S_c(\omega)) \, dP(\omega).$$

Therefore $\lambda(S_c(\omega)) = 0$ for almost every ω.

Section 4.4

1. It suffices to consider the "lim sup" statement. Let $Z(t) = tB(1/t), t > 0$; $Z(0) = 0$, as in Problem 4, Section 4.3. By 4.4.4,

$$\limsup_{t \to \infty} \frac{Z(t)}{(2\sigma^2 t \ln \ln t)^{1/2}} = 1 \quad \text{a.e.}$$

Set $h = 1/t$ to obtain

$$\limsup_{h \to 0^+} \frac{B(h)}{h(2\sigma^2 h^{-1} \ln \ln h^{-1})^{1/2}} = 1 \quad \text{a.e.,}$$

and the result follows for $t_0 = 0$. But $B(t_0 + t) - B(t_0), t \geq 0$, is also separable Brownian motion, and the result follows for any $t_0 \geq 0$.

Section 4.5

1. The "only if" part is immediate from 4.5.3 and the remark after 4.5.4, since $\mathscr{F}(X(r), r < t) \subset \mathscr{F}(X(r), r \leq t)$ and $\mathscr{F}(X(r), r > t) \subset \mathscr{F}(X(r), r \geq t)$.

Thus assume $\mathcal{F}(X(r), r < t)$ and $\mathcal{F}(X(r), r > t)$ are conditionally independent given $X(t)$. If $B \in \mathcal{S}$, $t' > t$, then

$$P\{X(t') \in B \mid \mathcal{F}(X(r), r \leq t)\}$$
$$= P\{X(t') \in B \mid \mathcal{F}(X(r), r < t) \vee \mathcal{F}(X(t))\}$$
$$= P\{X(t') \in B \mid X(t)\} \qquad \text{by conditional independence.}$$

2.   Let p, q, r be transition probabilities on $(S, \mathcal{S})$. Then

$$[p * (q * r)](x, B) = \int_S p(x, dy)(q * r)(y, B)$$

$$= \int_S p(x, dy)\left[\int_S q(y, dz)r(z, B)\right],$$

and

$$[(p * q) * r](x, B) = \int_S (p * q)(x, dz)r(z, B)$$

$$= \int_S \left[\int_S p(x, dy)q(y, dz)\right]r(z, B).$$

(This means that we integrate $r(\cdot, B)$ with respect to the measure μ, where
$\mu(A) = \int_S p(x, dy)q(y, A)$.) If we replace $r(z, B)$ by an indicator I_C, both of
the above integrals are equal to $\int_S p(x, dy)q(y, C)$. The usual passage from
indicators to nonnegative simple functions to nonnegative measurable functions completes the proof.

3.   By definition of composition,

$$[P_0 * (p * q)](B) = \int_S P_0(dx) \int_S p(x, dy)q(y, B)$$

$$[(P_0 * p) * q](B) = \int_S \left[\int_S P_0(dx)p(x, dy)\right]q(y, B).$$

Replace $q(y, B)$ by an indicator I_C and proceed as in Problem 2.

Section 4.6

1.   By Lévy's theorem, $Y(t)$ converges in distribution to 0 as $t \to 0$, hence
$Y(t) \overset{P}{\to} 0$ (see RAP, p. 335, Problem 8). But $|Y(t + h) - Y(t)|$ has the same
distribution as $|Y(|h|)|$, and it follows that $Y(t + h) \overset{P}{\to} Y(t)$ as $h \to 0$.

2. (a) By 4.1.6, we may use the set T_0 of dyadic rationals as the separating set. Since $Y(t + h) - Y(t)$ is Poisson, the sample functions are integer valued and increasing on T_0, and hence on $[0, \infty)$ by separability. (If $s < t$ and $Y(s, \omega) > Y(t, \omega)$, then $Y(s_1, \omega) > Y(t_1, \omega)$ for some $s_1, t_1 \in T_0$, with $s_1 < t_1$.)

(b) By (a), almost every sample function has only jump discontinuities; since $Y(t)$ is Poisson (λt) and hence finite a.s., there can be only finitely many jumps in any bounded interval, so the sample functions are constant between jumps. To show that only jumps of height 1 are possible, note that

$$\sup_{0 \le t \le 1} [Y(t^+) - Y(t^-)]$$

$$= \lim_{n \to \infty} \max_{1 \le k \le 2^n} [Y(k2^{-n}) - Y((k-1)2^{-n})],$$

and

$$P\left\{ \max_{1 \le k \le m} [Y(k/m) - Y((k-1)m)] \le 1 \right\} = [P\{Y(1/m) \le 1\}]^m$$

$$= [e^{-\lambda/m}(1 + \lambda/m)]^m$$

$$\to 1 \qquad \text{as} \quad m \to \infty.$$

Thus $\sup_{0 \le t \le 1} [Y(t^+) - Y(t^-)] \le 1$ a.e. Repeat this argument on the interval $[n, n + 1]$, $n = 1, 2, \ldots$, to conclude that

$$\sup_{t \ge 0} [Y(t^+) - Y(t^-)] \le 1 \quad \text{a.e.,}$$

as desired.

Section 4.7

1. Let $t_n \in T$, $t_n \downarrow t$. Choose a decreasing sequence $s_n \in T_0$, with $|s_n - t_n| < 1/n$ and $|f(s_n) - g(t_n)| < 1/n$. Then $s_n \downarrow t$, hence $f(s_n) \to g(t)$; therefore $g(t_n) \to g(t)$, proving right-continuity.

Let $t_n \in T_0$, $t_n \uparrow t$. Choose an increasing sequence $s_n \in T_0$, with $|s_n - t_n| < 1/n$ and $|f(s_n) - g(t_n)| < 1/n$. Then $s_n \uparrow t$, so $f(s_n)$ approaches a limit L, hence $g(t_n) \to L$.

2. Fix $t \in T$, and choose $t_n \in T_0$ with $t_n \downarrow t$. Then $X(t_n) \to Y(t)$ a.e., and $X(t_n) \to X(t)$ in probability, so that $X(t) = Y(t)$ a.e.

3. By 4.7.4, $X(t^+, \omega)$ and $X(t^-, \omega)$ exist for all t and almost every ω. Choose $t_1, t_2, \ldots$ in the separating set such that $t_n \to t$ and $X(t_n, \omega) \to X(t, \omega)$. But $\{t_n\}$ has either a subsequence $\{t_{n_i}\}$ decreasing to t or a subsequence $\{t_{m_j}\}$ increasing to t. Thus either $X(t_{n_i}, \omega) \to X(t^+, \omega)$ or $X(t_{m_j}) \to X(t^-, \omega)$. The result follows.

4. Let $\{t_n\}$ be a sequence in T with $t_n \uparrow t$. Since $X(t_n) \xrightarrow{P} X(t)$, there is a subsequence $\{t_{n_k}\}$ with $X(t_{n_k}) \to X(t)$ a.e. Since $X(t_{n_k}) \to X(t^-)$ by hypothesis, the result follows.

5. Let $A \in \mathscr{F}(X(0))$, and fix $t \geq 0$. Let T_n be the smallest number of the form $(j/n)t$, $n = 1, 2, \ldots, j = 0, 1, \ldots$, such that $T_n \geq T$. Then

$$\int_{A \cap \{T \leq t\}} X(T_n) \, dP = \int_{A \cap \{T_n \leq t\}} X(T_n) \, dP$$

$$= \sum_{j=0}^{n} \int_{A \cap \{T_n = jt/n\}} X(jt/n) \, dP$$

$$= \sum_{j=0}^{n} \int_{A \cap \{T_n = jt/n\}} X(t) \, dP$$

$$= \int_{A \cap \{T \leq t\}} X(t) \, dP.$$

As $n \to \infty$, $X(T_n) \to X(T)$ by right-continuity. By (a) and the dominated convergence theorem,

$$\int_{A \cap \{T \leq t\}} X(T) \, dP = \int_{A \cap \{T \leq t\}} X(t) \, dP.$$

But

$$\int_A X(0) \, dP = \int_A X(t) \, dP$$

$$= \int_{A \cap \{T \leq t\}} X(t) \, dP + \int_{A \cap \{T > t\}} X(t) \, dP$$

$$= \int_{A \cap \{T \leq t\}} X(T) \, dP + \int_{A \cap \{T > t\}} X(t) \, dP.$$

If we let $t \to \infty$ through an appropriate subsequence, we obtain, by (b) and (c),

$$\int_A X(0) \, dP = \int_A X(T) \, dP + 0,$$

and the result follows.

Chapter 5

Section 5.1

1. We must check the three conditions of 5.1.1.

(a) Set $g(x, t, \omega) = e^{-x} f(t - x/n, \omega)$; then g is $\mathscr{B}[0, \infty) \times \mathscr{B}[a, b] \times \mathscr{F}$-measurable. By Fubini's theorem, f_n is $\mathscr{B}[a, b] \times \mathscr{F}$-measurable.

(b) Fix t and set $h(x, \omega) = e^{-x} f(t - x/n, \omega)$; then h is $\mathscr{B}[0, \infty) \times \mathscr{F}(t)$-measurable. By Fubini's theorem, $f_n(t, \cdot)$ is $\mathscr{F}(t)$-measurable.

(c) This is immediate since f_n is bounded.

2. (a) By independence of increments,

$$E\left(\sum_{i=0}^{n-1} [B(t_{i+1}) - B(t_i)]^2\right) = \sum_{i=0}^{n-1} \sigma^2(t_{i+1} - t_i) = \sigma^2(b - a).$$

Thus

$$E\left\{\left|\sum_{i=0}^{n-1} [B(t_{i+1}) - B(t_i)]^2 - \sigma^2(b - a)\right|^2\right\}$$

$$= \text{Var} \sum_{i=0}^{n-1} [B(t_{i+1}) - B(t_i)]^2$$

$$= \sum_{i=0}^{n-1} \text{Var} \{[B(t_{i+1}) - B(t_i)]^2\}$$

$$= 2\sigma^4 \sum_{i=0}^{n-1} (t_{i+1} - t_i)^2$$

since $B(t_{i+1}) - B(t_i)$ is normal with mean 0 and variance $\sigma^2(t_{i+1} - t_i)$

$$\leq 2\sigma^4 |P| \sum_{i=0}^{n-1} (t_{i+1} - t_i)$$

$$= 2\sigma^4 |P| (b - a) \to 0 \qquad \text{as} \quad |P| \to 0.$$

(b) If $\varepsilon > 0$,

$$P\{|V(P_m) - \sigma^2(b - a)| > \varepsilon\} \leq \frac{\text{Var } V(P_m)}{\varepsilon^2} \qquad \text{by Chebyshev's inequality}$$

$$\leq \frac{2\sigma^4 |P_m| (b - a)}{\varepsilon^2}$$

by the calculation of part (a).

The result now follows from the Borel–Cantelli lemma.

(c)   If $\varepsilon > 0$,

$$P\{\,|\,S(P_m)\,| > \varepsilon\} \le \frac{1}{\varepsilon^2}\,\text{Var}\ S(P_m) \qquad \text{by Chebychev's inequality}$$

$$= \frac{1}{\varepsilon^2}\sum_{i=0}^{n-1}\text{Var}\ (\,|\,B(t_{i+1}) - B(t_i))^2 - (t_{i+1} - t_i)\,|\,)$$

by independent increments

$$= \frac{1}{\varepsilon^2}\sum_{i=0}^{n-1}C(t_{i+1} - t_i)^2$$

$$\le \frac{(b - a)C}{\varepsilon^2}\,|\,P_m\,| \to 0$$

where

$$C = \text{Var}\ \left(\left|\,\frac{(B(t_{i+1}) - B(t_i))^2}{t_{i+1} - t_i} - 1\,\right|\right)$$

$$= \text{Var}\ (\,|\,\sigma^2 Y - 1\,|\,) < \infty$$

where Y has the χ^2 distribution with one degree of freedom. The result follows as in (b).

3.   If f is of the form $f(t) = \sum_{i=0}^{n-1} a_i I_{[t_i,\, t_{i+1})}(t)$ where $a = t_0 < t_1 < \cdots < t_n = b$ and the a_i are constants, both definitions give $\sum_{i=0}^{n-1} a_i[B(t_{i+1}) - B(t_i)]$. In the general case, take f_n of the above form converging to f in $L^2([a, b], \mathscr{B}[a, b], m)$ where m is Lebesgue measure. But m is also the measure associated with the mov obtained from Brownian notion, so it follows that $\int_a^b f_n(t)\ dB(t)$, defined in the Chapter 2 sense, converges to $\int_a^b f(t)\ dB(t)$ in $L^2(\Omega, \mathscr{F}, P)$. On the other hand, $f_n \in \mathscr{C}$ and f_n (regarded as a function of two variables) converges to f in $L^2([a, b] \times \Omega, \mathscr{B}[a, b] \times \mathscr{F}, m \times P)$. By 5.1.4, $\int_a^b f_n(t)\ dB(t)$, defined in the Chapter 5 sense, converges to $\int_a^b f(t)\ dB(t)$ in $L^2(\Omega, \mathscr{L}, P)$. The result follows.

Section 5.2

1.   (a)   By the Chapman–Kolmogorov equation,

$$I(t + \Delta t)\,|\,x_0,\, t_0)$$

$$= \int_{-\infty}^{\infty} f(x)\left[\int_{-\infty}^{\infty} p(x, t + \Delta t\,|\,z, t)p(z, t\,|\,x_0,\, t_0)\ dz\right]\,dx.$$

By Taylor's formula we may write

$$f(x) = \sum_{k=0}^{2} \frac{f^{(k)}(z)(x-z)^k}{k!} + \frac{f^{(3)}(\xi)(x-z)^3}{3!}$$

where ξ is between z and x. By Fubini's theorem and also by the fact that $\int_{-\infty}^{\infty} p(x, t + \Delta t \mid z, t)\, dx = 1$, we have

$$\frac{I(t + \Delta t \mid x_0, t_0) - I(t \mid x_0, t_0)}{\Delta t}$$

$$= \sum_{k=1}^{2} \frac{1}{k!} \int_{-\infty}^{\infty} \left[\int_{-\infty}^{\infty} \frac{(x-z)^k}{\Delta t} p(x, t + \Delta t \mid z, t)\, dx \right]$$

$$\times p(z, t \mid x_0, t_0) f^{(k)}(z)\, dz$$

$$+ \frac{1}{3!} \int_{-\infty}^{\infty} \left[\int_{-\infty}^{\infty} \frac{(x-z)}{\Delta t} p(x, t + \Delta t \mid z, t) f^{(3)}(\xi)\, dz \right]$$

$$\times p(z, t \mid x_0, t_0)\, dx.$$

By the assumptions on the convergence of the M_i, we obtain

$$\frac{\partial I}{\partial t} = \sum_{k=1}^{2} \frac{1}{k!} \int_{-\infty}^{\infty} M_k(z, t) p(z, t \mid x_0, t_0) f^{(k)}(z)\, dz,$$

as desired.

(b) Integrate by parts and use $f^{(k)}(\pm \infty) = 0$ to obtain

$$\frac{\partial I}{\partial t} = \sum_{k=1}^{2} \frac{(-1)^k}{k!} \int_{-\infty}^{\infty} \frac{\partial^k}{\partial x^k} [M_k(x, t) p(x, t \mid x_0, t_0)] f(x)\, dx.$$

But

$$\frac{\partial I}{\partial t} = \frac{\partial}{\partial t} \int_{-\infty}^{\infty} f(x) p(x, t \mid x_0, t_0)\, dx$$

$$= \int_{-\infty}^{\infty} f(x) \frac{\partial}{\partial t} p(x, t \mid x_0, t_0)\, dx,$$

and the result follows.

(c) We have $E[B(t + \Delta t) - B(t)]^k \mid B(t) = x] = A_k(\Delta t)$ where $A_1(\Delta t) = 0$, $A_2(\Delta t) = \sigma^2 \Delta t$, $A_3(\Delta t) = o(\Delta t)$, so that $m(x, t) = 0$, $\sigma^2(x, t) = \sigma^2$. The forward equation is $\partial p/\partial t = \frac{1}{2}\sigma^2\, \partial^2 p/\partial x^2$ (the heat equation), and direct substitution shows that $p(x, t \mid x_0, t_0) = [2\pi\sigma^2(t - t_0)]^{-1/2}$ $\times \exp\left[-(x - x_0)^2/2\sigma^2(t - t_0)\right]$ is a solution.

2. (a) By Taylor's formula,

$$p(x, t \mid z, t_0 + \Delta t)$$

$$= \sum_{k=0}^{2} \frac{1}{k!} \frac{\partial^k}{\partial x_0^k} p(x, t \mid x_0, t_0 + \Delta t)(z - x_0)^k$$

$$+ \frac{1}{3!} \left[\frac{\partial^3}{\partial x_0^3} p(x, t \mid x_0, t_0) \right]_{x_0 = \xi} (z - x_0)^3.$$

By the Chapman–Kolmogorov equation,

$$p(x, t \mid x_0, t_0) = \sum_{k=0}^{2} \frac{1}{k!} \frac{\partial^k}{\partial x_0^k} p(x, t \mid x_0, t_0 + \Delta t)$$

$$\times \int_{-\infty}^{\infty} (z - x_0)^k p(z, t_0 + \Delta t \mid x_0, t_0) \, dz$$

$$+ \frac{1}{3!} \int_{-\infty}^{\infty} \left[\frac{\partial^3}{\partial x_0^3} p(x, t \mid x_0, t_0) \right]_{x_0 = \xi}$$

$$\times (z - x_0)^3 p(z, t_0 + \Delta t \mid x_0, t_0) \, dz.$$

Since $\int_{-\infty}^{\infty} p(z, t_0 + \Delta t \mid x_0, t_0) \, dz = 1$, we have

$$\frac{1}{\Delta t} [-p(x, t \mid x_0, t_0 + \Delta t) + p(x, t \mid x_0, t_0)]$$

$$= \sum_{k=1}^{2} \frac{1}{k!} M_k(x_0, t_0, \Delta t) \frac{\partial^k p}{\partial x_0^k}$$

$$+ \frac{1}{3!} \int_{-\infty}^{\infty} \frac{\partial^3 p(x, t \mid \xi, t_0)}{\partial x_0^3} (z - x_0)^3$$

$$\times p(z, t_0 + \Delta t \mid x_0, t_0) \, dz.$$

Let $t \to 0$ to obtain the desired result.

Section 5.3

1. (a)

$$P(|\Phi(f)| > \eta) = P(|\Phi(f)| > \eta \text{ and } \int_a^b |f(t)|^2 \, dt > \varepsilon^2)$$

$$+ P\left(|\Phi(f)| > \eta \text{ and } \int_a^b |f(t)|^2 \, dt \le \varepsilon^2 \right) \quad (*)$$

The first term of (*) is at most

$$P\left(\int_a^b |f(t)|^2\, dt > \varepsilon^2\right) = P\left(\frac{[\int_a^b |f(t)|^2\, dt]^{1/2}}{1 + [\int_a^b |f(t)|^2\, dt]^{1/2}} > \frac{\varepsilon}{1+\varepsilon}\right)$$

$$\leq ((1+\varepsilon)/\varepsilon)N\left(\left[\int_a^b |f(t)|^2\, dt\right]^{1/2}\right),$$

using respectively the fact that the function $\varepsilon \to \varepsilon/(1 + \varepsilon)$ is strictly increasing on $[0, \infty)$ and the Chebychev inequality. Now define $f_\varepsilon(t, \omega)$ to be $f(t, \omega)$ if $\int_a^b |f(t, \omega)|^2\, dt \leq \varepsilon^2$ and 0 otherwise; note that f_ε is in $\mathscr{C}$ so the stochastic integral of f_ε is defined. Then the second term of (*) is at most

$$P\left(\left|\int_a^b f_\varepsilon(t)\, dB(t)\right| > \eta\right) \leq \frac{1}{\eta^2} E\left(\left|\int_a^b f_\varepsilon(t)\, dB(t)\right|^2\right)$$

$$\leq \frac{1}{\eta^2} E\left(\int_a^b f_\varepsilon^2(t)\, dt\right)$$

$$\leq \varepsilon^2/\eta^2.$$

  (b) Put $Y = |X|/(1 + |X|)$. Then $|X| \leq \eta$ implies $Y \leq \eta$ and $0 \leq Y \leq 1$ always holds, so

$$N(X) = EY$$
$$= E(YI_{\{|X|\leq\eta\}}) + E(YI_{\{|X|>\eta\}})$$
$$\leq E(\eta) + E(I_{\{|X|>\eta\}})$$
$$= \eta + P(|X| > \eta).$$

Put $\eta = \varepsilon^{1/2}$ and use (a) to get

$$N(\Phi(f)) \leq \varepsilon^{1/2} + \varepsilon + ((1+\varepsilon)/\varepsilon)C$$

where $C = N(\int_a^b |f(t)|^2\, dt)$. If $C = 0$, let $\varepsilon \downarrow 0$ to get the desired inequality; if $C > 0$ take $\varepsilon = C^{2/3}$ and use the fact that $0 < C \leq 1$ to get

$$N(I(f)) \leq C^{1/3}(1 + C^{1/3} + 1 + C^{2/3}) \leq 4C^{1/3}.$$

  (c) From (b) we have $d_2(\Phi(f), \Phi(g)) \leq 4(d_1(f, g))^{1/3}$ which implies uniform continuity.

  If $f \in \mathscr{C}_1$, define $f_n(t, \omega)$ to be $f(t, \omega)$ if $|f(t, \omega)| \leq n$ and 0 otherwise; then $f_n \in \mathscr{C}$ and

$$\int_a^b |f(t) - f_n(t)|^2\, dt = \int_a^b |f(t)|^2 I_{[-n,\,n]^c}(f(t))\, dt \to 0 \quad \text{a.s.} \qquad \text{as} \quad n \to \infty$$

(since $\int_a^b |f(t)|^2\, dt < \infty$ a.s. the dominated convergence theorem applies for a.e. ω) and therefore in probability. Hence $d_1(f_n, f) \to 0$, so $\mathscr{C}$ is dense in $\mathscr{C}_1$.

2. (a)   By 5.1.3 and Problem 1 of 5.3 there exist step functions α_n, β_n satisfying (i) and (ii); we show that with an appropriate subsequence (iii) is also satisfied. Now

$$|X_n(t) - X(t)| \leq \left|\int_a^t (\alpha_n(u) - \alpha(u))\, du\right| + \left|\int_a^t (\beta_n(u) - \beta(u))\, dB(u)\right|.$$

The first term is at most

$$\int_a^b |\alpha_n(u) - \alpha(u)|\, du \leq \left\{(b - a)\int_a^b |\alpha_n(u) - \alpha(u)|^2\, du\right\}^{1/2} \to 0 \qquad \text{a.s.}$$

By Problem 1 we have for the second term

$$N\left(\int_a^t (\beta_n(u) - \beta(u))\, dB(u)\right)$$

$$\leq 4\left\{N\left(\left(\int_a^t |\beta_n(u) - \beta(u)|^2\, du\right)^{1/2}\right)\right\}^{1/3}$$

$$\leq 4\left\{N\left(\left(\int_a^b |\beta_n(u) - \beta(u)|^2\, du\right)^{1/2}\right)\right\}^{1/3}$$

$$= \eta_n$$

where the bound η_n does not depend on t. Thus $\int_a^t (\beta_n(u) - \beta(u)\, dB(u)$ converges to zero in probability at a rate that does not depend on t, $a \leq t \leq b$. It follows that there exists a subsequence $n_1 < n_2 < \cdots$ such that $\int_a^t (\beta_{n_k}(u) - \beta(u))\, dB(u)$ converges to zero uniformly in t for a.e. ω—just follow the standard proof that convergence in probability implies that some subsequence converges a.s. and use the fact that η_n does not depend on t.

   (b)   By 5.3.1 for step functions we have

$$f(X_n(t), t) - f(X_n(s), s)$$

$$= Y_n(t) - Y_n(s)$$

$$= \int_s^t [\alpha_n(u)f_x(X_n(u), u) + f_t(X_n(u), u)$$

$$+ \tfrac{1}{2}\beta_n^2(u)f_{xx}(X_n(u), u)]\, du$$

$$+ \int_s^t \beta_n(u)f_x(X_n(u), u)\, dB(u).$$

For a.e. ω the left side approaches $f(X(t), t) - f(X(s), s) = Y(t) - Y(s)$ since f is continuous and $X_n(t) \to X(t)$ a.s. We claim that the right side approaches (in probability) the corresponding expression with all n's omitted; we prove this only for the last integral since the first is similar. By Problem 1 it is sufficient to show that

$$\int_s^t |\beta_n^2(u)f_x(X_n(u), u) - \beta^2(u)f_x(X(u), u)|^2 \, du \to 0$$

in probability. The square root of this is at most

$$\sqrt{\int_a^b |\beta_n(u) - \beta(u)|^2 \, du} \sqrt{\int_a^b |f_x(X_n(u), u)|^2 \, du}$$

$$+ \sqrt{\int_a^b |\beta(u)|^2 \, du} \sqrt{\int_a^b |f_x(X_n(u), u) - f_x(X(u), u)|^2 \, du}.$$

The first term approaches zero a.s. by a(i) and the fact that f_x is bounded. The second term approaches zero a.s. by a(iii) and the fact that $\int_a^b |\beta(u)|^2 \, du < \infty$ a.s. Thus

$$Y(t) - Y(s) = \int_s^t [\alpha(u)f_x(X(u), u) + f_t(X(u), u)$$

$$+ \tfrac{1}{2}\beta^2(u)f_{xx}(X(u), u)] \, du + \int_s^t \beta(u)f_x(X(u), u) \, dB(u).$$

This is the integral form of 5.3.1.

3. By the Itô differentiation, formula 5.3.1

$$\psi(X(t), t) - \psi(X(a), a) = \int_a^t \left[\frac{\alpha(s)}{\sigma(X(s), s)} + \psi_t(X(s), s) \right.$$

$$\left. - \frac{1}{2}\frac{\beta^2(s)\sigma_x(X(s), s)}{\sigma^2(X(s), s)} \right] ds$$

$$+ \int_a^t \frac{\beta(s)}{\sigma(X(s), s)} \, dB(s).$$

Comparison with (4) gives

$$\frac{\alpha(s)}{\sigma(X(s), s)} - \frac{\beta^2(s)\sigma_x(X(s), s)}{2\sigma^2(X(s), s)} = \frac{m(X(s), s)}{\sigma(X(s), s)}$$

and

$$\frac{\beta(s)}{\sigma(X(s), s)} = 1,$$

from which the result follows.

4. Put $Y(t) = f(X(t), t)$ where $f(x, t) = \ln x - \ln X(0)$. Then $Y(0) = 0$ and 5.3.1 gives

$$Y(t) - Y(0) = \int_0^t [m(s) - \tfrac{1}{2}\sigma^2(s)] \, ds + \int_0^t \sigma(s) \, dB(s);$$

the first result follows.

$E(X^n(t) \mid X(0) = x) = E(x^n e^Z)$ where Z is normal with mean $a = n \int_0^t [m(s) - \tfrac{1}{2}\sigma^2(s)] \, ds$ and variance $b^2 = n^2 \int_0^t \sigma^2(s) \, ds$. The characteristic function of Z is $E(e^{iuZ}) = e^{iau - (b^2 u^2/2)}$; put $u = -i$ to get $E(e^Z) = e^{a + b^2/2}$. Thus

$$E(X^n(t) \mid X(0) = x) = x^n \exp \left\{ n \int_0^t [m(s) + (n - \tfrac{1}{2})\sigma^2(s)] \, ds \right\}.$$

5. (*Open-Loop Control*) By Problem 3, $\{X(t)\}$ satisfies the stochastic differential equation $dX(t) = Mc(t) \, dt + \sigma c(t) \, dB(t)$. We assume $X(0) = x$ and omit writing this condition in the following expressions. Thus

$$X(b) - x = \int_0^t Mc(t) \, dt + \int_0^t \sigma c(t) \, dB(t),$$

$$E(X^2(b)) = \text{Var} \, (X(b)) + (EX(b))^2$$

$$= \sigma^2 \int_0^b c^2(t) \, dt + \left[x + \int_0^b Mc(t) \, dt \right]^2$$

$$= F(c).$$

Using the usual variational technique, we minimize $F(c)$ by setting

$$0 = \frac{\partial}{\partial \varepsilon} F(c + \varepsilon \, \Delta c) \bigg|_{\varepsilon = 0}$$

$$= \int_0^b 2 \left[\sigma^2 c(t) + Mx + M^2 \int_0^b c(s) \, ds \right] \Delta c(t) \, dt.$$

Since this holds for all Δc, the expression in brackets is zero. It follows that c does not depend on t and that $c(t) = -Mx/(\sigma^2 + M^2 b)$, $0 \le t \le b$.

(*Linear Closed-Loop Control*) By Problem 3, $\{X(t)\}$ satisfies the stochastic differential equation

$$dX(t) = [M\alpha(t)X(t) + \tfrac{1}{2}\sigma^2\alpha^2(t)X(t)] \, dt + \sigma\alpha(t)X(t) \, dB(t).$$

By Problem 4 (assuming $X(0) = x$), we have

$$E(X^2(b)) = x^2 \exp \left\{ 2M \int_0^b \left(\sigma\alpha(s) + \frac{M}{2\sigma} \right) ds - \frac{bM^2}{2\sigma^2} \right\}$$

which is a minimum when $\alpha(s) = -M/2\sigma^2$, $0 \le s \le b$.

Index

A

Almost invariant function, 117
Almost invariant set, 117
Ash, R. B., 160, 248
Asymptotic equipartition property, 149
Autoregressive scheme, 74, 76

B

Bernoulli shift, 141, 160
Billingsley, P., 159, 160, 248
Blumenthal, R. M., 209, 248
Bochner's theorem, 22
Borel's normal number theorem, 140
Breiman, L., 160, 209, 248
Brownian motion, 24–26, 96, 175–186

C

Cantor function, 140
Chain rule for stochastic differentials, 226
Chapman–Kolmogorov equation, 194
Characteristic function, 6
Chung, K. L., 160, 248
Compound Poisson process, 27
Conditional entropy, 149
Conditional independence, 188
Continuity, L^2 sense, 32

Convergence to white noise, 92
Correlation function, 94
Covariance function, 14
Cramér, H., 112, 248
Cyclic model, 69
  generalized, 70

D

Decomposition of stationary processes, 57–68
Deterministic process, 63
Differentiation, L^2 sense, 32, 90
Doleans–Dade, C., 233, 248
Dyadic transformation, 140
Dynkin, E. B., 209, 248

E

Elementary measure with orthogonal values, 51
Ensemble average, 116
Entropy, 149, 151
Ergodic measure, 132
Ergodic theorem, 126, 129, 133
Ergodic theory, 113–160
Ergodic transformation, 117
Estimation, 43
Estimation error, 87

F

Fatou's radial limit theorem, 237
Fokker–Planck equation, 224
Fourier inversion theorem, 245
Fourier–Plancherel theorem, 92, 246
Fourier transform, 242
Freedman, D., 209, 248

G

Gallager, R. G., 159, 160, 248
Garcia, A. M., 159, 248
Gaussian distribution, multivariate, 16
Gaussian process, 19
Gaussian white noise, 97
Getoor, R. K., 209, 248
Gikhman, I. I., 112, 209, 248
Green's function, 105

H

Halmos, P. R., 119, 159, 248
Hannan, E. J., 112, 248
Hardy–Lebesgue space H^2, 237
Herglotz's theorem, 21
Hermitian symmetric, 18
Hitting time, 174

I

Impulse response, 105
Independent increments, 10, 195
Index set, 2
Initial distribution, 190
Integration, L^2 sense, 34–36
Integration by parts, stochastic, 110
Invariant function, 117
Invariant set, 117
Itô's differentiation formula, 227

J

Jacobs, K., 159, 249
Joint entropy, 149

K

Kalman filter, 46
Karhunen–Loève expansion, 37, 38
Kieffer, J., 160, 249
Kolmogorov extension theorem, 4

Kolmogorov shift, 147
Kolmogorov's backward equation, 226
Kolmogorov's forward equation, 224

L

L^2 stochastic process, 2
L^2 theory, 14
Langevin velocity process, 110
Law of the iterated logarithm, 181
Law of large numbers, 60, 131
Leadbetter, M. R., 112, 248
Leibniz's rule, stochastic, 110
Linear filter, 70, 89
Linear operation, 70, 89

M

McKean, H. P. Jr., 233, 248, 249
McMillan, B., 160, 249
Markov chain, 141
Markov process, 186
Markov property, 12, 186
Markov subsequence, 193
Martingale, 200
Maximal ergodic theorem, 126
Mean ergodic theorem, 133, 134
Measurable process, 169
Measurable space, 1
Measure associated with an mov, 51, 53
Measure with orthogonal values (mov),
    51–56
Measure-theoretic isomorphism, 137
Mercer's theorem, 38
Meyer, P. A., 233, 248
Mixed scheme, 76
Mixing transformation, 119
Moving average, 73

N

Negligible set, 162
Nondeterministic process, 63
Nonnegative definite, 18
Nonuniform Poisson process, 28
Normal number, 140

O

Ornstein, D., 160, 249

P

Papoulis, A., 112, 249
Pointwise ergodic theorem, 124, 129
Poisson kernel, 235
Poisson process, 9, 27, 28
Prediction, 44, 50, 63, 77, 81, 97, 100, 110, 111
Prediction error, 64
Probability space, 1
Progressive measurability theorem, 170
Progressively measurable process, 169
Purely nondeterministic process, 63

R

Random function, 2
Random sequence, 2
Random variable, 1
Random vector, 1
Rational spectral density, 75
Recurrence time, 145
Reflection principle, 175
Renewal process, 9
Rozanov, Yu. A., 112, 249

S

Sample function, 2
Second order calculus, 30–36
Separability theorem, 166
Separable process, 162
Separating set, 162
Shannon, C. E., 160, 249
Shannon–McMillan theorem, 148, 157
Shift operator, 63
Shift transformation
    Bernoulli, 141, 160
    Kolmogorov, 147
    one-sided, 114
    two-sided, 115
Shot noise process, 29
Skorokhod, A. V., 112, 209, 248
Smoothing, 44
Spectral density, 57
Spectral distribution function, 57
Spectral measure, 57
Spectral representation, 58
Spectral theory, 50
Spectrum, 58

Standard modification, 165
State space, 2
Stationary process, 15
Stochastic differential, 226
Stochastic differential equation, 91, 104, 210, 219
Stochastic integral, 50, 210
Stochastic mov, 55
Stochastic process, 2
Stopping time, 172
Strong Markov property, 205
Submartingale, 200
Supermartingale, 200
Symmetric, 18
Szëgo–Kolmogorov–Krein theorem, 87

T

Tail σ-field, 141
Time average, 116
Transfer function, 70, 89
Transformation
    conservative, 124
    dyadic, 140
    ergodic, 117
    incompressible, 124
    infinitely recurrent, 124
    measurable, 113
    measure-preserving, 113
    mixing, 119
    one-sided shift, 114
    recurrent, 124
    tail trivial, 147
    two-sided shift, 115
Transition probability, 190, 192

U

Uhlenbeck–Ornstein process, 110
Uncertainty, 149

W

White noise, 69, 92, 97
Wold decomposition, 65
Wong, E., 112, 249

Y

Yaglom, A. M., 112, 249

Probability and Mathematical Statistics

A Series of Monographs and Textbooks

Editors   **Z. W. Birnbaum**   **E. Lukacs**
            University of Washington   *Bowling Green State University*
            Seattle, Washington   *Bowling Green, Ohio*

1. Thomas Ferguson. Mathematical Statistics: A Decision Theoretic Approach. 1967
2. Howard Tucker. A Graduate Course in Probability. 1967
3. K. R. Parthasarathy. Probability Measures on Metric Spaces. 1967
4. P. Révész. The Laws of Large Numbers. 1968
5. H. P. McKean, Jr. Stochastic Integrals. 1969
6. B. V. Gnedenko, Yu. K. Belyayev, and A. D. Solovyev. Mathematical Methods of Reliability Theory. 1969
7. Demetrios A. Kappos. Probability Algebras and Stochastic Spaces. 1969
8. Ivan N. Pesin. Classical and Modern Integration Theories. 1970
9. S. Vajda. Probabilistic Programming. 1972
10. Sheldon M. Ross. Introduction to Probability Models. 1972
11. Robert B. Ash. Real Analysis and Probability. 1972
12. V. V. Fedorov. Theory of Optimal Experiments. 1972
13. K. V. Mardia. Statistics of Directional Data. 1972
14. H. Dym and H. P. McKean. Fourier Series and Integrals. 1972
15. Tatsuo Kawata. Fourier Analysis in Probability Theory. 1972
16. Fritz Oberhettinger. Fourier Transforms of Distributions and Their Inverses: A Collection of Tables. 1973
17. Paul Erdös and Joel Spencer. Probabilistic Methods in Combinatorics. 1973
18. K. Sarkadi and I. Vincze. Mathematical Methods of Statistical Quality Control. 1973
19. Michael R. Anderberg. Cluster Analysis for Applications. 1973
20. W. Hengartner and R. Theodorescu. Concentration Functions. 1973
21. Kai Lai Chung. A Course in Probability Theory, Second Edition. 1974
22. L. H. Koopmans. The Spectral Analysis of Time Series. 1974
23. L. E. Maistrov. Probability Theory: A Historical Sketch. 1974
24. William F. Stout. Almost Sure Convergence. 1974

25. E. J. McShane. Stochastic Calculus and Stochastic Models. 1974
26. Z. Govindarajulu. Sequential Statistical Procedures. 1975
27. Robert B. Ash and Melvin F. Gardner. Topics in Stochastic Processes. 1975
28. Avner Friedman. Stochastic Differential Equations and Applications, Volumes 1 and 2. 1975
29. Roger Cuppens. Decomposition of Multivariate Probabilities. 1975

In Preparation

Eugene Lukacs. Stochastic Convergence, Second Edition

Harry Dym and Henry P. McKean. Gaussian Processes: Complex Function Theory and the Inverse Spectral Method

A 5
B 6
C 7
D 8
E 9
F 0
G 1
H 2
I 3
J 4